Lecture Notes in Computer Sc

Edited by G. Goos and J. Hartmanis

A. Nakamura   M. Nivat   A. Saoudi
P. S. P. Wang   K. Inoue (Eds.)

Parallel Image Analysis

Second International Conference, ICPIA '92
Ube, Japan, December 21-23, 1992
Proceedings

Springer-Verlag

Berlin Heidelberg New York
London Paris Tokyo
Hong Kong Barcelona
Budapest

Series Editors

Gerhard Goos
Universität Karlsruhe
Postfach 69 80
Vincenz-Priessnitz-Straße 1
W-7500 Karlsruhe, FRG

Juris Hartmanis
Cornell University
Department of Computer Science
4130 Upson Hall
Ithaca, NY 14853, USA

Volume Editors

Akira Nakamura
Meiji University, Faculty of Science, Dept. of Information Science
Higashimita, Tama-ku, Kawasaki-shi, 214 Japan

Maurice Nivat
University Paris VII, LITP
2, place de Jussieu, F-75221 Paris, France

Ahmed Saoudi
University Paris XIII, Institut Galilée L.I.P.N.
Av. J. B. Clement, F-93430 Villetaneuse, France

Patrick S. P. Wang
College of Computer Science, Northeastern University
161 Cullinane Hall, Boston, MA 02115, USA

Katsushi Inoue
Yamaguchi University, Dept. of Computer Science and Systems Engineering
Ube, 755 Japan

CR Subject Classification (1991): I.4-5, G.1, F.1-2, F.4

ISBN 3-540-56346-6 Springer-Verlag Berlin Heidelberg New York
ISBN 0-387-56346-6 Springer-Verlag New York Berlin Heidelberg

Typesetting: Camera ready by author/editor
Printing and binding: Druckhaus Beltz, Hemsbach/Bergstr.
45/3140-543210 - Printed on acid-free paper

Preface

This volume contains the papers selected for presentation at the Second International Conference on Parallel Image Analysis (ICPIA '92), held in Ube, Japan, December 21–23, 1992.

The conference topics are data structures, parallel algorithms and architectures, neural networks, computational vision, syntactic generation and recognition, and multidimensional models. The first meeting (International Colloquium on Parallel Image Processing) with the same topics took place in Paris, France, June 17–19, 1991. Selected papers in this meeting are included in a 1992 special issue of the International Journal of Pattern Recognition and Artificial Intelligence. The aim of our meeting is to bring together specialists from various countries who are interested in the above topics and to stimulate both theoretical and practical research in the very important field of parallel image processing and analysis.

The proceedings consist of invited papers, a summary of a tutorial paper, and communications. The following Program Committee members took part in the evaluation and selection of submitted papers: M. Brady (Univ. of Oxford, England), L. Davis (Univ. of Maryland, U.S.A.), M.J.B. Duff (Univ. of London, England), O. Faugeras (INRIA, France), S. Grossberg (Boston Univ., U.S.A.), K. Inoue (Yamaguchi Univ., Japan), S. Levialdi (Univ. La Sapienza di Roma, Italy), K. Morita (Yamagata Univ., Japan), M. Nivat (Univ. Paris VII, France), A. Rosenfeld (Univ. of Maryland, U.S.A.), H. Samet (Univ. of Maryland, U.S.A.), A. Saoudi (Univ. Paris XIII, France), O. Shirai, (Osaka Univ., Japan), R. Siromoney (Madras Christian College, India), P.S.P. Wang (Northeastern Univ., U.S.A.).

We would like to thank all the Program Committee members for their meritorious work in evaluating the submitted papers as well as the following referees, who assisted the Program Committee members: K. Aizawa, M. Crochemore, D. Dreyfus, K. Fukushima, E. Gelenbe, K. Krithivasan, G.M. Landau, A. Restivo, U. Vishkin. We would also thank the Local Arrangement Committee members (Y. Hamamoto, A. Ito and Y. Yamamoto) for their efforts.

We gratefully acknowledge the following organizations which supported the conference:

Yamaguchi Industrial Technology Development Organization (Japan)
The Kajima Foundation (Japan)
International Communications Foundation (Japan)
Department of Computer Science, Meiji University (Japan)
Faculty of Engineering, Yamaguchi University (Japan)
The Institute of Electronics, Information and Communication Engineers, Technical Groups on Computation, Pattern recognition and Understanding (Japan)

Thanks are due also to Professor K. Kobayashi (Tokyo Institute of Technology) for helping us organize the conference. Last but not least, we want to thank Springer-Verlag for excellent co-operation in the publication of this volume.

Tokyo, Paris, Boston and Ube
October 1992

Akira Nakamura
Maurice Nivat
Ahmed Saoudi
Patrick S.P. Wang
Katsushi Inoue

Invited Speakers:
K. Fukushima (Osaka Univ., Japan)
T.S. Huang (Univ. of Illionois at Urbana-Champaign, U.S.A.)
S. Tanimoto (Univ. of Washington, U.S.A.)
P. Quinton (CNRS-IRISA, France)

Tutorial:
P.S.P. Wang (Northeastern Univ., U.S.A.)

General Chairman:
A. Nakamura (Meiji Univ., Japan)

Program Chairman:
M. Nivat (Univ. Paris VII, France)

Program Vice-Chairmen:
K. Inoue (Yamaguchi Univ., Japan)
A. Rosenfeld (Univ. of Maryland, U.S.A.)
A. Saoudi (Univ. Paris XIII, France)
P.S. Wang (Northeastern Univ., U.S.A.)

Program Committee Members:
M. Brady (Univ. of Oxford, England)
L. Davis (Univ. of Maryland, U.S.A.)
M.J.B. Duff (Univ. of London, England)
O. Faugeras (INRIA, France)
S. Grossberg (Boston Univ. U.S.A.)
S. Levialdi (Univ. La Sapienza di Roma, Italy)
K. Morita (Yamagata Univ., Japan)
H. Samet (Univ. of Maryland, U.S.A.)
Y. Shirai (Osaka Univ., Japan)
R. Siromoney (Madras Christian College, India)

Local Arrangement:
Y. Hamamoto (Yamaguchi Univ., Japan)
A. Ito (Yamaguchi Univ., Japan)
Y. Yamamoto (National Museum of Ethnology, Japan)

Table of Contents

From equations to hardware. Towards the systematic mapping of algorithms onto parallel architectures*

François Charot, Patrice Frison, Eric Gautrin, Dominique Lavenier, Patrice Quinton and Charles Wagner

IRISA-CNRS and IRISA-INRIA, Campus de Beaulieu, 35042 Rennes Cedex, France
e-mail : quinton@irisa.fr

1 Introduction

Advances in VLSI technology make it possible to realize systems of very high complexity in a small volume of hardware using integration. In many application fields, it is necessary to implement certain algorithms, or even complete information processing systems, directly in silicon. Application domains which are likely to benefit are in the field of signal processing and scientific computing, with applications to telecommunications, medical imaging, speech processing, image analysis and compression, radar and sonar, etc.

The design of an application specific architecture is a long and difficult task, which spans many different steps, ranging from functional specification, simulation, to chip partitioning, design and assembly, making use of different techniques and tools.

During the design process, the application considered for hardware implementation is subject to many transformations providing different views of the same objects. Let us symbolize the space of evolution of the design as a triangle, as shown in figure 1. The vertices of this triangle are *specification*, *simulation* and *silicon*. When moving downwards, the design gets a higher degree of particularization, and therefore, becomes more dependent on the target technology. The left region represents the final target of the designer, that is to say, a system which is completely finalized in terms of VLSI. The right region represents forms of the design which are suitable for simulation at various levels. We have sketched inside this triangle the positioning of several tools that we have been using at IRISA, and that will be described later on in this paper.

The main goal of the design process can be stated simply: try to obtain a VLSI system which meets the performance and specification requirements of the project, at as low a cost as possible. To reach this, several major goals must be reached :

- the system must be safe and reliable. This means that the functional behaviour of the system should be correct, and if possible, proved to be correct.
- the design time should be as short as possible. Indeed, the cost of the total system depends heavily on this parameter, not only because man-power is costly, but also since a short time-to-market is a crucial factor of success especially when the technology is changing so fast.
- the system must be efficient, i.e., optimal in hardware cost and speed. In fact, the goal is most often to reach a system which just meets some given speed requirement, while minimising real-estate in term of silicon.

There exist no general accepted methods for specifying and implementing a complete hardware system[Hay88]. The different design steps are most often performed using a set of heterogeneous tools, which do not guarantee the coherence of the design through its various representations.

---

* This work is funded by the French Coordinated Research Program C^3 of the Ministère de la Recherche et de la Technologie, and by the ESPRIT Basic Research Action nr. 3280.

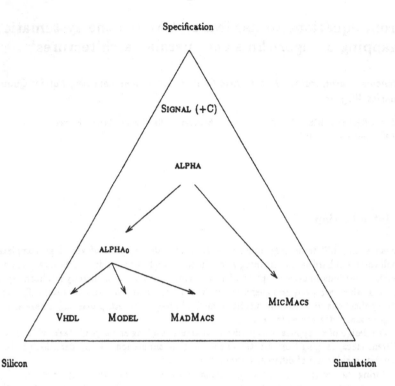

Fig. 1. The design space of a special-purpose system together with the tools currently experimented at IRISA. The SIGNAL language is used for specification purpose. Specification and derivation of regular parts is done using ALPHA. The result of the derivation is an ALPHA0 program, which can be translated in VHDL or MODEL or given to the MADMACS system for layout generation. MICMACS is a lego architecture for real-time simulation.

Several formalisms have been studied and applied for the modeling of systems – algebraic specifications, functional languages, temporal logic, Petri nets, STATECHARTS, etc. [HGdR88] – , with the goal of studying their properties and reasoning about their behaviour. However, the proof power of these formalisms is currently orders of magnitude away from the needs of applications.

The state-of-the art approach consists of using a single framework such as VHDL[ABOR90] for both simulation and description. But even if the use of VHDL represents significant progress over the previously prevailing methodology, it can hardly be said that all the mentioned problems are solved. The most severe limitation of VHDL is its lack of formal semantics, which prevents its use for doing formal verification.

In this paper, we consider several steps which we believe to be essential in the design path of a special purpose architecture, and we present methodologies for achieving design requirements. These solutions are based on experience gathered in the Parallel VLSI Architecture group of IRISA. Our effort is guided by three key ideas:

- the most permanent part of a system is its high-level specification; indeed, the lifetime of a system spans a period of time which goes far beyond one particular implementation as a special-purpose architecture. The architecture is therefore technology dependent, and as the technology changes, so does the structure of the implemented system. We believe that the use of a formal specification language, supporting synthesis techniques and verification, will eventually provide a great benefit for designers;

- in many situations, a real-time simulation of a system is needed; however, even supercomputers do not have the computing power needed to simulate in real-time systems such as those we are currently considering for implementation. We advocate a *lego* approach based on programmable hardware elements and software components that would make it possible to quickly emulate a desired architecture.

- finally, we think that methods for synthesizing regular parts of systems are becoming mature, and that they will eventually provide a great saving in design time, as well as more independence from the technology.

In section 2, we consider the problem of specifying a system. Our approach relies upon the premise that most embedded systems are a mix of computation intensive but regular parts glued together by highly complex control mechanisms. We present the use of the *synchronous language* SIGNAL for the specification and the functional simulation of a system, and we explain why such a language is useful for learning about the synchronization difficulties of the application early in the design process.

In section 3, we describe an approach to the real-time simulation of parallel systems, corresponding to the bottom right region of the design space triangle. Our claim is that *building blocks* for rapidly assembling a real-time simulator for a system can be designed both at the hardware and the software level. We illustrate this concept on the design of a special-purpose chip for doing spelling correction with application to the optical reading of postal addresses. The machine upon which our experiment is based is named MICMACS.

Section 4 is concerned with the design of regular architectures. We describe the use of a language named ALPHA which was designed especially for the description and the synthesis of systolic architectures - but has application beyond this class of architectures. In our approach, ALPHA serves as a formal specification language for regular algorithms. It is also a framework for the formal derivation of parallel architectures which are described as a subset of the language named ALPHA$_0$. This language can then be translated into hardware languages such as VHDL or MODEL[Sol90]. ALPHA$_0$ may also provide input to a program named MADMACS, which does automatic assembly. MADMACS is presented in section 5.

2 Specifications

Most embedded processing systems, at least among those which are candidate for VLSI or ULSI implementation, contain computation intensive but regular parts. This parts are amenable to highly parallel hardware implementations such as systolic, SIMD or pipeline architectures. The specification and implementation of these regular parts will be covered in section 4. One of the main difficulties faced by a system designer is to take into account the interaction between these regular parts, in such a way that the global utilization of resources such as memory, input/output, and silicon can be optimized. The control of the system as a whole can therefore become extremely complex. Our opinion is that this problem is often underestimated, and this greatly increases the probability of residual error in the final design. Moreover, this problem is even more severe when one has to implement the specification on complex parallel architectures. Indeed parallelization transformations on the initial specification can dramatically affect the control of the whole system, therefore leading to intractable situations.

Another problem of design is that an application has to be represented by different descriptions, at different stages in the design process. The initial specification has to be augmented with technology related information, has to be transformed to fit the constraints of the implementation. It is essential that these descriptions be checked against one another, either by *synthesis*, or by *formal verification*. To this end, a specification language should meet several requirements:

- the execution should conform to the specification and be independent of the underlying architecture;
- transformations applied on the description should be valid, and if possible, formally proved;
- the language must allow for parallel and hierarchical descriptions, for the obvious reason of modularity;
- the formalism should permit both the description of algorithms and architectures;
- the ease in specification should not come at the cost of final efficiency of simulation.

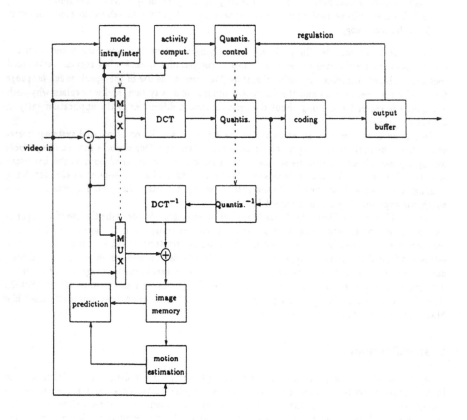

Fig. 2. Structure of a typical video compression system

The SIGNAL language

The so-called *synchronous language* approach[BB91] has been defined to address some of the above mentioned problems, mainly, the mastering of synchronization of real-time systems. Synchronous languages are based upon the hypothesis that calculations or reaction of the program take no time. This makes it possible to express and check logical properties of an application, before taking into account the physical properties of a particular hardware realization. The implementation of a specific process is obtained when one associates an execution time to each elementary action, which should be checked to have a bounded duration. Examples of synchronous languages are ESTEREL, LUSTRE, and SIGNAL[BB91].

In the following, we report our experience in using the SIGNAL language[LGLL91], developed at IRISA, to specify systems. A SIGNAL process is a system of equations whose variables are signals, i.e., infinite sequences of values each associated with a discrete time. The set of time instants when the signal is defined is called the *clock* of the signal. The kernel of the SIGNAL language is made of basic operators which are use to define elementary SIGNAL processes. The specification of more complex processes is obtained by composition of elementary processes. There are four elementary processes: component-wise signal arithmetic operators, delay, subsampling of a signal by a condition, and shuffle of two signals. External processes can be written in another language, provided they can be considered as instantaneous operations.

```
process   counter =
      ( integer N )                            (* 1 *)
      { ? logical h                            (* 2 *)
      ! integer nmod }                         (* 3 *)
      (| synchro { nmod, h }                   (* 4 *)
      | znmod:= nmod $1                         (* 5 *)
      | nmod:= ( 1 when( znmod=N )) default(znmod+1 ) (* 6 *)
      |) !! znmod                              (* 7 *)
         where
         integer znmod init. 0
end;
```

Fig. 3. Description of a counter using SIGNAL

As an example, consider the modulo counter described in figure 3. The counter is described as a process *counter* parameterized by N (line 1). The input is a boolean signal h (line 2), and the output is an integer signal $nmod$ (line 3) whose value is the number of ticks of h modulo the parameter N. The instruction of line 4 specifies that h and $nmod$ are synchronous, that is to say, they have the same clock. Line 5 gives the definition of an intermediate signal $znmod$ which is a delayed version of $nmod$. The definition of $nmod$ in line 5 can be read as follows : $nmod$ has value 1 whenever its past value $znmod$ reaches value N (this is the reset of the molulo function). Otherwise, (default operator), $nmod$ receives the value of $znmod$ increased by 1. Finally, line 7 indicates that the signal $znmod$ is local to the process.

A SIGNAL program can be easily associated to a conventional signal flow graph. A graphical SIGNAL editor can be used to input programs, using a mix of graphic and text representation. A typical representation of a SIGNAL process is shown in figure 4. Boxes represent SIGNAL processes, connection ports correspond to signals, and links between boxes figure flow of data between the processes. Processes can be organized into a hierarchy.

Application to image compression

We have experimented SIGNAL on an application of animated image sequences compression. This example is interesting for several reasons. First, compression algorithms are now normalized (for example, MPEG[Gal91]) and thus are "real life" examples. Second, they are good examples of complex signal processing systems.

The compression algorithms we consider belong to the class of hybrid coding schemes. Inter-image predictive coding exploits the temporal redundancy of successive images, whereas intra-image coding takes benefit of the spatial correlation of luminance and chrominance amplitude around a

6

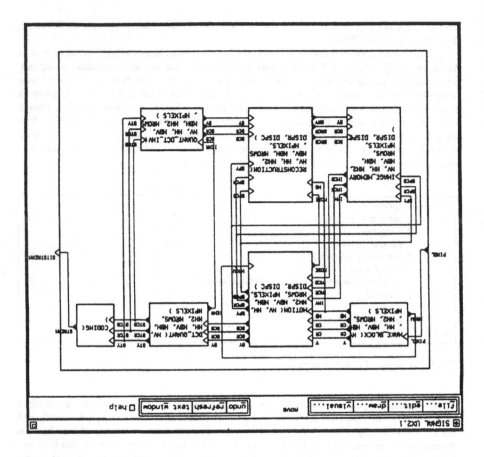

Fig. 4. Graphical view of the image coding process which is composed of a block segmentation (*MAKE_BLOCK*), a motion estimation module (*MOTION*), transformation processes (*DCT_QUANT, QUANT_DCT_INV*) and image memory management processes (*RECONSTRUCTION, IMAGE_MEMORY*)

given pixel. Transformations are applied on subblocks of the image, – typically 8 × 8 –, as shown in figure 2.

The functional specification of such an application using SIGNAL is a direct translation of the structure of its block diagram. A SIGNAL process is associated with each block of the diagram, and the data transfers between processes consist in blocks of pixels.

The main advantages of SIGNAL for representing specifications are the following ones:

– the compiler can checks many coherence properties of programs. For example, the SIGNAL compiler checks that there exists no deadlock situations resulting from cycles in the definition of signals. More generally, the compiler carries out a detailed analysis of the clocks, and checks the synchrony of signals. This puts some constraints on the way programs have to be written, which results in a better specification of events.

– SIGNAL expressions obey algebraic rules which can be used either to derive programs, or to check the equivalence of programs. This property is well suited to architecture design, because synthesis and formal verification are pervasive in this field.

3 Simulation and real-time emulation

The simulation of specifications is important for two reasons. First, behavioral specifications cannot be proved, and they must checked by simulation. Second, during the design process, specifications are often refined. This may affect the behavior of the system, and in many cases, real-time simulation is the only way of seeing the effects of these refinements. This is in particular true for image compression application. Even normalized compression coding schemes leave the designer with the freedom to choose the representation of data and the best architectural organization for algorithms, for example.

Functional simulation can often be done directly from the specification language. SIGNAL, for example, can be executed. However, the simulation of such a language is unsatisfactory, because it is much too slow to attain real-time, even when executed on a super computer.

Accelerating the simulation may rely upon various techniques:

1. one can execute in parallel the specification language. However, this is possible only if a parallel compiler is available for the specification language. This is seldom the case.
2. one can rewrite the most time consuming parts of the specification in another language which can be executed more efficiently on a general-purpose parallel architecture.
3. finally, on can execute the time consuming parts on a special-purpose programmable architecture.

Although solution 2 seems to be reasonable, we believe that in practice, only solution 3 can lead to real-time. Indeed, the computing power needed to simulate systems such as thouse we consider here is much higher than the one available even on super computers. Moreover, general purpose often lack the input/output capabilities needed for high performance signal processing applications applications.

Our proposition is to build a *lego* of hardware and software elements partly tailored to the domain of application. The hardware elements needed to reach this goal must be easy to interconnect so that parallel architectures can be quickly assembled, and their communication bandtwidth and computing power must be balanced, in order to efficiently execute fine grain parallel algorithms. These requirements exclude the use of Digital Signal Processing chips or general purpose processors available on the shelf[2]. Our approach is supported by the experience we report here, which concerns the design of special-purpose architectures for string matching problems, with an application to correction of optical reading.

The MicMacs machine

The MicMacs machine is a VLSI programmable systolic array which was designed at IRISA[FL91a] to become the building block of various systolic architectures. The MicMacs machine can be seen as a peripheral for high speed systolic-like data processing. It is organized into two modules as shown in figure 5. The first module, called MICS, is the systolic array properly speaking, while the second one, called MACS, is in charge of supplying the array with instructions and data. The systolic architecture – the MICS module – is a network of locally interconnected processing elements operating in SIMD mode. The processing element itself is a single chip named API15C which was also designed at IRISA.

To support systolic applications, the processing element was designed along the following ideas :

---

[2] One of the few processors which would meet theses specification is the IWARP[BCC*90], but to our knowledge, it is not available as an independent component.

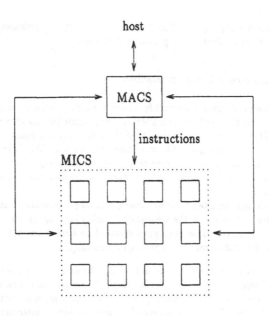

Fig. 5. MicMacs architecture

- it is a *single chip* processor. This is essential to meet the high speed requirements of applications while keeping the hardware volume reasonably low,
- it is *programmable*, in order to cover as large a class of applications as possible,
- it follows a SIMD *execution mode*, in order to simplify the control of the machine and its programming,
- its instruction set is *dedicated* to systolic computations. In particular, local reinitializations and boundary conditions which are present in almost all systolic algorithms can be handled efficiently by special conditional instructions,
- the chip has *several parallel I/O ports*. Thus, different interconnection topologies – linear or bi-dimensional – can be supported, and the communications do not slow down computations when fine grain parallelism is required.

These requirements resulted in the design of the API15C systolic processor.

The MACS control module has two functions : it generates instructions and handles data transfers to/from the systolic network. It can be observed that these functions are executed by programs which have roughly the same structure: the program which generates the flow of instructions is *data-independent* whenever the program handling the data *data-dependent*. As observed in [AAG*87], merging both programs in a single one often results in inefficiencies. In MICMACS, these functions are implemented as two separate controllers running independent processes. These controllers have to be synchronized when data are supplied to the network. Synchronization operations are implemented directly by a hardware mechanism to keep throughput performance high.

Two different machines were build upon this model of architecture. The first one is a linear array of 18 API15C processors [Lav89]. It can be considered as a general-purpose architecture (cf figure 6a) for regular computation, and as such, can be used to program efficiently video processing algorithms. The second machine is an accelerator for string matching algorithms and was developed as a vehicle to study automatic recognition of typewritten postal addresses. The address is read by an Optical Character Recognizer (OCR) device. This operation is not fault-free and, therefore, it has to be followed by a correction algorithm. To meet the reading speed of the OCR, the correction

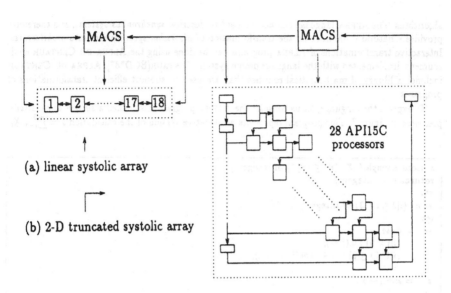

Fig. 6. systolic arrays

algorithm must be able to do the comparison of 2 million words in one second. These performances can be reached only with a 2-D systolic array. The topology of this architecture is shown in figure 6b: it consists of three diagonals of a rectangular array.

A single-chip version

The 28 processor version was used also to study a single chip architecture for spelling correction named API69[Lav92]. The idea was to integrate on silicon the same 2-D structure after processors were customized to the application. The resulting chip is an array of 69 processors containing 300 000 transistors which can process more than 2 million words per second.

Our approach, which consisted of designing a special purpose parallel machine using a flexible building block processor and a powerful control module, allowed us to test very rapidly the architecture before integrating it in silicon. In the case of spelling correction, this step was fundamental, since the possible variants of the string comparison algorithm had to be simulated in real-time.

4 Derivation of parallel regular algorithms

As noticed earlier, the regular parts of applications are the ones which perform most of the computations, and are candidates for parallel VLSI implementation. These recent years, much work has been devoted to automatic techniques for designing regular algorithms (see [Thi92] for a survey). The motivation for using such techniques was the observation that systolic algorithms can be represented by space time reindexing of systems of linear recurrence equations (see [Qui92]). The work we report here is based on the ALPHA DU CENTAUR design environment developed in our laboratory.

The ALPHA language

The ALPHA language[DVQS91, DGL*91, LMQ91] is based on the recurrence equation formalism. It is therefore an *equational language* whose constructs are well-suited for the expression of regular algorithms. The ALPHA language can also be used to describe *synchronous systems*, and therefore, provides a natural framework for the transformation of *algorithm specifications* into *architectures*. Interactive transformations of ALPHA programs can be done using the ALPHA DU CENTAUR environment, implemented with the language design system CENTAUR[BCD*87]. ALPHA DU CENTAUR includes a library of mathematical routines that are used to support efficient transformations of programs.

To explain the language, let us consider the ALPHA program, also called a *system of equations*, presented in figure 7. This program represents an iterative version of the calculation $s = \sum_{i=1}^{3} X_i$.

```
system example ( X : {i|1 ≤ i ≤ 3} of integer )
returns ( s : integer );
var
sum : {i|0 ≤ i ≤ 3} of integer ;
let
sum = case
        {i|i = 0} : 0.(i →);
        {i|1 ≤ i ≤ 3} : X + sum.(i → i − 1);
      esac;
s   = sum.(→ 3);
tel ;
```

Fig. 7. Example of ALPHA program

It takes an input variable X, indexed on the set $\{i|1 \le i \le 3\}$, and returns an integer s. The program makes use of a local variable *sum*, defined on the set $\{i|0 \le i \le 3\}$. Between the keywords **let** and **tel** are the definitions of *sum* and *s* that we now explain in more detail.

In ALPHA, variables and expressions are functions from a set of integral points of $\mathbf{Z}^n$ called the *spatial domain* of the variable or the expression, to a set of values of a given type (boolean, integer, real). ALPHA expressions are obtained by combining variables (or recursively, expressions) together with two sorts of operators : *motionless operators* and *spatial operators*.

Motionless operators are simply point-wise generalization of classical arithmetic operators to ALPHA expressions. As an example, given one-dimensional variables X and Y, the expression $X + Y$ represents a function defined on the intersection of the domains of X and Y whose value at index i is $X_i + Y_i$.

Spatial operators operate explicitly on spatial domains. The *dependence operator* combines *dependence functions* and expressions. Dependence functions are affine mapping between spatial domains, and are denoted $(i, j, ... \rightarrow f(i, j, ..))$ where f is an affine expression. Given an expression E and a dependence function *dep*, $E.dep$ denotes the composition of functions E and *dep*. As an example, the expression $sum.(i \rightarrow i-1)$ denotes the expression whose i-th element is sum_{i-1}. Note that constants are defined on $\mathbf{Z}^0$, and $(\rightarrow i)$ denotes a mapping from $\mathbf{Z}^0$ to $\mathbf{Z}$: the definition $s = sum.(\rightarrow 3)$ in figure 7 says that s is sum_3 (the value of *sum* at index 3.)

The *restriction* operator restricts the domain of an expression by means of linear constraints. In figure 7, the expression $\{i|1 \le i \le 3\} : X + sum.(i \rightarrow i - 1)$ restricts the domain of $X + sum.(i \rightarrow i - 1)$ to the segment $[1, 3]$. The case operator combines expressions defined on disjoint domains into a new expression, as shown by the definition of the variable *sum* in figure 7.

Spatial operators allow recurrence equations to be expressed. In figure 7, the value of the variable *sum* is the sequence of partial sums of the elements of X and is defined by means of a case, whose first branch specifies the initialization part and the second one the recurrence itself.

Finally, ALPHA includes *reduction* operators. They can be used for writing directly expressions containing $\sum$ or Π operators. With a reduction operator, the definition of *s* in the above program could be simply written $s = red(+, (i \to), X)$, which means "sum all element of variable X over the i index".

Transformations

ALPHA follows the substitution principle : any variable can be substituted by its definition, without changing the meaning of the program. For example, substituting *sum* in the definition of s in program of figure 7, gives the program shown in figure 8. One can prove that any ALPHA expression

```
system example ( X : {i|1 ≤ i ≤ 3} of integer )
returns ( s : integer );
var
sum : {i|0 ≤ i ≤ 3} of integer ;
let
sum = case
        {i|i = 0} : 0.(i →);
        {i|i > 0} : X + sum.(i → i − 1);
      esac;
  s   = ( case
        {i|i = 0} : 0.(i →);
        {i|i > 0} : X + sum.(i → i − 1);
      esac).(→ 3);
tel ;
```

Fig. 8. Program 7 after substituting sum by its definition

can be rewritten as an equivalent expression, called its *normal form*, whose structure consists of one single case expression. The normalization process serves to simplify expressions obtained by complex transformations. It can be used, together with the substitution, to do a symbolic simulation of an ALPHA program. For example, the definition of s in program (8) becomes after normalization :

$$s = X.(\to 3) + sum.(\to 2) .$$

By repeating this substitution-normalization process, we obtain

$$s = X.(\to 3) + (X.(\to 2) + (X.(\to 1) + 0.(\to))) ,$$

which is just the mathematical definition of s.

A *change of basis* can be applied to the index space of any local variable, using a straightforward syntactic transformation of the equations [LMQ90]. The change of basis transformation is used for *space-time reindexing*, which is the most fundamental transformation of the synthesis of regular array. The idea behind this transformation is to reindex the variables of a program in such a way that one dimension of the new index space can be interpreted as the *time*, while the remaining indexes represent a processor number.

A motion estimator

An effective algorithm for video-compression uses the idea of motion compensation. In this algorithm, the previous position of part of the current image is searched. Only the distance (or motion vector) is then transmitted instead of all the pixels. This results in a large compression of the video-signal.

The current image is divided into small neighbourhoods of $m \times n$ pixels (current window) whose location in the previous image is to be determined. In order to limit the number of comparisons, this search is restricted to a window of appropriate size in the previous image.

The kernel of the motion detection algorithm lies in the computation of a difference measure S between two windows. Let the reference window be represented by variable $O(k', l')$ and the new window by variable $N(k, l)$. Let (i, j) denote the distance vector between the center of the reference window and the center of one position of the new window. Then the difference associated with this motion vector is $S(i, j)$ given by

$$S(i, j) = \left\{ \sum_{k,l} \mid N(k, l) - O(k', l') \mid \right\}. \tag{1}$$

The minimum value of $S(i, j)$ is

$$\Delta = \min_{i,j}(S(i, j)). \tag{2}$$

The motion vector to be transmitted is the pair (i, j) for which the minimum in (2) is reached.

Figure 9 shows the ALPHA description of this algorithm. It combines both equations into a single line of ALPHA code. From this initial specification, the final architecture is derived through

```
system motion_estimator ( N : {k, l|k ≥ 1; l ≥ 1; 8 ≥ l; 8 ≥ k} of integer;
                          O : {ip, jp|ip ≥ 1; jp ≥ 1; 23 ≥ jp; 23 ≥ ip} of integer)
returns ( Delta : integer ) ;

let
 Delta = red( min , (i, j →) ,
         red( + , (i, j, k, l → i, j) ,
          | N.(i, j, k, l → k, l) - O.(i, j, k, l → i + k - 1, j + l - 1) | ) );
tel ;
```

Fig. 9. Initial description of the motion estimation application in ALPHA

formal transformations which lead to an ALPHA description of about 100 lines. Figure 10 presents a fragment of this description. One can see that *Delta* is now defined by means of two variables $S1$ and $S2$. Each one of these variables corresponds to the serialization of one reduction in the initial system. One can also notice that these variables are indexed by $t1$, $t2$, x, and y. The first two indexes represent a multidimensional time, and the last two ones must be interpreted as a processor number. Therefore, this final description can be interpreted as the description of a 2-D architecture for solving the problem.

5 Automatic processor array generation

In this last section, we consider the problem of generating layouts from the description of a regular architecture. Our research is motivated by the observation that available circuit compilers – gate arrays, standard cells, datapath generators, array compilers, etc.–, consist most often of a unique placement and wiring strategy. They produce dense layouts only if this strategy fits the circuit topology. For instance, the so-called datapath approach[MLB*86] is only efficient for the computational part of a chip, but is unacceptable for generating control logic. Past experience in the design of regular arrays with available compilers have shown poor results[DGL*91] compared to designs made by hand.

```
system motion_estimator ( N : {k, l|l ≥ 1; k ≥ 1; 8 ≥ k; 8 ≥ l} of integer;
                          O : {ip, jp|jp ≥ 1; ip ≥ 1; 23 ≥ ip; 23 ≥ jp} of integer )
returns ( Delta : integer );
var
S2 : {t1, t2, z, y|16 ≥ t1; t1 ≥ 1} , {t1, t2, z, y|t1 = 0} of integer;
S1 : {t1, t2, z, y|z ≥ 1; t1 ≥ 1; 16 ≥ t1; 16 ≥ z} , {t1, t2, z, y|16 ≥ t1; z = 0; t1 ≥ 1} of integer ;
let
Delta = S2.(→ 16, 40, 16, 8);
S1    = case
           {t1, t2, z, y|16 ≥ t1; z = 0; t1 ≥ 1} : eltn( min ).(t1, t2, z, y →);
           {t1, t2, z, y|z ≥ 1; t1 ≥ 1; 16 ≥ t1; 16 ≥ z} :
           min ( S1.(t1, t2, z, y → t1, z + 23, z − 1, 8) , D2.(t1, t2, z, y → t1, z + 24, z, 8) );
           esac;
S2    = case
           {t1, t2, z, y|t1 = 0} : eltn( min ).(t1, t2, z, y →);
           {t1, t2, z, y|16 ≥ t1; t1 ≥ 1} :
           min ( S2.(t1, t2, z, y → t1 − 1, 40, 16, 8) , S1.(t1, t2, z, y → t1, 40, 16, 8) );
           esac;
tel ;
```

Fig. 10. Final version of the motion estimator

There is a wide range of possible array structures – linear, bidimensional, triangular, etc.–, and it is not practical to define a unique place and route strategy nor to develop a single compiler for each conceivable array structure. The alternative chosen was to implement an environment for the development of specific generators, named MADMACS.

A regular array consists of active elements (processors, memories, latches, etc.) interconnected with neighbours. The layout of a regular array can be generated in two steps, as follows :

Active element generation : first, the layout of active elements is generated by classical compilers. However, this generation must be constrained in order to retain the routing regularity at the array level.

Array assembly : the array is then produced by abutment of the active elements.

The MADMACS system[GP92] is an environment for the development of generators. Basically, MADMACS is a LISP language based design tool, coupled with a graphical front-end. Mixing language and graphics is interesting for two reasons. On the one hand, the use of a language gives a great flexibility. Parameters and conditional generation instructions can be used for example. On the other hand, a graphical front-end gives a great interactivity. With the "WYSIWYG" macro-command mechanism, one can develop functions for repetitive tasks such as routing. The most important feature of MADMACS is the possibility that it offers of using coordinate free movements and functions. It is similar to the EMACS text editor in which some commands are independent of word or line size. A designer can therefore interactively produce a symbolic layout by using only object size independent operations. Such procedures can be easily generalized into a program. The MADMACS system has been used to develop several generators. In particular, the layout of the API69 chip presented in section 3[FL91b] was designed in order to test the concepts. The floorplan of this chip is shown in figure 11.

6 Conclusion

We have presented some of the research directions followed by our group regarding methods for designing special-purpose architectures. These directions are based on several ideas :

14

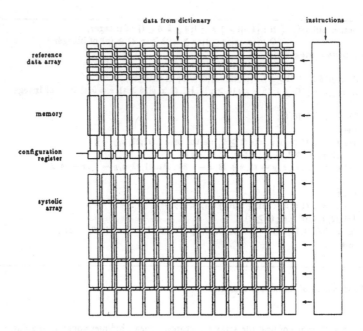

data from dictionary   instructions

reference
data array

memory

configuration
register

systolic
array

Fig. 11. Systolic spelling co-processor layout organization.

- formal specifications will play a prominent role in the future, as they permit the synthesis and verification of designs,
- real-time simulation is very often needed before a special-purpose architecture is committed to silicon. A hardware and software *lego* approach is a good way to reach the performances needed while keeping the development effort at a reasonable level,
- regular parallel architectures will be the key to the design of VLSI computationally intensive parts of systems. The design of these architectures can make benefit of new powerful methods such as those which are studied for systolic arrays.

We have illustrated these principles through various examples. The use of the SIGNAL synchronous language for the specification of a video compression algorithm was presented. We have then described the MICMACS *lego* architecture. The principles of the ALPHA DU CENTAUR environment for the design of regular architectures was explained and illustrated on the example of a motion estimation algorithm. Finally we have explained the concepts of the MADMACS layout generator for regular arrays.

In the future, it is very likely that approaches such as the ones mentioned in this paper will become more commonly used, either directly, or more likely embedded in integrated CAD frameworks.

The main impediment to the spreading of such methods is certainly their current esoteric notations which are very close to the formal model they come from. A considerable effort has thus to be done in order to shorten the distance between these languages and tools and the way designers are used to work.

Acknowledgements

The authors would like to thank Doran Wilde for his careful reading of the paper and his thoughtful comments.

References

[AAG*87]   M. Annaratone, E. Arnould, T. Gross, H.T. Kung, M. Lam, O. Menzilcioglu, and J.A. Webb. The warp computer : architecture, implementation, and performance. *IEEE tr. on Computers*, C-36(12):1523–1538, December 1987. Systolique, Architecture parallele generale.

[ABOR90]   R. Airiau, J.-M. Bergé, V. Olive, and J. Rouillard. *VHDL. Du langage à la modélisation.* Collection Informatique, Presses Polytechniques et Universitaires Romandes, 1990.

[BB91]   A. Benveniste and G. Berry. The Synchronous Approach to Reactive and Real-Time Systems. *PIEEE*, 9(79):1270–1281, sep 1991.

[BCC*90]   S. Borkar, R. Cohn, G. Cox, T. Gross, H.T. Kung, M. Lam, M. Levine, B. Moore, W. Moore, C. Peterson, J. Susman, J. Sutton, J. Urbanski, and J. Webb. *Supporting Systolic and Memory Communication in iWarp.* Technical Report, Carnegie Mellon University, 1990.

[BCD*87]   P. Borras, D. Clément, Th. Despeyroux, J. Incerpi, G. Kahn, B. Lang, and V. Pascual. *CENTAUR : the System.* Technical Report 777, INRIA, 1987.

[DGL*91]   C. Dezan, E. Gautrin, H. Le Verge, P. Quinton, and Y. Saouter. Synthesis of systolic arrays by equation transformations. In *ASAP'91*, IEEE, Barcelona, Spain, September 1991.

[DVQS91]   C. Dezan, H. Le Verge, P. Quinton, and Y. Saouter. The ALPHA DU CENTAUR environment. In P. Quinton and Y. Robert, editors, *International Workshop Algorithms and Parallel VLSI Architectures II*, North-Holland, Bonas, France, June 1991.

[FL91a]   P. Frison and D. Lavenier. Experience in the design of parallel processor arrays. In *International Workshop on Algorithms and Parallel VLSI Architectures II*, Bonas (France), jun 1991.

[FL91b]   P. Frison and D. Lavenier. A fully integrated systolic spelling co-processor. In *VLSI91 : International Conference on Very Large Scale Integration*, August 1991.

[Gal91]   D. Le Gall. MPEG: A Video Standard for Multimedia Applications. *Communications of the ACM*, 34(4):46–58, April 1991.

[GP92]   E. Gautrin and L. Perraudeau. Madmacs : a tool for the layout of regular arrays. In *IFIP Workshop on Synthesis, Generation and Portability of Library Blocks for ASIC Design*, March 1992.

[Hay88]   John P. Hayes. *Computer Architecture and Organization.* Mc Graw Hill, New York, 1988.

[HGdR88]   C. Huizing, R. Gerth, and W. P. de Roever. Modelling statecharts behaviour in a fully abstract way. In M. Dauchet and M. Nivat, editors, *13th Colloquium on Trees in Algebra and Programming CAAP'88, Lecture Notes in Computer Science*, pages 271–294, Springer Verlag, Nancy, France, March 1988. Volume 299.

[Lav89]   Dominique Lavenier. *MicMacs : un réseau systolique linéaire programmable pour le traitement des chaines de caractères.* PhD thesis, Université de Rennes 1, jun 1989.

[Lav92]   D. Lavenier. A high performance systolic chip for spelling correction. In *Euro Asic ' 92*, pages 381–384, IEEE computer Society Press, jun 1992.

[LGLL91]   Paul Le Guernic, Thierry Gautier, Michel Le Borgne, and Claude Le Maire. Programming real-time applications with SIGNAL. *Proceedings of the IEEE*, 79(9):1321–1336, septembre 1991.

[LMQ90]   H. Le Verge, C. Mauras, and P. Quinton. A language-oriented approach to the design of systolic chips. In *International Workshop on Algorithms and Parallel VLSI Architectures*, Pont-à-Mousson, June 1990. To appear in the Journal of VLSI Signal Processing, 1991.

[LMQ91]   H. Le Verge, C. Mauras, and P. Quinton. The ALPHA language and its use for the design of systolic arrays. *Journal of VLSI Signal Processing*, 3:173–182, 1991.

[MLB*86]   T. Mashburn, I. Lui, R. Brown, D. Cheung, G. Lum, and P. Cheng. Datapath: a cmos data path silicon assembler. In IEEE, editor, *23rd Design Automation Conference*, pages 722–729, 1986.

[Qui84]   P. Quinton. *The Systematic Design of Systolic Arrays.* Technical Report, Microelectronics Center of North Carolina Research Report, July 1984.

[Qui92]   P. Quinton. Systems of recurrence equations. In *Conpar 92 - VAPP V Tutorial*, Lyon (France), September 1992.

[Sol90]   *SOLO 1400 Reference Manual.* ES2 Publication Unit, European Silicon Structures Limited, Berkshire, United Kingdom, 1990.

[Thi92]   L. Thiele. Compiler techniques for massive parallel architectures. In P. Dewilde, editor, *State of the Art in Computer Science*, Kluwer Academic Publisher, 1992.

Visual Pattern Recognition
with Neural Networks

Kunihiko Fukushima

Faculty of Engineering Science, Osaka University
Toyonaka, Osaka 560, Japan

Abstract. The "neocognitron" was first proposed as a hierarchical neural network model for the mechanism of visual pattern recognition in the brain. It is capable of deformation-resistant pattern recognition. Various experiments have demonstrated its powerful ability to recognize visual patterns. For example, the authors have designed several systems which recognize hand-written characters, such as, a system recognizing ten numerals, and a system recognizing alphanumeric characters. This paper also discusses recent advances in the neocognitron. The network has been modified to have an architecture closer to that of the real biological brain, and a new learning algorithm has been introduced.

The "selective attention model" has not only forward but also backward connections in a hierarchical network. It has the ability to segment patterns, as well as the function of recognizing them. The principles of this selective attention model can be extended to be used for several applications: for example, the recognition and segmentation of connected characters in cursive handwriting of English words, and the recognition of Chinese characters.

1 Introduction

The author has proposed various models for visual pattern recognition in the brain. The idea of these models is useful in obtaining new design principles for visual pattern recognition systems. This paper introduces two of such models: a "neocognitron" model, which is capable of deformation-resistant pattern recognition, and a "selective attention" model, which has the ability to segment patterns, as well as the function of recognizing them.

2 Neocognitron

The neocognitron [1],[2],[3] can acquire the ability to recognize patterns by learning, and can be trained to recognize any set of patterns. After learning, it can recognize input patterns robustly, with little effect from deformation, changes in size, or shifts in position. Since the neocognitron has the ability to generalize, it is not necessary to teach all the deformed versions of the training patterns. It is even able to correctly recognize a pattern which has not been presented before, provided it resembles one of the training patterns.

2.1 Network Architecture of the Neocognitron

The neocognitron is a hierarchical neural network consisting of many layers of neuron-like cells. The lowest stage of the network is the input layer U_0, and consists of a two-dimensional array of receptor cells. Each of the succeeding stages has a layer U_S consisting of "S-cells" followed by another layer U_C of "C-cells". Thus, in the whole network, layers of S-cells and C-cells are arranged alternately, as shown in Fig. 1. Each rectangle in the figure represents a two-dimensional array of cells. Notation U_{Sl}, for example, is used to indicate the layer U_S of the lth stage. There are forward connections between cells in adjoining layers. A cell receives its input connections from only a limited number of cells situated in a small area in the preceding stage. The size of the area covered by the input connections of a single cell becomes larger in higher stages.

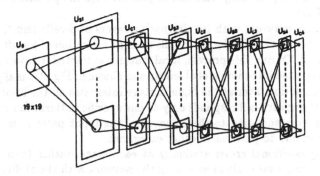

Fig. 1. Hierarchical network structure of the neocognitron.

The S-cells are feature-extracting cells. They have variable input connections, and can acquire the ability to extract features by learning (or training). After learning is finished, S-cells, with the aid of the subsidiary V-cells (which are not drawn explicitly in Fig. 1), can extract features from the input pattern. In other words, an S-cell is activated only when a particular feature is presented in a certain position in the input layer. The features extracted by the S-cells are determined during the learning process. Generally speaking, local features, such as a line at a particular orientation, are extracted in the lower stages. More global features, such as part of a training pattern, are extracted in the higher stages.

C-cells are put into the network to allow for positional error in the features extracted by S-cells. The connections from S-cells to C-cells are fixed and invariable. Each C-cell receives signals from a group of S-cells which extract the same feature, but from slightly different positions. The C-cell is activated if at least one of these S-cells is active. Even if the stimulus feature is shifted and another S-cell is activated instead of the first one, the same C-cell keeps responding. Hence, the C-cell's response is less sensitive to shifts of the input pattern. We can also say that the spatial response of a layer of S-cells is blurred at the succeeding layer of C-cells.

The layer of C-cells at the highest stage is the recognition layer, representing the final result of the pattern recognition by the neocognitron.

The density of cells in each layer is designed to decrease with the order of the stage, because each cell in a higher stage usually has a larger receptive field, and neighboring cells come to receive similar signals owing to the overlapping of their receptive fields.

Each layer of S-cells or C-cells is divided into subgroups, called "cell-planes", according to the features to which they respond. Each rectangle drawn with heavy lines in Fig. 1 represents a cell-plane. The connections converging to the cells in a cell-plane are homogeneous and topographically ordered. In other words, the connections have a translational symmetry: all the cells in a cell-plane receive input connections of the same spatial distribution, in which only the positions of the preceding cells shift in parallel with the positions of the cells in the cell-plane.

In the entire network, with its alternating layers of S-cells and C-cells, the process of feature-extraction by the S-cells and the toleration of a shift by the C-cells is repeated. During this process, local features extracted in lower stages are gradually integrated into more global features as shown in Fig. 2. Finally, each C-cell of the recognition layer at the highest stage integrates all the information of the input pattern; each cell responds only to one specific pattern. In other words, only one cell, corresponding to the category of the input pattern, is activated. Other cells respond to patterns of other categories.

Tolerating positional errors gradually at each stage, rather than all in one step, plays an important role in endowing the network with the ability to recognize not only shifted but also distorted patterns. Let an S-cell in an intermediate stage of the network have already been trained to extract a global feature, which consists of a combination of local features. The S-cell usually responds even to a distorted version of the global feature, in which the locations of the local features are deviated from their original positions. This is because a small amount of deviation of the locations of the local features is tolerated by the blurring operation of the C-cells placed before the S-cell.

Thus, the network has the ability to generalize, and to correctly recognize a pattern which has not been presented before, provided it resembles one of the training patterns.

2.2 Training the Neocognitron

The neocognitron can be trained to recognize patterns through either unsupervised or supervised learning.

In the case of unsupervised learning, the self-organization of the network is performed using two principles. The first principle is a kind of "winner-take-all" rule: among the cells situated in a certain small area, only the one responding most strongly has its input connections reinforced. The amount of reinforcement of each input connection to this maximum-output cell is proportional to the

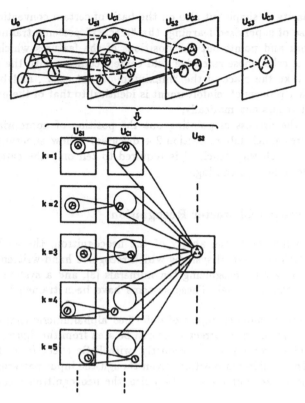

Fig. 2. Illustration of the process of pattern recognition in the neocognitron [1].

intensity of the response of the cell from which the relevant connection leads. Thus, the maximum-output cell comes to acquire the ability to extract a feature of the stimulus which is now presented to the input layer.

Once a cell is thus selected and reinforced to respond to a feature, the cell usually loses its responsiveness to other features. When a different feature is presented, a different cell usually yields the maximum output and has its input connections reinforced. Thus, a "division of labor" among the cells occurs automatically.

The second principle is introduced in order that the connections being reinforced always preserve the homogeneity (or translational symmetry) of the network. The maximum-output cell not only grows by itself, but also controls the growth of neighboring cells, working, so to speak, like a seed in crystal growth. To be more specific, all of the other S-cells in the cell-plane, from which the "seed cell" is selected, follow the seed cell, and have their input connections reinforced by having the same spatial distribution as those of the seed cell.

Supervised learning is useful when we want to train a system to recognize, for instance, hand-written characters, which should be classified not only on the basis of similarity in shape but also on the basis of certain conventions.

In the case of supervised learning, the "teacher" presents training patterns to the network and points out the positions of the features which should be extracted. The cells whose receptive field centers coincide with the positions of the features take the place of the "maximum-output cells", and become seed cells. The other process of reinforcement is identical to that of the unsupervised learning, and occurs automatically.

However, the process of pointing out the position of appropriate features usually requires much labor. Section 2.4 discusses a new supervised learning algorithm, in which the "teacher" is required to tell only the category of the training pattern he is presenting.

2.3 Handwritten Character Recognition

In order to demonstrate the ability of the neocognitron, the author and his group have designed several systems which recognize hand-written characters: for example, a system recognizing ten numerals [3], and a system recognizing alphanumeric characters [4]. These systems have been trained by supervised learning.

Figure 3 shows some examples of deformed alphanumeric characters which the system has recognized correctly. As can be seen from the figure, the neocognitron recognizes input patterns robustly, with little effect from deformation, changes in size, or shifts in position. Even though the input patterns have some parts missing or are contaminated by noise, the neocognitron recognizes them correctly.

The present system, which can recognize thirty-five alphanumeric characters (where the alphabetic character 'O' and numeral '0' are treated as the same pattern) [4], is somewhat larger in scale than the previous system recognizing ten numerals [3]. However, the increase in the size of the network required is not proportional: the number of cells in the present system is 70,045 while that of the previous system is 36,321. The number of characters to be recognized increased 3.5 times, but the number of cells increased only around 1.9 times. This is because the same local features are commonly contained in many patterns of different categories. Hence, the number of local features to be extracted in the lower stages does not increase substantially. Although the number of features to be extracted in the higher stages increases with an increase in the number of characters to be recognized, it does not substantially affect the total number of cells in the network either, because the density of the cells is small in the higher stages. Therefore, the total number of cells in the whole network does not increase as much when compared to the increase in number of categories.

2.4 Improvement of the Neocognitron

Many approaches are now in progress to improve the neocognitron. We discuss some of them in this section.

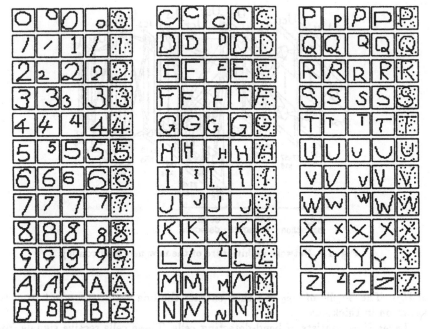

Fig. 3. Some examples of deformed alphanumeric characters which the neocognitron recognized correctly [4].

As has been discussed above, the conventional supervised learning can endow the neocognitron with the ability of robust pattern recognition, if a good training pattern set is skillfully chosen. However, a great amount of labor is required to construct a good training pattern set: the "teacher" is required to tell not only the categories of the training patterns, but also the positions at which important features exist. On the other hand, the conventional algorithm for unsupervised learning [2], by which all of the training progresses automatically, produces a somewhat lesser ability to recognize deformed patterns.

In order to construct a neocognitron which can be easily trained to have a powerful ability to recognize patterns, improvements have been made from two sides [5]. One is the improvement of the network architecture [6], and the other is the development of a new learning algorithm which does not require much labor [7].

The improved neocognitron [6] has built-in edge-extracting cells, line-extracting cells, and bend-detecting cells. Figure 4 shows the network architecture.

Layer U_{S1} consists of edge-extracting S-cells. The S-cells of layer U_{S2} extract line components using the edge information extracted in U_{S1}. Although a change in thickness of a line causes a shift in position of both edges, the effect of the positional shift of the edges is absorbed by the blurring operation in the C-cells

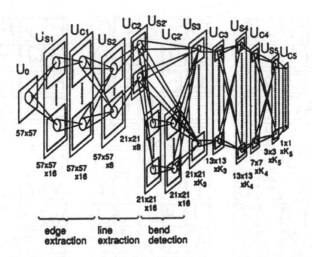

edge   line   bend
extraction   extraction   detection

Fig. 4. Network architecture of the new neocognitron [6].

of U_{C1}. The S-cells of U_{S2} can then extract the line with little effect from the variation in thickness.

Layer $U_{S2'}$ consists of bend-detecting cells. These cells receive signals from U_{C2}, which responds to line segments, and detect bend-points and endpoints of the lines. More specifically, an S-cell of $U_{S2'}$ receives input connections antagonistically from two groups of C-cells located at slightly different positions in the same cell-plane of U_{C2}. Since these C-cells respond optimally to line segments of the same orientation, their outputs become different if the line is curved or terminated. The S-cell detects this unbalance of the outputs of these C-cells. Therefore, when a straight line is presented to the input layer U_0, the response of the S-cells of $U_{S2'}$ appears at an endpoint of the line. When a curved line is presented, the S-cells' response represents bend-points of the curved line. The structure of this part of the network is the same as the one used in the curvature-extracting network proposed previously by Fukushima [8],[9].

Layer U_{S3} receives input connections not only from U_{S2} but also from $U_{S2'}$. The S-cells in layers U_{S3}, U_{S4}, and U_{S5} have variable input connections, which are modified through learning.

In contrast to the conventional neocognitron, the new network shows considerable robustness even when trained by unsupervised learning. The amount of accepted deformation is quite large compared to the conventional neocognitron trained by unsupervised learning.

Supervised learning is still required, however, if we want to construct a multi-font recognition system which can recognize not only deformed patterns, but also various patterns of completely different styles of writing.

The new learning algorithm [7] combines unsupervised and supervised learning. Variable connections are reinforced by unsupervised learning at first, and are modified afterward by supervised learning if the result of the recognition is wrong. This algorithm does not require much labor from the "teacher", who is required to tell only the category of the training pattern he is presenting.

Backward signal paths similar to the ones used in the "selective attention model" discussed below have been added to the neocognitron to be used only for the training. The cells in the backward paths are arranged in a mirror image of the cells in the forward paths. The forward and the backward connections also make a mirror image to each other, but the directions of signal flow through the connections are opposite.

After finishing the unsupervised learning, learning by error-correction is begun. Test patterns of various styles of writing are presented to the network, and the response of the recognition layer U_{C5} is observed. If the result of recognition is incorrect, the desired output is fed to the highest stage of the backward paths. To be more specific, the "teacher" activates the w_{C5}-cell which corresponds to the category of the training pattern.

Backward signals from the desired recognition cell are compared with the forward signals from the training pattern, as shown in Fig. 5. If discrepancies between the two signals are detected, new feature-extracting cells are automatically generated in the network. Operations which used to be manually performed can now progress automatically.

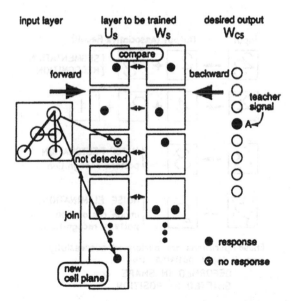

Fig. 5. Supervised learning using backward signals [7].

The new neocognitron trained by the improved learning algorithm has recognized handwritten numerals robustly, with little effect from deformation, shifts in position, or changes in thickness of the lines.

3   Selective Attention Model

3.1   Original Model of Selective Attention

Although the neocognitron has considerable ability to recognize deformed patterns, it does not always recognize patterns correctly when two or more patterns are presented simultaneously. The selective attention model has been proposed to eliminate these defects [10],[11],[12].

The model has not only forward (i.e., afferent or bottom-up) but also backward (i.e., efferent or top-down) connections between the hierarchically connected layers of cells. Figure 6 summarized the functions of the model. When a composite stimulus, consisting of two patterns or more, is presented, the model focuses its attention selectively to one of the patterns, segments it from the rest, and recognizes it. After the identification of the first segment, the model switches its attention to recognize another pattern. The model also has the function of associative recall. Even if noise or defects affect the stimulus pattern, the model can recognize it and recall the complete pattern from which the noise has been eliminated and defects corrected. These functions can be successfully performed even for deformed versions of training patterns, which have not been presented during learning.

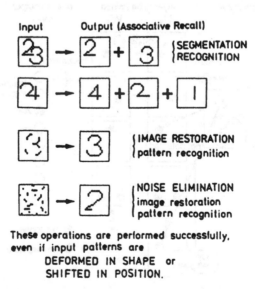

Fig. 6.   Summary of the functions of the selective attention model [13].

The principles of this selective attention model can be extended to be used for several applications. Two examples are offered in this section. One is the recognition and segmentation of connected characters in cursive handwriting of

English words. The second example is the recognition of Chinese characters (or *kanji* characters).

3.2 Connected Character Recognition

Generally speaking, machine recognition of connected characters in cursive handwriting of English words is a difficult problem. It cannot be successfully performed by a simple pattern matching method, because each character changes its shape by the effect of the characters before and behind. In other words, the same character can be scripted differently when it appears in different words, in order to be connected smoothly with the characters in front and in the rear.

Although the original model of selective attention already has the ability to recognize and segment patterns, it does not always work well when too many patterns are presented simultaneously. The model was then modified and extended to be able to recognize connected characters in cursive handwriting [14],[15],[16]. A search controller was added to the original model, in order to restrict the number of patterns to be processed simultaneously: The new model processes the patterns contained in a small "search area", which is moved by a search controller. The position control of the search area does not need to be accurate, as the original model, by itself, has the ability to segment and recognize patterns, provided the number of patterns present is small.

Basic Structure of the Model. The connected character recognition system is a hierarchical multilayered network, which has backward as well as forward connections between layers. Figure 7 shows how the different kinds of cells, such as u_S, u_C, w_S and w_C, are interconnected in the network. Each circle in the figure represents a cell. Notation u is used to represent the cells in the forward paths, and w the cells in the backward paths. Although the figure shows only one of each kind of cell in each stage, numerous cells actually exist, arranged in a two-dimensional array. We will use notation u_{Cl}, for example, to denote a u_C-cell in the lth stage, and U_{Cl} to denote the layer of u_{Cl}-cells. The highest stage of the network is the Lth stage ($L = 4$).

The signals through forward paths manage the function of pattern-recognition. If we consider the forward paths only, the model has almost the same structure and function as the neocognitron. Cells u_S, u_{SV} and u_C correspond respectively to S-, V- and C-cells in the neocognitron, and layer U_{CL} at the highest stage works as the recognition layer.

In the selective attention model used for the connected character recognition, feature-extracting cells u_S in the first stage have fixed input connections and extract line components of various orientations. In all other stages, input connections of u_S-cells are variable and reinforced by unsupervised learning.

The cells and connections in the backward paths are arranged in the network in a mirror image of those in the forward paths. The signals through the backward paths manage the function of selective attention and associative recall.

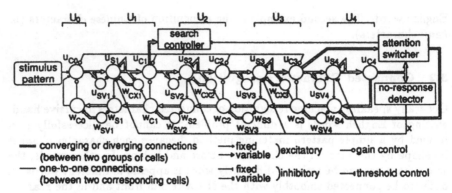

Fig. 7. Network architecture of the selective attention model illustrating the interconnections between different kinds of cells [15].

Segmentation. The output signal of the recognition layer U_{CL} is sent through the backward paths, and reaches the recall-layer W_{C0} at the lowest stage. The routes of the backward signals are controlled by the gate signals, which are given to w_S-cells in the backward paths from the corresponding u_S-cells in the forward paths. Guided by the gate signals, the backward signals reach exactly the same positions at which the input pattern is being presented.

When two patterns or more are simultaneously presented to the input layer U_{C0}, usually only one cell corresponding to the category of one of the stimulus patterns is activated in the recognition layer U_{CL}. This is partly because of the lateral inhibition between u_S-cells, and also because of the attention focusing by gain control signals from the backward paths, which will be discussed later. Since the backward signals are sent only from the activated recognition cell, only the signal components corresponding to the recognized pattern reach the recall-layer W_{C0}. Therefore, the output of the recall-layer can be interpreted as the result of segmentation. Even if the stimulus pattern which is now recognized is a deformed version of a training pattern, the deformed pattern is segmented and emerges with its deformed shape.

Threshold Control. Take, for example, a case in which the stimulus contains a number of incomplete patterns which are contaminated with noise and have several parts missing. Even when the pattern recognition in the forward path is successful, and only one cell is activated in the recognition layer U_{CL}, it does not necessarily mean that the segmentation of the patten is also completed in the recall-layer W_{C0}.

When some part of the input pattern is missing and the feature which is supposed to exist there fails to be extracted in the forward paths, the backward signal flow is interrupted at that point and cannot proceed any further because no gate signals are received from the forward paths. This situation is detected by w_{CX}-cells. The w_{CX}-cells send threshold-control signals to u_S-cells around that area, and decrease the threshold for feature-extraction. Thus, the model is forced to extract even vague traces of the undetected feature.

Once a feature is thus extracted in the forward paths, the backward signal can then be further transmitted to lower stages through the path unlocked by the gate signal from the newly activated forward cell. Hence, a complete pattern, in which defective parts are interpolated, emerges in the recall-layer W_{C0}.

A threshold-control signal is also sent from the no-response detector shown at far right in Fig. 7. If all of the recognition cells are silent, the no-response detector sends the threshold-control signal to the u_S-cells in all stages, and lowers their threshold for feature-extraction until at least one recognition cell becomes activated.

Focusing Attention. When a backward cell w_C is activated, it sends a gain control signal to the corresponding u_C-cell and increases the gain of the cell. Thus, only the forward signal flow in the paths in which backward signals are flowing is facilitated. Since the backward signals are usually sent back only from one activated recognition cell, only the forward paths relevant to the pattern which is now recognized are facilitated. This means that attention is selectively focused on only one of the patterns in the stimulus.

In the model for connected character recognition, (and not in the model for Chinese character recognition), a search controller is introduced in order to restrict the number of patterns to be processed simultaneously. The model mainly processes the patterns contained in a small "search area", which is moved by the search controller.

In all stages except U_{CL}, the u_C-cells receive gain control signals not only from the w_C-cells but also from the search controller. The gain control signal from the search controller produces the search area by decreasing the the gain of the u_C-cells situated outside of the search area.

The position of the search area is shifted to the place in which a larger number of line extracting cells u_{C1} are activated. The search area has a size somewhat larger than the size of one character. The boundary of the search area is not sharply restricted: the gain of the u_C-cells is controlled to decrease gradually around the boundary. It is not necessary to control the position and the size of the area accurately because the original selective attention model has the ability to segment and recognize patterns by itself, provided the number of patterns present is small. The only requirement is that the search area covers at least one pattern. It does not matter if it covers a couple of patterns simultaneously.

Switching Attention. The fatigue of the cells is effectively used in the model for switching attention to another pattern. A u_C-cell is fatigued if it receives a strong gain control signal. It can maintain high gain only when it is receiving a large gain control signal. Once the gain control signal disappears, the gain of the u_C-cell decreases rapidly, and cannot recover for a long period of time. This is effective in preventing the model from recognizing the same character twice.

Once a pattern has been recognized and segmented, the attention is automatically switched to recognize another pattern. To be more exact, there is a detector in the network which determines the timing of attention switching. The detector sends a command to switch attention when the following two conditions are simultaneously satisfied: 1) the number of activated recognition cells u_{CL} is only one, and 2) the total activity of layer U_{CL-1} has nearly reached a steady state.

Once a command to switch attention is given to the network, the backward signal flow is cut off for a short period. Since the gain control signal from the w_C-cells disappears, the gain of the u_C-cells falls to the level determined by the degree of fatigue of the cells. The effect of the threshold control signal is also reset at this moment.

The search controller again seeks a place in which a larger number of line extracting cells u_{C1} are activated, and shifts the search area to the new place. If all of the responses from the cells of layer U_{C1} are small enough because of the fatigue, however, the model stops working, assuming that all characters in the input string have already been processed.

Computer Simulation. This section offers some results of a preliminary experiment with computer simulation to check the capability of the model to recognize connected characters. In this experiment, the model was taught only five characters, instead of the whole set of 26 alphabetical characters. The input layer of the model has a rectangular shape, and consists of 57×19 cells.

The network was trained with unsupervised learning in a similar way as for the original model [11],[12]. The five training patterns shown in Fig. 8(a) were repeatedly presented to the network during the training phase. The size of each training pattern was a 19×19 pixel array. These characters were presented only in this shape, and anything like a deformed version of them was not presented during the training.

Figure 8(b) shows how the response of layer W_{C0} of the network changes with time, when a connected character string (shown at the top of the figure) is presented to the input layer. Time t after the first presentation of the character string is indicated in the figure. As can be seen from the figure, character 't' is first recognized and segmented, then 'o', 'a', and 'c', follow. Although the characters in the input string are different in shape from the training characters shown in Fig. 8(a), recognition and segmentation of the characters have been successfully performed.

Figure 8(c) shows some examples of input character strings which have been successfully recognized and segmented. It can be seen from the figure that the input strings are processed correctly, even if the spacing between the characters changes. A string which contains two of the same characters with somewhat different shapes can also be processed successfully.

(a) Five training patterns used for learning.

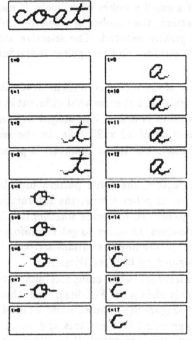

(b) Time course of the response of layer W_{C0}, in which the result of segmentation appears. A character string presented to the input layer is shown at the top.

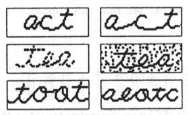

(c) Some examples of character strings which have been successfully recognized and segmented.

Fig. 8. Performance of the model for connected character recognition [15].

3.3 Chinese Character Recognition

Recognition of Chinese characters is also difficult, but for another reason. There are an enormous number of different characters, and each of these characters has a very complicated shape. However, it is fortunate that most of the Chinese characters can be decomposed into a couple of fundamental parts, such as left-hand radical (i.e., the side, or called *hen* in Japanese), right-hand radical (i.e., the body, or *tsukuri*), upper radical (i.e., the crown part, or *kanmuri*), and so on. In other words, a large number of different characters can be composed by a simple combination of a small number of radicals. If each of these radicals is recognized as a single pattern, the number of categories, to which patterns have to be classified, can be greatly reduced. The selective attention model can be successfully used for this purpose and recognizes individual radicals in a single character [14].

The network structure of the model is fundamentally the same as the one for connected character recognition, but somewhat different in details. The model for Chinese character recognition is also a four-staged network, but has a square-shaped input layer consisting of 29 × 29 cells. In the recognition phase after finishing the self-organization, one input character is presented at a time to the input layer.

In some Chinese characters, the same pattern may be a radical placed at several different positions. In other words, the information on the shape of a radical, only, is not always enough to decide whether it is in the left-hand, right-hand or upper radical position. In order to get the information on the position of the radical in a character, the model contains a circuit to detect the center of gravity of the pattern segmented in layer W_{C0}.

In a preliminary experiment with computer simulation, the network was taught only eight radicals shown in Fig. 9(a), and then the performance of the network was tested. Figure 9(b) shows some examples of Chinese characters which have been segmented into radicals and recognized correctly. When a largely deformed character is presented, there are some cases where patterns segmented in layer W_{C0} are incomplete in shape and have some parts missing, but the character is still recognized correctly. Although the scale of the simulation is still very small, we can see that this approach is promising for Chinese character recognition.

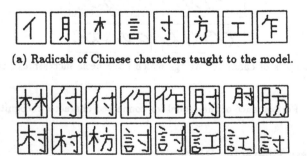

(a) Radicals of Chinese characters taught to the model.

(b) Some examples of Chinese characters which have been segmented into radicals and recognized correctly.

Fig. 9. A preliminary experiment for Chinese character recognition [14].

Acknowledgement: This work was supported in part by Grant-in-Aid #02402035 for Scientific Research (A), and #04246105 for Scientific Research on Priority Areas on "Higher-Order Brain Functions", both from the Ministry of Education, Science and Culture of Japan.

References

1. K. Fukushima: "Neocognitron: A self-organizing neural network model for a mechanism of pattern recognition unaffected by shift in position". *Biological Cybernetics* 36[4], 193–202 (April 1980).
2. K. Fukushima, S. Miyake: "Neocognitron: A new algorithm for pattern recognition tolerant of deformations and shifts in position". *Pattern Recognition* 15[6], 455–469 (1982).
3. K. Fukushima: "Neocognitron: A hierarchical neural network capable of visual pattern recognition". *Neural Networks* 1[2], 119–130 (1988).
4. K. Fukushima, N. Wake: "Handwritten alphanumeric character recognition by the neocognitron". *IEEE Trans. on Neural Networks* 2[3], 355–365 (May 1991).
5. N. Wake, K. Fukushima: "Improved learning method for the neocognitron", (in Japanese). *IEICE Transactions, D-II* (Inst. Electronics, Information, Commun. Engineers, Japan) J75-D-II, to appear (1992).
6. K. Fukushima, N. Wake: "Improved neocognitron with bend-detecting cells". *IJCNN'92-Baltimore*, Baltimore, MD, U.S.A., Vol. IV, 190–195 (June 1992).
7. K. Fukushima, N. Wake: "An improved learning algorithm for the neocognitron". In I. Aleksander, J. Taylor (eds.): *Artificial Neural Networks, 2*, Amsterdam: North-Holland, 1992, pp. 497–505.
8. K. Fukushima: "A feature extractor for curvilinear patterns: A design suggested by the mammalian visual system". *Kybernetik* 7[4], 153–160 (Sept. 1970).
9. K. Fukushima: "A feature extractor for a pattern recognizer — A design suggested by the visual system", (in Japanese). *NHK Technical Journal* 23[5], 351–367 (Sept. 1971).
10. K. Fukushima: "A neural network model for selective attention in visual pattern recognition". *Biological Cybernetics* 55[1], 5–15 (Oct. 1986).
11. K. Fukushima: "A neural network model for selective attention in visual pattern recognition and associative recall". *Applied Optics* 26[23], 4985–4992 (Dec. 1987).
12. K. Fukushima: "A neural network for visual pattern recognition". *Computer* (IEEE Computer Society) 21[3], 65–75 (March 1988).
13. K. Fukushima: "Neural network models for visual pattern recognition". In R. Eckmiller, G. Hartmann, G. Hauske (eds.): *Parallel Processing in Neural Systems and Computers*, Amsterdam: North-Holland, 1990, pp. 351–356.
14. K. Fukushima, T. Imagawa, E. Ashida: "Character recognition with selective attention". *IJCNN-91-Seattle*, Seattle, WA, U.S.A., pp. I-593–I-598 (1991).
15. T. Imagawa, K. Fukushima: "Character recognition in cursive handwriting with the mechanism of selective attention", (in Japanese). *Trans. IEICE* J74-D-II[12] 1768–1775 (Dec. 1991).
16. K. Fukushima, T. Imagawa: "Recognition and segmentation of connected characters with selective attention". *Neural Networks* 5, in print (1992).

Object Recognition by a Self-organizing Neural Network which Grows Adaptively

J. Weng, T. S. Huang, N. Ahuja

Beckman Institute
University of Illinois at Urbana-Champaign

Abstract. We describe a new type of neural network for object recognition which we call a Cresceptron. The term "Cresceptron" was coined from Latin *cresco* (grow) and *perceptio* (perception). The primary objective of the Cresceptron framework is to automatically handle manually intractable tasks: such as constructing a network that can recognize many objects from real world images. The Cresceptron uses a hierarchical structure, and the network adaptively and incrementally grows through learning. For recognition, the network is made largely translationally invariant by using the same neuron at all the positions of each neural plane. Scale invariance is achieved through a multi-resolution representation with the framework of visual attention. Limited orientational invariance is obtained by variation tolerance. Complete orientational invariance is not sought here since the recognition should report also the orientation. It is interesting to note that psychophysical studies have demonstrated that the human vision system does not have perfect invariance in either translation, scale, or orientation.

1 Structure

The Cresceptron has a pyramidal structure of several layers or modules. Each module contains a set of neural planes. When one goes up the pyramid, the resolution (or number of neurons) decreases but the complexity of the concepts (represented by the neurons) increases. In our current experiments, the network has 7 modules and the fovea resolution is 64 x 64 pixels. The inputs at the lowest module are 8 directional edge elements. As one goes up the pyramid, these edge elements are grouped into larger and larger structures.

2 Learning

In learning, the network is initialized to be empty. Then, a series of images are presented sequentially. From each image, the human operator selects the object to be learned from the input image by drawing a polyhedron that follows the outline of the object. Then, the network learns the object hierarchically. When a new concept is detected in a module, a new neuron with the synaptic connections is created. The input from the corresponding active neural planes is copied to the synaptic connections. A sigmoidal function determines an upper threshold such that if the same concept is presented, the output is saturated above. Since all the neurons at different positions of a neural plane are the same, only one neuron (with connections) needs to be created for each neural plane. Each training sample can cause many neurons to be created. Finally, if the sample is not recognized, a new node is created at the output layer of the highest level. A label that identifies the object is attached to this node so that later the node reports the label when it responds.

Over the entire network, knowledge sharing occurs naturally among different samples, since the same feature may appear in different samples. Therefore, the size increase of the network will gradually slow down, while sufficient low and intermediate level knowledge has been learned to cover most of the training samples and thus further new nodes will be mostly needed for only high levels. This implies that this network can learn a large number of objects. When the space reaches the equipment limit, the network "forgets" less frequently used or very "old" knowledge by deleting the corresponding nodes and thus keeps its size manageable.

3 Recognition

Once the network has been trained, it can be presented with new inputs for recognition. The result can be one of the following: (1) No output neuron responds: the new input is not recognized. (2) Only one output neuron responds: the input is recognized uniquely. (3) Two or more output neurons respond. The last case occurs when the input is similar to several learned samples. For example, the input contains a face that is taken at an orientation between those of the two learned samples. All the recognitions should be reported together with the response values as the confidence for further interpretation. For example, if the recognized objects indicate two different orientations of a face, a confidence weighted orientation sum can be used to approximately predict the actual viewing angle of the current face.

4 Segmentation

Once an object is recognized, the network can identify the location of the recognized object from the image. This is done by back tracking the response paths of the network from top level down to the lowest level. All the edges that have contributed to the recognition are marked in the input image. A closure can be easily computed from these marked edges to give the segmented region.

5 Implementation and Results

The objective of the network is to locate and recognize general objects in real world scenes, including manmade objects such as buildings, furniture and machine parts, and natural objects such as trees, animals, and human faces. The final implementation of the system will consist of a front-end graphics-supported workstation and a fast massively parallel neurocomputer which is connected with the front end. The neurocomputer should be able to perform recognition tasks in real time.

For the theoretical and algorithmic development, this system is currently being simulated on a SUN SPARC workstation, and an interactive user interface has been developed which allows effortless training and examination of the network. The system digitizes video signals into (512 x 512)-pixel images or directly accepts digital images of resolution up to 512 x 512. As mentioned earlier, the first version of the system uses the directional edges in the image as the input to the network. From each input image, the neural network accepts 8 edge images, each of which records the zero-crossings of the directional derivative of the Gaussian smoothed image along one of the 8 possible directions.

A few dozen sample images digitized from live TV programs have been used for training. A neural network has been automatically created through learning of these images. It has a total of about 70,000 synaptic connections. The system successfully recognizes the trained human figures in the sample images and can distinguish different objects. A few figures have been scaled slightly differently and been positioned at different places in the fovea, and they can still be recognized correctly. When more and more instances have been learned, the system is able to recognize objects more precisely by confidence-weighted interpretation. Work is underway to use a much larger set of sample images and collect statistical data about the performance and the size of the network.

Acknowledgement: This work was supported by National Science Foundation Grant IRI-89-02728.

INTELLIGENT PATTERN RECOGNITION and APPLICATIONS
— A Tutorial

P. S. P. Wang
College of Computer Science
Northeastern University, Boston, MA 02115

How do people learn and recognize things? These amazing capability has been taken for granted for years. It is until recently, when one tries to use computers or machines to do things like recognizing handwritten characters or faces, it becomes clear that such seemingly trvial tasks by human being turns out extremely difficult, if not impossible, by mechanical means such as computers. After decades of rigorous attacks, such research is still as fresh as ever, and such mystery as for how human beings can do it remains largely unknown but just began to unfold. In a sense, "human brain" is the "smartest" or the "most intelligent" mechanism than any computer can provide. While no one knows exactly the detailed, sophisticated organism in human brains, the best way one can do, is perhaps to "simulate" what "might" be going on. In a way, the study of *pattern recognition* and *artificial intelligence* serves the purpose. This lecture intends to get some inside views of pattern recognition techniques using artificial intelligence (AI) and neural network (NN) methodologies, and its applications to a lot of interesting and important problems including optical character recognition (OCR), zipcode recognition, bank check recognition, industrial parts inspection, scene analysis and image understanding. computer vision, and learning.

Introduction

For the past few decades, there is a growing interest in the study of artififcial intelligence and rule-based expert systems. Pattern recognition plays an important role in such systems. In fact, there is now much interaction between expert systems and pattern analysis. It is interesting to see that the core of pattern recognition, in cluding "learning techniques" and "inference" also plays an important and central role in artificial intelligence. Visual perception, scene analysis, and image understanding are also essential to robotic vision. On the other hand, the methods in artificial intelligence such as knowledge representation, semantic networks, and heuristic searching algorithms can also be applied to improve the pattern representation and matching techniques in many pattern recognition problems – leading to "smart" pattern recognition. Moreover, the recognition and understanding of sensory data like speech or images, which are major concerns in pattern recognition, have always been considered as important subfields of artificial intelligence.

This lecture covers the following subtopics:

(1) Overview of Pattern Recognition (PR) [including work done at MIT]

(2) Overview of Artificial Intelligence (AI) [including wrok done at MIT]

(3) The Relation Between PR and AI
Concentrating on Learning

(4) The Concepts of Learning and Inferencing

Supervised vs Non-supervised

(5) The Four Main Approaches to PR
Statistical (Classical, Decision-Theoretical)

Syntactical (Linguistical, Grammatical)
Structural, and
Histogram

(6) Parallelism and Some Multi-Dimensional Models for PR
Examples and Applications of Parallel Array Grammars and Others

(7) Degrees of Recognizability, Learnability, Understandability and Ambiguity

(8) Knowledge Representation and Semantic Networks for PR

(9) Neural Networks and Character Recognition

(10) An Example : Line-Drawing PR and 3-D Object Recognition
BM Method vs Extended Freeman Chain Code(EFCC) vs Improved EFCC(IEFCC);
3-dimensional vs 2-dimensional

(11) Another Example: Knowledge Pattern Representation of Chinese Characters
Hierarchical Structure, Induced Knowledge, Syntax-Semantics Correlation, Com-
mon Patterns and Logical Relations Between Characters, and New Character Principle

The following bibliography is useful in understanding this lecture.

1. C.H.Chen (ED), "Special Issues of IEEE Workshop on Expert Systems and Pattern Analysis", *IJPRAI*, v1, n2, 1987

2. K.S.Fu, *Syntactic Pattern Recognition and Applications*, Prentice-Hall, 1982

3. P. Winston, *Artificial Intelligence*, Addison-Wesley, 1992 (3rd edition)

4. S.L.Tanimoto, *The Elements of Artificial Intelligence*, CSP, 1991

5. P.S.Wang, "Knowledge Pattern Representation of Chinese Characters", *IJPRAI*, v2,n1, 161-179, 1988

6. H.Bunke and P.S.Wang (eds), "Editorial", *IJPRAI*, v1,n1, 1987

7. H.Lu and P.S.Wang, "A Comment on "A Fast Parallel Algorithm for Thinning Digital Patterns" ", *CACM*, v29,n3, 239-242, 1986

8. P.S.Wang and Y.Y.Zhang, "A Fast and Flexible Thinning Algorithm", *IEEE Trans. on Computers*, 1989

9. P.S.Wang, "Hierarchical Structures and Complexities of Parallel Isometric Patterns", *IEEE-PAMI*, v5,n1, 92-99, 1983

10. P.S.Wang, "An Application of Array Grammars to Clustering Analysis for Syntactic Patterns", *Pattern Recognition*, v17,n4, 441-451, 1984

11. P.S.Wang, "A New Character Recognition Scheme with Lower Ambiguity and Higher Recognizability", *Pattern Recognition Letters*, 3, 431-436, 1985

12. P.S.Wang, "On-Line Chinese Character Recognition", *Proc. IGC 6th International Conference on Electronic Imaging*, 209-214, 1988

13. P.S.Wang, "A More Natural Approach for Recognition of Line-drawing Patterns", *SPIE*

Image Pattern Recognition, v755, 141-160, 1987

14. C.Cook and P.S.P.Wang, "A chomsky hierarchy of isotonic array grammars and languages", *Computer Graphics and Image Processing*, v8, 144-152 (1978)

15. S.Edelman, H. Bulthoff, D.Weinshall, *Stimulus familiarity determines recognition strategy for novel 3-D*, MIT AI Lab. Memo. 1138, July 1989.

16. T. Marill, *Computer perception of three-dimensional objects.* MIT AI Lab Memo. 1136, August 1989.

17. T. Marill, *Recognizing three-dimensional objects without the use of models.* MIT AI Lab. Memo. 1157, September 1989.

18. T. Marill. "Emulating the human interpretation of line-drawings as 3-d objects", IJCV, v6-2, (1991) 147-161

19. A. Rosenfeld, " Preface ", *Array grammars, patterns and recognizers*, P.S.P. Wang(Ed.), World Scientific Publishing Co. (WSP), 1989.

20. A. Rosenfeld, *Picture languages: formal models for picture recognition*, Academic Press, New York, 1979

21. R. N. Shepard and J.Metzler, "Mental rotation of 3-D objects" *Science* 171, pp.701-703(1971).

22. R.Siromoney, "Array language and Lindenmayer systems- a survey", *The Book of L*, G. Rozenberg and A. Salomma (ed), Springer Verlag, 1986

23. S.Ullman, *An approach to object recognition: aligning pictorial descriptions.* MIT AI Lab. Memo. 931, Dec. 1986.

24. P.S.P. Wang(Ed.), *Array grammars, patterns and recognizers World Scientific Publishing Co. (WSP)*, 1989.

25. P.S.P.Wang, "3-D Object Representation by Array Grammars", *SPIE Intelligent robots and computer vision IX: algorithms and techniques(1990)*, v.1381, 210-216 (1990)

26. P. Winston with S. Shellard (ed), *Artificial Intelligence at MIT - Expanding Frontiers*, MIT Press, 1990

27. K. Sugihara, *Machine interpretation of line drawings*, MIT Press, Cambridge (1986)

28. C.Y.Suen, "Foreword", *Character and Handwriting Recognition*, P.S.P. Wang (ed), WSP, 1991

29. H.Bunke and A. Sanfeliu(ed), *Syntactic and Structural Pattern Recognition - Theory and Applications*, WSP (1990)

30. P.S.P.Wang(ed), *Character and Handwriting Recognition - Expanding Frontiers*, WSP (1991)

31. S.Impedovo, L.Ottaviano and S. Occhinegro, *Optical Character Recognition - A Survey*, IJPRAI, WSP, v5 n1&2, (1991)

32. S.Impedovo, J.S.Simon and H.L.Teulings (ed), *Proceedings of International Workshop on Frontiers in Handwriting Recognition (IWFHR)*, ELSVIER (1992)

33. P.S.P.Wang, M.V.Nage and A.Gupta, *A New Neural Net Based Hybrid Approach to Handwritten Numeral Recognition*, Proc. IWFHR'91, (1991)

34. I. Guyon, *Applications of Neural Nets to Character Recognition*, IJPRAI v5, n1&2, (1991)

35. I.Guyon and P.S.P.Wang (ed), *Neural Networks and Character Recognition*, WSP, to appear in 1993

36. A. Bischoff and P.S.P.Wang, "Handwritten digit recognition using neural networks" , SPIE Conf on Intelligent Robotics and Computer Vision, 1991, 436-447

37. L.Baird and P.S.P.Wang, "3-D Object Recognition using Gradient Descent and Array Grammars", SPIE Conf on Intelligent Robotics and Computer Vision, 1991, 711-719

38. P.S.P.Wang, C.H.Chen and L. Pau,(ed) *Handbook of Computer Vision and Pattern Recognition.* WSP, NJ (1993) to appear

39. P.S.P.Wang and I. Gyuon,(ed), *Advances in Neural Network and Pattern Recognition*, WSP, NJ, (1993) to appear

Path-Controlled Graph Grammars
for
Syntactic Pattern Recognition

Kunio Aizawa[1] and Akira Nakamura[2]

[1]Department of Applied Mathematics, Hiroshima University
Higashi-Hiroshima, 724 Japan

[2]Department of Computer Science, Meiji University
Kawasaki, Kanagawa, 214 Japan

Abstract. The graph structure is a strong formalism for representing pictures in syntactic pattern recognition. Many models for graph grammars have been proposed as a kind of hyper-dimensional generating systems, whereas the use of such grammars for pattern recognition is relatively infrequent. One of the reason is the difficulties of building a syntax analyzer for such graph grammars. In this paper, we define a subclass of nPCE graph grammars and present a parsing algorithm of O(n) for both sequential and parallel cases.

1 Introduction

The graph structure is a strong formalism for representing pictures in syntactic pattern recognition. Many models for graph grammars have been proposed as a kind of hyper-dimensional generating systems (see e.g., [1-3]), whereas the use of such grammars for pattern recognition is relatively infrequent. One of the reason are the difficulties of building a syntax analyzer for such graph grammars. In [4], Flansinski introduce a parsing algorithm of time complexity $O(n^2)$ for a subclass of edNLC graph grammars.

In this paper, we define a subclass of nPCE graph grammars and present a parsing algorithm of O(n) for both sequential and parallel cases. Originally, nPCE graph grammars are introduced in [5] for describing uniform structures. These grammars utilize partial path groups to define their embedding function. By using such embedding mechanism, nPCE graph grammars can generate various uniform structures, e.g., rectangular-, hexagonal-, triangular-arrays, trees, hypercubes, etc., with only change the properties of path group used in the embedding functions. So they are useful for describing pictures on digital spaces.

In Section 2, some basic definitions of IE graphs are reviewed. In Section 3, a subclass of nPCE graph grammars, $L(1)$-$nPCE_4$ graph grammars, which generate IE-graphs is defined and its generative power is investigated. In Section 4, a parsing algorithm of time complexity $O(n)$ for $L(1)$-$nPCE_4$ graph grammars is introduced, where n is the number of nodes in the parsed IE-graph. In Section 5, a parallel version of $L(1)$-$nPCE_4$ graph grammars is introduced and its generative power is investigated. It is also shown that the parsing algorithm for $L(1)$-$nPCE_4$ graph grammars can be used for their parallel version.

2 Basic Definitions

In this section, we review the definitions of EDG-graphs and its indexed version (IE-graphs) which are introduced in [6] and [4].

Definition 2.1. A *directed node- and edge-labelled graph (EDG-graph)* over Σ and Γ is a quintuple $H = < V, E, \Sigma, \Gamma, \varphi >$, where V is the finite, nonempty set of nodes, Σ is the finite, nonempty set of node labels, Γ is the finite nonempty set of edge labels, E is the set of edges of the form $< v, \lambda, w >$, where u, w $\in$ V, $\lambda \in \Gamma$, φ: $V \to \Sigma$ is the node labelling function.

Let us take a set of edge labels as shown in Fig. 1 representing "RIGHT", "LEFT", "UP", and "DOWN", and ordered -h $\leq$ -v $\leq$ h $\leq$ v.

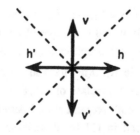

Fig. 1. The set of edge labels.

Definition 2.2.   An EDG-graph H is called an *OS-graph* if

(1)   for each $\lambda \in \Gamma$ there exists an inverse edge label $\lambda^{-1} \in \Gamma$,

(2) Γ is simply ordered by a relation $\leq$,

(3)   for each $v \in V$, if there exists $< v, \lambda, w > \in E$ then there does not exist $< v, \gamma, z >$ such that $\lambda = \gamma$ or $< z, \beta, v > \in E$ such that $\lambda^{-1} = \beta$.

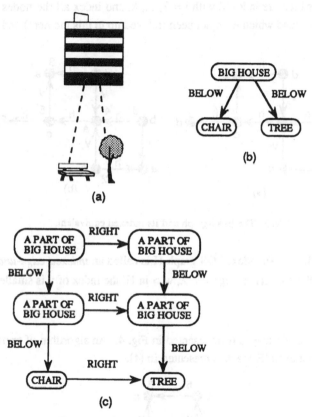

Fig. 2. The scene representation by EDG-graphs.

Note that the third condition means that a node of OS-graph cannot be connected with two other nodes by edges with the same labels. It may be omitted when we consider multi-resolution graph grammars. For example, the picture shown in Fig. 2a can be represented by a graph in Fig. 2b. In this case, two edges having the same label "BELOW" emerge from the node labelled "BIG HOUSE" to two nodes labelled "TREE" and "CHAIR". If we use higher resolution graph, Fig. 2a can be represented by an OS-graph in Fig. 2c. So, it can be considered that the graph in Fig. 2b is obtained by contracting four nodes labelled with "A PART OF BIG HOUSE" to one node labelled with "BIG HOUSE".

Now we discriminate the so-called *S-node* which corresponds tp the lowest leftmost object of a picture represented by a OS-graph. Then we index an S-node v_0 of a graph with 1, and we index all the nodes which are adjacent to v_0 with the help of a relation $\leq$ according to an increasing order $i = 2, ..., k$. After that, we successively choose nodes which are indexed with $i = 2, ..., k$, and index all the nodes which are adjacent to them (and which have not been indexed up to this moment) and so on (see Fig. 3a-b).

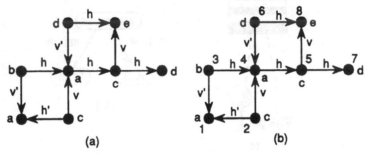

(a)   (b)

Fig. 3. The EDG-graph and its indexed equivalent.

Definition 2.3.   An indexed OS-graph H is called an *indexed edge-unambiguous (IE-graph)*, if there exists an edge $< v, \lambda, w >$ in H, the index of v is smaller than that of w and $\lambda^{-1} \leq \lambda$.

An example of the IE-graphs is represented in Fig. 4. An algorithm of transformation of an OS-graph into an IE-graph is presented in [4].

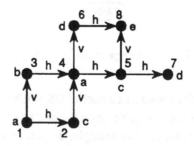

Fig. 4. The indexed edge-unambiguous OS-graph

At the end of this section we introduce a string description on an IE-graph. This string description for a graph is introduced in [7].

Definition 2.4. Let n_k be a node (having index k and a label n) of an IE-graph. A *characteristic description of the node* n_k is a quadruple $<n_k, r, (e_1...e_r), (i_1...i_r)>$, where r is the out-degree of n_k, $(i_1...i_r)$ is the string of node's indices, to which edges going out from n_k come into, $(e_1...e_r)$ is the string of edge labels ordered in such a way that the edge labelled with e_j comes into the node indexed with i_j.

Definition 2.5. For a given IE-graph $H = < V, E, \Sigma, \Gamma, \varphi >$, where $V = \{n_1, ..., n_k\}$, and for each I_i (i = 1, ..., k) which is a characteristics description of a node n_i, the string $I_1...I_k$ is called a *characteristic description of a graph H*.

a_1	c_2	b_3	a_4	c_5	d_6	d_7	e_8	
2	1	1	2	2	1	0	0	is the characteristic description
h v	v	h	h v	h v	h	—	—	
2 3	4	4	5 6	7 8	8	—	—	

of the graph shown in Fig. 4.

3 Generative Power of PCE Grammars Generating IE-Graphs

In this section, we review the definitions of nPCE graph grammars and define their restricted versions. These definitions of nPCE graph grammars are slightly different from the original version [5] since the definitions of the path group are different. Then we examine their powers for generating IE-graphs. At first we review the definitions of the path groups describing the square grid [8].

Definition 3.1. A *discrete space* is a finitely presented abelian path group $\Gamma = (X/D)$, where X has 2n generators $s_1, s_2, ..., s_n, s_1^{-1}, s_2^{-1}, ..., s_n^{-1}$, and D contains all relations other than the commutativity $(s_i s_j s_i^{-1} s_j^{-1} = 1)$ and the inverse iterations $(s_i s_i^{-1} = 1)$. The *square grid* is a discrete space described by a four generators $s_1 = $ (above), $s_2 = $ (right), $s_1^{-1} = $ (below), $s_2^{-1} = $ (left), and $D = \emptyset$.

Note that the path groups defined above can also be defined on a graph generated by a graph grammar by regarding the edge labels of the generated graph as the generators.

Definition 3.2. For any given OS-graph H, its node P, and a string $\pi = \{c_1 c_2...c_i\}$ of its edge labels, $P\pi$ is *realizable on H* if and only if there exists a set of nodes $\{P_0, P_1, ..., P_i\}$ such that $P_0 = P$ and $< P_j, c_j, P_{j-1} >$ or $< P_{j-1}, c_j^{-1}, P_j >$ is an edge of H $(1 \le j \le i)$.

We review briefly the definition of nPCE grammars using abelian path groups to control embedding mechanism.

Definition 3.3.   *A node-replacement graph grammar using path controlled embedding with 4 generators abelian path groups, ($nPCE_4$ grammar), is a* construction $G = < \Sigma_N, P, \psi, Z, \Delta_N, \Delta_E >$, where Σ_N is a finite nonempty set of node labels, Δ_N is a finite nonempty subset of Σ_N, called terminal node labels, $\Delta_E = \{h', v', h, v\}$, called terminal edge labels, P is a finite set of productions of form (v_a, β), where v_a is a graph consisting of only one node labelled with a $\in \Sigma_N$, β is a connected OS-graph, ψ is a mapping from Δ_E^+ into Δ_E provided that for any $\pi \in \Delta_E^+$, ψ maps π into c, the first lebel of π, i.e., there exists a $\sigma \in \Delta_E^*$ such that $\pi = c\sigma$, Z is a connected OS-graph over (Σ_N, Δ_E) called *the axiom.*

A direct derivation step of a $nPCE_4$ grammar G, $\Rightarrow_G$, is performed as follows:

Let H be an OS-graph.   Let $p = (v_a, \beta)$ be a production in P.   Let β' be isomorphic to β (with h an isomorphism from β' into β), where β' and $H-v_a$ have no common nodes.   Then the result of the application of p to H (by using h) is obtained by replacing v_a with β' and adding edges $< u, \lambda, w >$ between every nodes u in β' and every w in $H-v_a$ such that if the path from node 1 to node u on β' is σ then w is the node of H defined by $v_a c\sigma$ or its equivalent path under abelian path group with four generators and if $\psi(c\sigma) = h$ or v then the added edges are $< u, \psi(c\sigma), w >$, otherwise $< w, \psi(c\sigma)^{-1}, u >$, where c is an element of Δ_E.

We will denote the reflexive and the transitive closure of $\Rightarrow_G$ by $\Rightarrow_G^*$ and the transitive closure of $\Rightarrow_G^+$. The language of G, denoted as L(G), is defined by $L(G) = \{ H \mid H$ is an OS-graph over (Δ_V, Δ_E) and $Z \Rightarrow_G^* H \}$.

We here present an example of the derivations of $nPCE_4$ grammars.

Example 3.1. The application of the production in Fig. 5a to the graph in Fig. 5b results the graph in Fig. 5c.

One of the advantages to introduce the partial path groups into the path controlled embedding mechanism is flexibility of describing various structures. As mentioned above, two-dimensional square arrays correspond to abelian group with 4 generators,

43

and other types of arrays (triangular, hexagonal) correspond to other types of abelian groups. For detail discussions of those structures, see [8]. In [9], Rosenfeld defined a different type of group structure on a graph which are used to define the original version of nPCE graph grammars. The groups structure for more complicated structures are discussed in [10]. We define here the restricted version of nPCE$_4$ grammar in almost the same way of L(1) grammars in [11].

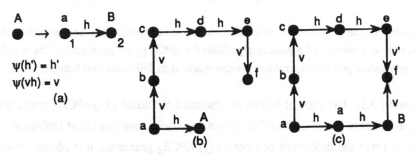

Fig. 5. An example of a direct derivation step of nPCE$_4$ graph grammars.

Definition 3.4. Let H = < V, E, Σ, Γ, φ > be an IE-graph. S-node is called the node of the *first level*. Node v is called the node of *n-th level* if
(1)   there exists < w, λ, v > ∈ E such that w is the node of the (n-1)-th level,
(2)   for each < u, λ, v > ∈ E or < v, λ, u > ∈ E, u is a node at least of (n-1)-th level.

Definition 3.5. Let G = < Σ$_N$, P, ψ, Z, Δ$_N$, Δ$_E$ > be an nPCE$_4$ grammar. Then G is called the L(1)-nPCE$_4$ grammar, if
(1)   P is a finite set of production of the form $(v_a, β)$, where β is an IE-graph

having a characteristic description:
$$\begin{array}{cccc} Y_1 & Y_2 & \cdots & Y_m \\ r_1 & r_2 & \cdots & r_m \\ E_1 & E_2 & \cdots & E_m \\ I_1 & I_2 & \cdots & I_m \end{array} \quad or \quad \begin{array}{c} Y_1 \\ 0 \\ - \\ - \end{array} \quad , where \quad \begin{array}{c} Y_i \\ r_i \\ E_i \\ I_i \end{array}$$
is a characteristic description of the node Y_i, i = 1, ..., m, Y_1 is a terminal node, Y_i, i = 2, ..., m is the node of the second level.
(2)   There are no two production rules having the same left-hand side and the same label of the node indexed by 1.
(3)   Z is an IE-graph having a characteristic description which satisfies conditions of (1).

Definition 3.6. An nPCE$_4$ grammar is called *closed* if, for each derivation processes of this grammar, Z = g_0 ⇒ g_1 ⇒ ... ⇒ g_n, a graph g_i, i = 0, ..., n is the IE-graph.

Definition 3.7. Let a derivation of a closed $L(1)$-$nPCE_4$ grammar be given by: $Z = g_0 \Rightarrow g_1 \Rightarrow ... \Rightarrow g_n$. This derivation is called the *regular left-hand side derivation*, if

(1)   for each $i = 0, ..., n-1$, a production rule is applied to the node having the least index in the graph g_i.

(2)   node indices are not changed during the derivation.

As we will see in the next section, the restrictions defined above dramatically reduce the time complexity of a parsing algorithm for nPCE graph grammars. These effects on generative power of grammars are summarized as following two lemmas.

Lemma 3.1. The class of IE-graphs generated by closed $L(1)$-$nPCE_4$ grammars is equal to that of closed $L(1)$-$nPCE_4$ grammars without condition (1) of Definition 3.5.

Proof: From the definitions of closed $L(1)$-$nPCE_4$ grammars, it is obvious that the class of IE-graphs generated by closed $L(1)$-$nPCE_4$ grammars is contained in the class generated by closed $L(1)$-$nPCE_4$ grammars without condition (1). Therefore we will show that for any given closed $L(1)$-$nPCE_4$ grammars $G = <\Sigma_N, P, \psi, Z, \Delta_N, \Delta_E>$ without condition (1), we can construct closed $L(1)$-$nPCE_4$ grammars $G' = <\Sigma_N', P', \psi', Z', \Delta_N', \Delta_E'>$ such that $L(G) = L(G')$. First of all, we define $\Delta_N' = \Delta_N$, $\Delta_E' = \Delta_E$, $\psi' = \psi$. For any production rule $p = (v_a, \beta)$ of G, let β contain the nodes of the n-th level. All of these nodes must be connected directly to the nodes of the (n-1)th level. For each such the (n-1)th level node v_i, determine the graph α_i consisting of v_i and the n-th level nodes which are directly connected to v_i. For a node of the n-th level, if there exist more than one such node of the (n-1)th level, one of them is picked out arbitrarily. Because the ignored edges at this step are added by ψ automatically in the derivation process of G'. For each such graph α_i, we introduce new nonterminal node label X_i, and replace α_i in β by one node labelled with X_i. Thus, such β should contain at most the nodes of the (n-1)th level. Then for each α_i, the production rule (v_{X_i}, α_i) is added to P'. By repeating this process for n-2 times, rule p of G is decomposed to rules whose right-hand sides have the nodes of at most the second level. Next, for each such decomposed rule, let the node of right-hand side indexing by 1 be labelled with a nonterminal node label A. Then each such node is rewritten at most four times except A→B type rules since G' generates the set of IE-graphs as its language. Moreover such nonterminal nodes must finally be rewritten to terminal nodes. So it is not so difficult that $L(G) = L(G')$.

Example 3.2. The production rule of Fig. 6a is decomposed to the three rules of Fig. 6b.

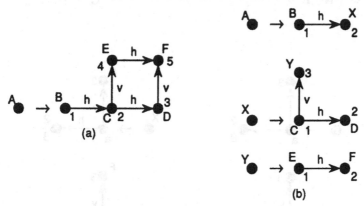

(a)

(b)

Fig. 6. An example of rule decomposition.

Example 3.3. The production rules of Fig. 7a are translated to two rules of Fig. 7b satisfying the Condition (1) of Definition 3.5.

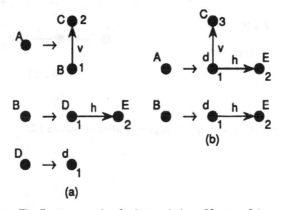

(a)

(b)

Fig. 7. An example of rule translation of Lemma 3.1.

To examine the effect of Condition (2) of Definition 3.5, we need one more definition of the finite axiom grammars.

Definition 3.8. A *finite axiom $L(1)$-$nPCE_4$ grammars, $(FL(1)$-$nPCE_4$ grammars)*, is a construction $G = < \Sigma_N, P, \psi, F, \Delta_N, \Delta_E >$, where $\Sigma_N, P, \psi, \Delta_N, \Delta_E$ are the same as in the definition of $L(1)$-$nPCE_4$ grammars, F is the finete set of IE-graphs having characteristic descriptions which satisfy Condition (1) of Definition 3.5.

The *language* of an FL(1)-nPCE$_4$ grammar G is defined as L(G) = { H | H is an IE graph over (Δ_N, Δ_E) and there exist a Z in F such that Z $\Rightarrow_G^*$ H }.

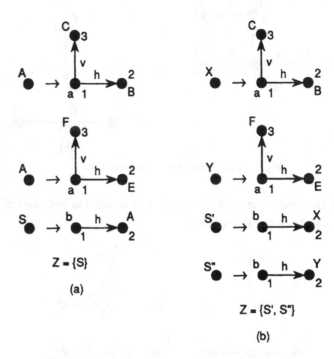

Fig. 8. An example of edge translation of Lemma 3.2.

Lemma 3.2. The class of IE-graphs generated by closed FL(1)-nPCE$_4$ grammars is equal to that of closed L(1)-nPCE$_4$ grammars without condition (2) of Definition 3.5.
Proof: For any given finite set of axiom { Z_1, ..., Z_n }, if we add the new production rules (v_S, Z_1), ..., (v_S, Z_n) and set the single node graph labelled with S as the axiom, it is obvious that the class of FL(1)-nPCE$_4$ grammars is contained in the class of L(1)-nPCE$_4$ grammars without Condition (2). Therefore we will show that for any given closed L(1)-nPCE$_4$ grammars G = < Σ_N, P, ψ, Z, Δ_N, Δ_E > without condition (2), we can construct closed FL(1)-nPCE$_4$ grammars G' = < Σ_N', P', ψ', Z', Δ_N',

Δ_E' > such that $L(G) = L(G')$. Assume here that for each production rule p of G, there exists no nonterminal symbol in the right-hand side which is same as that of left-hand side. By making use of additional nonterminal symbols, it is not so difficult to construct such grammar from given one. Firstly, we define $\Delta_N' = \Delta_N$, $\Delta_E' = \Delta_E$, $\psi' = \psi$. Then new nonterminal node label S is added to Σ_N' and new rule (v_S, Z) is added to P'. For any given rules $p_i = (v_a, \beta_i)$, i = 1, ..., n, of G which have the same left-hand side and the same label of the node indexed by 1, new nonterminal node labels $X_1, ..., X_n$ are added to Σ_N' and rule p_i is replaced by $p_i' = (v_{X_i}, \beta_i)$. For each rule q_j which has the right-hand side containing v_a, new nonterminal node label Y is added to Σ_N' and new rule $q_j' = (v_Y, \gamma_{j_i})$ is added to P', where γ_{j_i} is obtained by replacing v_a by v_{X_i}. By applying these steps to every set of rules which have the same left-hand side and the same label of the node indexed by 1, and adding each newly introduced nonterminal node label for S to the set of axiom, the language of G' is equal to that of G.

Example 3.4. The production rules of Fig. 8a and an axiom S are translated to four rules satisfying Condition (2) of Definition 3.5 and the set of axioms { S', S" }.

From Lemma 3.1 and 3.2, the next theorem is straightforward.

Theorem 3.1. The class of IE-graphs generated by closed FL(1)-nPCE$_4$ grammars is equal to that of closed nPCE$_4$ grammars.

4 A Parsing Algorithm for L(1)-nPCE$_4$ Grammars

In this section, an algorithm for parsing closed L(1)-nPCE$_4$ grammars which has time complexity of O(n) are presented, where n is the number of nodes of the parsed graph. In general, the connectivity of nPCE grammars defined by the embedding function ψ cannot be determined locally. However, in the case of closed L(1)-nPCE$_4$ grammars, the situation is entirely different. To show this, we need one more property of IE-graphs generated by closed L(1)-nPCE$_4$ grammars. By using almost the same arguments as in Theorem 4.1 of [11], it is easy to see the next lemma.

Lemma 4.1. Let G be a closed L(1)-nPCE$_4$ grammar. Then at any step $H_i \Rightarrow_G H_{i+1}$ of its derivation, node Y_1 of applied production rule

$$\begin{matrix} Y_1 & Y_2 & \cdots & Y_m \\ r_1 & r_2 & \cdots & r_m \\ E_1 & E_2 & \cdots & E_m \\ I_1 & I_2 & \cdots & I_m \end{matrix}$$

$A \rightarrow$ (the second row) receives the same index as the node A_k in the graph H_i.

Now, we define a parsing algorithm for closed $L(1)$-$nPCE_4$ grammars.

G is an analyzed graph,

H is a graph derived during parsing,

Z is an axiom of a closed $L(1)$-$nPCE_4$ grammar,

$v_G(i)$ is a label of the node of the graph G indexed by i,

$v_H(i)$ is a label of the node of the graph H indexed by i,

n is the number of nodes of the graph G.

Procedure: CHOOSE(A, a, k)

> if there exists a production rule such that A is the label of the left-hand side,
> and a is a label of the first level node of the right-hand side,
> then k := the number of such a production,
> else k := 0

Procedure: PRODUCTION(H, i, k)

> the application of k-th production rule for i-th node of the graph H.

> if H has j nodes, characteristic description of i-th node is $\begin{matrix} v_H(i) \\ r_H(i) \\ E_H(i) \\ I_H(i) \end{matrix}$, and the

> right-hand side of k-th rule is $\begin{matrix} Y_1 & Y_2 & \cdots & Y_m \\ r_1 & r_2 & \cdots & r_m \\ E_1 & E_2 & \cdots & E_m \\ I_1 & I_2 & \cdots & I_m \end{matrix}$, then

$v_H(i) := Y_1$,

$r_H(i) := r_H(i) + r_1$,

$E_H(i) :=$ concatenation of E_1 and $E_H(i)$,

$I_H(i) :=$ concatenation of I_1 and $I_H(i)$,

nodes $Y_2, ..., Y_m$ become the node indexed by $j+1, ..., j+m-1$ of H.

Function: INCLCON(G, H, i)

> the Boolean function checking whether the following conditions hold for the characteristic description of node i of H or not:

$v_H(i) = v_G(i)$,

$r_H(i) \le r_G(i)$,

each index of $I_H(i)$ is included in $I_G(i)$,

$E_H(i)$ is the restriction of $E_G(i)$ to $I_H(i)$.

Algorithm 4.1.   The parsing algorithm for $L(1)$-$nPCE_4$ grammars.

    $H := Z$;

    err := false;

    for i := 1 **to** n **do**

            if err = false **then**

                begin

                    if $v_H(i)$ is nonterminal node **then**

                        begin

                            CHOOSE($v_H(i)$, $v_G(i)$, k);

                            if k = 0 **then** err := true

                                else PRODUCTION(H, i, k)

                      end;

                    if not INCLCON(G, H, i) **then** err := true

                end;

Before we show the validity of Algorithm 4.1, we examine the following example.

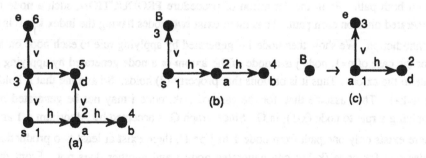

Fig.9. An example of a parsing step of Algorithm 4.1.

Example 4.1. Let us assume that we analyze the graph G in Fig. 9a. We assume here that in some step of a parsing we have obtained the graph H shown in Fig. 9b. Then after the application of rule in Fig. 9c to the node indexed by 3, we get a characterristic description

s_1	a_2	c_3	b_4	d_5	e_6
2	1	2	0	0	0
h v	h	h v	—	—	—
2 3	4	5 6	—	—	—

and G is accepted by Algorithm 4.1 because all characteristic description of nodes in H satisfy the condition of procedure INCLCON. However, H is not equal to G since an edge from a_2 to d_5 is missing. Such missing edges are not an edge of the right-hand side of a production rule but are inserted by a embedding function ψ according to derivation steps. So they can be ignored in parsing processes.

Lemma 4.2. Let an IE-graph G be accepted by Algorithm 4.1. Then the characteristic description $\begin{array}{c} Y_1 Y_2 \cdots Y_n \\ r_1 \ r_2 \cdots r_n \\ E_1 E_2 \cdots E_n \\ I_1 \ I_2 \cdots I_n \end{array}$ of graph H has the following properties:

(1)   For each node j, ($2 \leq j \leq n$), there exists one and only one path π of H from node 1 to j. It is also a path of G.

(2)   Let $p_{0_j}(=Y_1)$, p_{1_j}, p_{2_j}, ..., $p_{n_j}(=Y_j)$ be nodes on π, then node j is obtained if and only if a production rule of the considered grammar is applied to each node on π, i.e., there exists no other path in G generating node j.

Proof: From the definition of procedure PRODUCTION, it is obvious that H is connected. So there exist a path from node 1 to j, and from the definition of procedure INCLCON, it is also a path of G. Assume that there exists another path of H from 1 to j. Then there exists at least one node k_j other than 1 (it may be node j itself) which is on both path. From the definition of procedure PRODUCTION, such a node is generated once on each path. Thus there exist two nodes having the index k_j. It is a contradiction. We show then node j is generated by applying rule to each node on π. For the case of n=1, node j is a node of the axiom or a node generated by applying a rule to the axiom. Thus it is obvious that property (2) holds. So assume that it holds to n=k-1. Then assume that, for the case of n=k, node j may not be generated by applying a rule to node $(k-1)_j$ in G. Since graph G is accepted by Algorithm 4.1 and there exists only one path from node 1 to j on H, there exist at least two production rules one for node $(k-1)_j$: one generates node j and another does not. From the assumption that node $(k-1)_j$ is generated from node $(k-2)_j$, these two rule must have same left-hand side and same label of the node indexed by 1. It is not allowed for L(1)-nPCE$_4$ grammars (see Condition (2) of Definition 3.5).

From the arguments of Lemma 4.1 and 4.2, it is not so difficult to see that the following theorem for the validity of Algorithm 4.1.

Theorem 4.1.   For any given IE-graph G and closed $L(1)$-$nPCE_4$ grammar with the regular left-hand side derivation, G is an element of $L(S)$ if and only if G is accepted by Algorithm 4.1.

The parsing algorithm presented in this section is a deterministic top-down syntax analyzer for closed $L(1)$-$nPCE_4$ grammars. It has time complexity of $O(n)$, where n is the number of nodes of considered graph because each procedure used in Algorithm 4.1 has time complexity of constant and there exists only one for-loop repeats n times.

A parsing algorithm for closed $FL(1)$-$nPCE_4$ grammars is quite straightforward since it is sufficient to apply Algorithm 4.1 for each axiom of given grammar. So it also has time complexity of $O(n)$ for large n.

5 Parallel Generation on $L(1)$-$nPCE_4$ Grammars

In this section, we extend nPCE grammars to the parallel systems, i.e., graph L-systems or GL-systems. Then a relationship between closed $L(1)$-$nPCE_4$ GL-systems and closed $L(1)$-$nPCE_4$ grammars is investigated.

Definition 5.1.   An extended *nPCE graph L-system with 4 generators abelian path groups*, (*EnPCE_4 GL-system*), is a construction $G = <\Sigma_N, P, \psi, Z, \Delta_N, \Delta_E>$, where all of its components are same as in the definition of $nPCE_4$ grammars.

The derivation steps of $EnPCE_4$ GL-system proceed in the almost same way as any other L-systems, i.e., all nodes are rewritten in parallel within a direct derivation step. A direct derivation step in an $EnPCE_4$ GL-system S, $\Rightarrow_{S_p}$, is performed as follows:

Let H be an OS-graph. Let $\{v_1, v_2, ..., v_n\}$ be the set of nodes of H and $p_1, p_2, ..., p_n$ are production rules applicable to $v_1, v_2, ..., v_n$, respectively, where $p_i = (a_i, \beta_i, \psi_i)$, $1 \leq i \leq n$. Let β_i' be isomorphic to β_i (with h_i an isomorphism from β_i' into β_i), where β_i's have no common nodes. Then the result of the application of p_is to H (by using h_is) is obtained by first removing v_is from H, then replacing v_i with β_i' and finally adding edges $< u, \lambda, w >$ between every nodes u in β_i' and every w in β_j', $i \neq j$, such that if the path from node 1 to node u of β_i' is σ_i, then w is either a node defined by $v_i c \sigma_i$ under abelian path group with four generators or a node defined by node 1 of β_j' defined by $v_i c$ and a path σ from node 1 to w in β_j' under abelian path group with four generators and if $\psi(c\sigma)$ = h or v then the added edges are $< u, \psi(c\sigma), w >$, otherwise $< w, \psi(c\sigma)^{-1}, u >$, where c is an element of Δ_E.

The L(1)-, closed properties for $EnPCE_4$ GL-systems can be defined in the same way as $nPCE_4$ grammars.

As mentioned in the statement of the following theorem, the generative power of closed L(1)-$EnPCE_4$ GL-systems is equal to closed L(1)-$nPCE_4$ grammars with regular left-hand side derivation. So, by introducing parallelism in L(1)-$nPCE_4$ grammars, the conditions of regular left-hand side derivation can be omitted.

Theorem 5.1. The class of IE-graphs generated by closed L(1)-$EnPCE_4$ GL-systems is equal to that of closed L(1)-$nPCE_4$ grammars with regular left-hand side derivation.

Proof: We first show that for any given closed L(1)-$EnPCE_4$ GL-systems S there exists a closed L(1)-$nPCE_4$ grammars G with regular left-hand side derivation such that L(S)=L(G). We will show that at each derivation step of S, the levels of all nonterminal nodes equal to each other. For the case of the first and the second levels, obviously this property holds because these nonterminal nodes are either nodes of axiom or generated by applying a rule to the axiom. Assume that it holds for the (k-1)th level. Obviously each node generated by applying production rules to nonterminal nodes of the (k-1)th level is at most k-th level. Assume that there exist a node v of less than k-th level. Then there must exist an edge e from a node w of less than (k-1)th level to v. From the assumption of induction, w is a terminal node and no rule is applied to w at this step. So such edge must be defined by the embedding function ψ rather than a production rule. Since w is a node of at most (k-2)th level, the length of the path from v to w other than e is at least two and the path is defined by e.g., $\alpha^{-1}\beta^{-1}$ for some α and β such that $\alpha^{-1}\leq\alpha$ and $\beta^{-1}\leq\beta$ in the case of two. Thus β is the path from node 1 to v in the right-hand side of a rule and from the definition of ψ, the path defined by $\alpha^{-1}\beta$ must be a pth from v to w. However from the definition of abelian path group, $\alpha^{-1}\beta^{-1}$ is not equivalent to $\alpha^{-1}\beta$. This is a contradiction. Therefore, it is easy to see that G can simulate a direct derivation step of S by regular left-hand side derivation. Conversely, for any step of derivation of G and any level n, G must apply rules to every nonterminal node of n-th level before it applies a rule to a nonterminal node of (n+1)th level, because of the definition of regular left-hand side derivation. So it is not so difficult to see that S can simulate the derivation steps of G for the nonterminal nodes of a level by single step.

From the results of Theorem 4.1 and 5.1, the next corollary is straightforward.

Corollary. For any given IE-graph G and closed $L(1)$-EnPCE$_4$ GL-system S, G is an element of $L(S)$ if and only if G is accepted by Algorithm 4.1.

References

[1] Ehrig, H., M. Nagl and G. Rozenberg (ed.): *Graph-Grammars and Their Application to Computer Science*, Lecture Notes in Computer Science, 153, Springer-Verlag, Berlin, 1983

[2] Ehrig, H., M. Nagl, G. Rozenberg and A. Rosenfeld (ed.): *Graph-Grammars and Their Application to Computer Science*, Lecture Notes in Computer Science, 291, Springer-Verlag, Berlin, 1987

[3] Ehrig, H., H.-J. Kreowski and G. Rozenberg (ed.): *Graph Grammars and Their Application to Computer Science*, Lecture Notes in Computer Science, 532, Springer-Verlag, 1991

[4] Flasinski, M.: Parsing of edNLC-graph grammars for scene analysis, *Pattern Recognition*, 21, pp. 623-629, 1988.

[5] Aizawa, K. and A. Nakamura: Graph grammars with path controlled embedding, *Theoretical Computer Science*, 88, pp. 151-170, 1991.

[6] Janssens, D., G. Rozenberg and R. Verraedt: On sequential and parallel node-rewriting graph grammars, *Computer Graphics and Image Processing*, 18, pp. 279-304, 1982.

[7] Shi, Q. Y. and K. S. Fu: Parsing and translation of attributed expansive graph languages for scene analysis, *IEEE Transaction on Pattern Analysis and Machine Intelligence*, PAMI-5, pp. 472-485, 1983.

[8] Mylopoulos, J. P. and T. Pavlidis: On the topological properties of quantized spaces, *Journal of the Association for Computing Machinery*, 18, pp. 239-254, 1971.

[9] Rosenfeld, A.: Partial path groups and parallel group constructions, in G. Rozenberg and A. Salomaa (ed.), *The Book of L*, Springer-Verlag, Berlin, pp. 369-382, 1986.

[10] Melter, R. A.: Tessellation graph characterization using rosettas, *Pattern Recognition Letters*, 4, pp. 79-85, 1986.

[11] Flasinski, M.: Characteristics of edNLC-graph grammar for syntactic pattern recognition, *Computer Vision, Graphics, and Image Processing*, 47, pp. 1-21, 1989.

Path-Controlled Graph Grammars
for
Syntactic Pattern Recognition

Kunio Aizawa[1] and Akira Nakamura[2]

[1]Department of Applied Mathematics, Hiroshima University
Higashi-Hiroshima, 724 Japan

[2]Department of Computer Science, Meiji University
Kawasaki, Kanagawa, 214 Japan

Abstract. The graph structure is a strong formalism for representing pictures in syntactic pattern recognition. Many models for graph grammars have been proposed as a kind of hyper-dimensional generating systems, whereas the use of such grammars for pattern recognition is relatively infrequent. One of the reason is the difficulties of building a syntax analyzer for such graph grammars. In this paper, we define a subclass of nPCE graph grammars and present a parsing algorithm of $O(n)$ for both sequential and parallel cases.

1 Introduction

The graph structure is a strong formalism for representing pictures in syntactic pattern recognition. Many models for graph grammars have been proposed as a kind of hyper-dimensional generating systems (see e.g., [1-3]), whereas the use of such grammars for pattern recognition is relatively infrequent. One of the reason are the difficulties of building a syntax analyzer for such graph grammars. In [4], Flansinski introduce a parsing algorithm of time complexity $O(n^2)$ for a subclass of edNLC graph grammars.

In this paper, we define a subclass of nPCE graph grammars and present a parsing algorithm of $O(n)$ for both sequential and parallel cases. Originally, nPCE graph grammars are introduced in [5] for describing uniform structures. These model representations may be considered as guidelines for emulating the focus of attention typical of biological systems.

The goal of this paper is to present a general method for 2-D object recognition using different resolution levels, where the object at each level is described by means of a syntactic approach: in practice the image variations from a coarse to a finer level are coded by formal grammar production rules. A running example on a limited set of printed characters illustrates the recognition method yet a number of open questions that require further research are provided in the last section.

2. Modeling the Visual System

In biological vision systems, particularly referring to primates, broadly speaking the light sensitive elements on the retina are distributed in a non homogenous way so that a central area is densely covered while the periphery has a smaller number of sensitive elements. Many authors [7,8,9] argued that this acuity distribution on the retina provides a hardwired implementation of scale invariant analysis for object recognition. Moreover, in this way the subject perceives a global view and then only the interesting/relevant components of such view are observed by redirecting the eye so as to expose the fovea (central area having the maximum resolution) to analyze those components.

This mechanism may also switch between different resolution levels (excluding the extremes); such switching also reflects a hypothesis-testing modality which is typical of a biological reasoning process that maximizes the use of limited physical resources. At each analyzed level, the total amount of processed information is constant since the number of picture elements which are handled is approximately the same regardless of the image resolution.

The crucial problem is how to allocate the visual system resources (retinal area and visual cortex) in order to explore and analyze the regions of interest within the image, i.e. the attention of the viewer is gradually centered on those parts of the image which convey significant information for object recognition. The eye movements during this process have been studied in the past, (see for instance [10]) proving that after a preliminary random exploration of the full image the eye fixation points correspond to the "interesting" subparts of the object.

In our system, which follows a similar paradigm, the recognition process is guided from a starting low resolution level, which is chosen as a minimal one to enable object detection, proceeding towards higher levels by collecting cues in a hypothesis-driven mechanism until sufficient evidence has been gathered.

3. Multi-Resolution Object Representations

In this section we review two related hierarchical data structures: the Pattern Tree and the Models Feature Graph; that support fast coarse-to-fine search procedures for dynamic vision analysis and object recognition.

In a pattern tree, subpatterns (having significant relevance for the overall recognition process) are organized along nodes of the tree, each descendant node contains a fraction (at a higher resolution level) of the sub-pattern present in its parent node. A full pyramid of images, taken at various levels of resolution, is first constructed (for instance a Gaussian Pyramid) and then subarrays are selected in each image representing object subpatterns. The size of the subarrays is roughly constant along the pyramid thus progressively smaller subpatterns are represented moving away from the root node. Subarrays considered at a given level of the pyramid are not necessarily disjoint since distinctive subpatterns may overlap.

The search for a given object is performed through a sequence of simple matching steps; the results of each step guide the search at the next step. The links in the tree help to locate the wanted subpatterns.

If an object is partially occluded it is still possible to identify it provided that a distinctive subpattern, represented by a path in the tree, is visible. Nevertheless, for an unambiguous representation of occluded or noisy objects, the next data structure is more adequate.

The second data structure, the Models Feature Graph (MFG), is a directed acyclic graph where nodes representing local features or subpatterns are organized into levels corresponding to different resolutions and nodes at a given level may be reached via different paths from the previous level. This is particularly useful for the cases of occlusion: if a feature represented by a parent node is occluded, but there are other parent nodes, there still remains a possible path that allows to exploit the feature of the child node.

An interesting aspect of this representation is the specification of the Match Set; that is a subset of nodes of the graph belonging to a connected path that allows to discriminate between the given object and all other possible objects. To this end, each node is associated with a weight (smaller than 1) specifying the significance of the feature allocated to that node. The weights are selected in such a way that any connected path with a total weight equal or greater than 1 unambiguously identifies the object. A Match Set strongly reduces the search time since it alone provides evidence of the presence of the object in the observed image.

In the Model Feature Graph presented in [6], the features represented in the graph nodes are sets of unconnected edge points stored in a generalized Hough transform R-table [11]. The matching processes performed at each level of the search are based on the voting results of the generalized Hough transform.

4. The Syntactic Description

The well known structural approach to pattern recognition [12] represents patterns as strings where the patterns are considered at a single resolution level and the decision making process for recognition is based on a parsing procedure of the string obtained in the first stage of the analysis. Each object class is represented by a formal grammar which accepts the string representing the object to be recognized. The unknown pattern is coded in a sentence which will be parsed, providing a yes/no answer according to whether the object belongs or not to that class. Each grammar may be defined as a quadruple $G=(V_n, V_t, P, S)$ where V_n (V_t) is the symbol vocabulary for non-terminal (terminal) symbols, P the production rules and S the starting symbol of the sentence to be parsed.

This approach, based on a single level representation of the object, implies that the V_n symbols do not always have a direct physical counterpart. Our approach, which is parallel and context-sensitive, implies the use of higher dimensional pattern grammars and, more particularly, we refer to the work of Kirsch and Dacey [13-14], where there is no difference between terminal and non-terminal symbols: they all correspond to some specified topological property of a boundary segment.

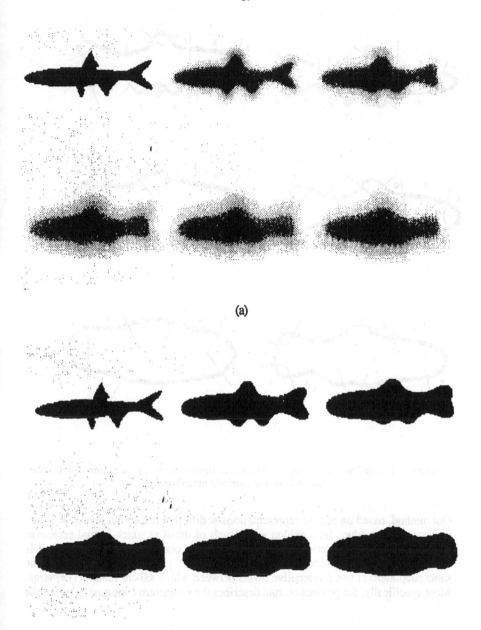

(a)

(b)

Figure 1.(a)   A six level Gaussian Pyramid representing fish. (b) Thresholded binary versions of figure (a)

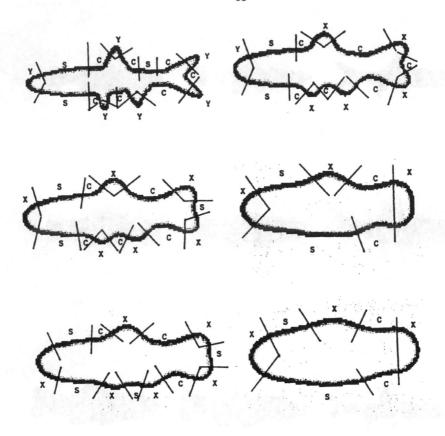

Figure 2. Labeled silhouettes of the content of figure after contour segmentation (see Table I for the symbols meanings).

Our method, based on pattern representations at different resolution levels, attempts to iteratively introduce (in a guided way) levels of detail at which formal sentences will be produced. Production rules formalize the process through which detail is augmented as the resolution is increased: each single production rule refers to the same subpattern at two consecutive levels between which salient feature(s) emerge. More specifically, the production rule describes the subpattern evolution between the low to high levels resolution. An initial symbol S corresponds, according to this view, to the coarsest representation of the pattern, which may be seen as a blob.

Terminal Symbols V_t	Labels
Very Concave	W
Concave	C
Straight	S
Convex	X
Very Convex	Y

Table I

L	Starting String	Production Rules
5	$x^1,s^1,c^1,x^2,c^2,x^3,s^2$	$\rightarrow$ $x^1,s^1,c^1,x^2,c^2,x^3,s^2$
4	$x^1,s^1,c^1,x^2,c^2,x^3,s^2$	
3	$x^1,s^1,x^2,s^2,x^3,c^1,x^4,s^3,x^5,c^2,x^6,c^3,s^4$	$s^1 \rightarrow s^1,x^2,s^2,x^3;$ $x^2 \rightarrow x^4,s^3,x^5; s^2 \rightarrow c^3,s^4$
2	$x^1,s^1,c^1,x^2,c^2,x^3,c^3,x^4,s^2,x^5,c^4,x^6,c^5,s^3$	$s^1 \rightarrow s^1,c^1; s^2 \rightarrow c^2$
1	$x^1,s^1,c^1,x^2,c^2,x^3,c^3,x^4,c^4,x^5,c^5,x^6,c^6,s^2$	$s^2 \rightarrow c^4$
0	$y^1,s^1,c^1,y^2,c^2,y^3,c^3,y^4,c^4,y^5,c^5,s^2,c^6,y^6,c^7,s^3$	$x^* \rightarrow y^*; c^5 \rightarrow c^5,s^2,c^6$

Table II: the first column indicates resolution levels, the second one the strings describing (following Table I) the silhouette of the object, the third column shows the production rules when a new detail appears. (Note that the implicant label refers to the the level above, while the implied label is the one of the current level). If a label (new detail) is originated from two adjacent subpatterns it will appear in both consequent production rules.

Figure 1a illustrates the pattern of a fish represented at 6 different resolution levels, with the corresponding descriptive sentences. This pattern, where only its silhouette will be analyzed, has been digitized on a 128x128 array. Next, following [15] the pattern at different resolution levels is produced by means of the so called Gaussian Pyramid since a conventional subsampling would introduce noise being context free. Note that the final presentation of each version is expanded to a standard common array size in order to allow comparisons to be made. Each version is then binarized as can be seen on figure 1b, to be next labeled in terms of the topological properties of the contour. In this example, five different levels of "curvature" have been defined so obtaining the corresponding figure 2.

For each different resolution level, a string may be constructed after associating each topological property to a symbol as shown in Table I.

By comparing the generated strings at the different consecutive levels, production rules describing the contour evolution may be extracted. Table II reports, for each resolution level, both the contour descriptive string and the corresponding used production rules.

An overview of the labeling process may be seen on the graph of figure 3 which corresponds to the Models Feature Graph.

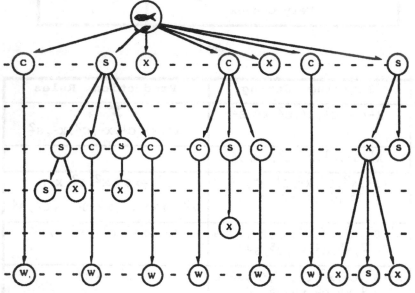

Figure 3.Model Feature Graph corresponding to the Production Rules of Table II.(The full strings can be easily obtained by replicating leaf nodes downwards until the base).

5. Recognizing Objects

The Model Feature Graph and the Production Rules in addition to providing a description of the object at different resolution levels also result to be an explicit representation of the focus of attention mechanism. The global recognition process may be subdivided in this way by a node-to-node series of elementary partial matches in which each match (mismatch) leads into the successive step in the process. This mechanism models the eye movements performed in humans when observing objects to be recognized, as described in section 2.

The basic strategy of this approach is to start from a coarse version of the object and, gradually, in a guided manner, reach the finer resolution levels where the significant details may be found. Three important issues must be considered for a correct implementation of this strategy:

- the coarsest level should be chosen so as to reliably establish which branch to be expanded from this node downwards;
- the finest level should contain enough detail to discriminate the object to be recognized among all possible objects that may be present in the scene;
- the possibility of verifying a correct recognition should be available starting from each level of the tree: for levels above the basic one by simply going in coarse to fine mode and for the basic level by observing the iso-level details.

As in all problems where the solution has to be found expanding a node (root of a tree) towards the leaves besides the depth-first node expansion, breadth-first or other parallel search strategies may be used (see for instance [16]).

Our method is based on the linguistic approach which relies on a formal description of the object components. This description is constructed using production rules which will then be compared with the MFG for each object. A recognition will be obtained when a isomorphism exists between the MFG and the list of the production rules.

An example is shown (in figure 3 representing the rules of figure 2), where top to bottom and left to right precedences have been chosen (for a different ordering of the rules another, correspondent, MFG exists).

In an object recognition process, under the constraint of a completely known closed world, the optimal strategy is obtained on the basis of the grammars of all the models of the possible objects present in the scene. To this aim, the techniques to detect the "minimal discriminant description" and the "maximal conjunctive generalization" [17] may be used.

6. An Implementation Example

We will now show, as an example, the recognition process of four printed characters (D B A and R) which have been chosen due to their common topological features and therefore to highlight the capabilities of our method. Let us firstly consider letter D (refer to figure 4); where figure 4a represents a six level Gaussian Pyramid of the input letter A starting from the finest resolution and figure 4b shows the corresponding binary versions. Table III (in analogy with Table II) illustrates, for each resolution level, the descriptive string and the associated production rule. At the coarsest level, the symbol corresponding to the letter to be recognized is provided.

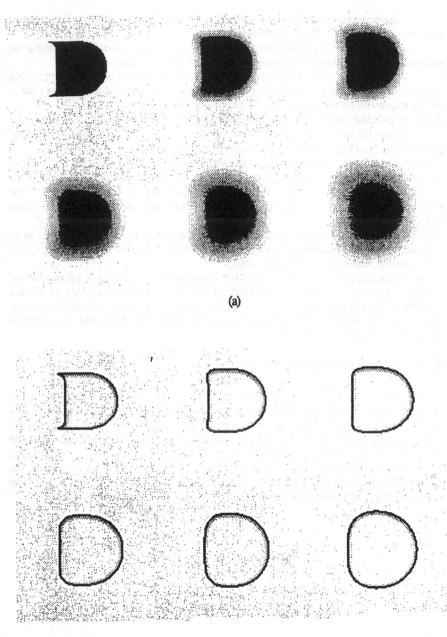

(a)

Figure 4.(a) A six level Gaussian Pyramid of the character D. (b) Contours of the
thresholded binary versions.

Every descriptive string starts with a straight line segment and has circular symmetry (the last symbol is the predecessor of the first symbol on the string since the object has a closed contour). The first production relative to the letter D (table III) generates a straight segment (S) and a convex one (X). The second production shows that the convex segment at the next level is split into straight segments separated by convex segments.

The last production rules reveal the formation of finer details at the upper and lower corners on the left side of the letter: moreover two concave parts connected to the straight segment are produced.

We now turn to letter B (refer to figure 5 and table IV). At level 4 we have the same description as for letter D except that the uniform curvature here is present on the right side instead of on the upper part as in D. At level 3 the production rule describes the typical generation of two convex segments separated by a straight segment from a convex segment: these two convex segments have higher curvature than the previous one.

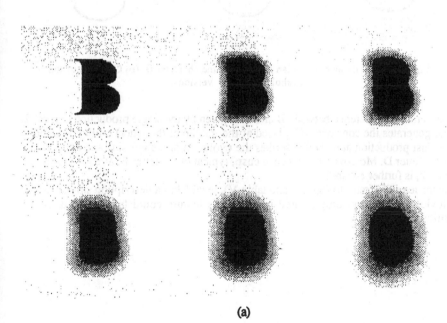

(a)

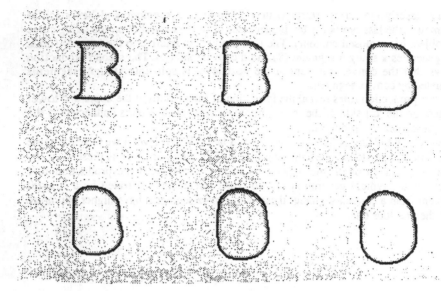

(b)

Figure 5.(a) A six level Gaussian Pyramid of the character B. (b) Contours of the
thresholded binary versions.

The semantic difference between B and D is given by the single production at level 2
that generates the concavity (C, placed at the middle of the letter) on the right side.
The last production at level 0 describes the evolution of the left side and is the same
as for letter D. Moreover, the concave cusp, typical of the letter B, already present at
level 2, is further enhanced.
As for letter A, (refer to figure 6 and table V), level 5 has a description similar to the
previous two letters only rotated by 90°: this feature could be described by an
attributed grammar.

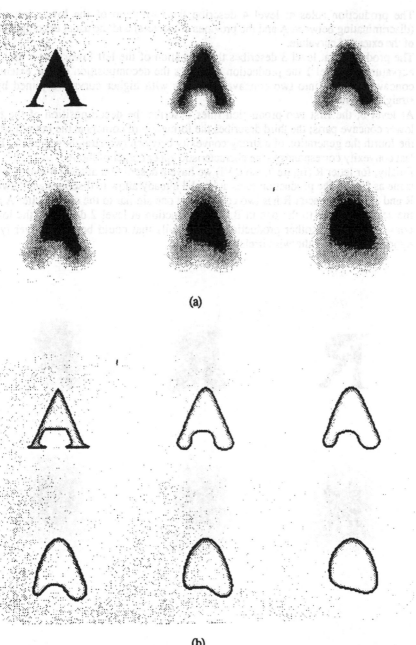

(a)

(b)

Figure 6.(a) A six level Gaussian Pyramid of the character A. (b) Contours of the thresholded binary versions.

The production rules at level 4 describe the formation of the lower intrusion (discriminating between A and the previous characters) and the usual decomposition of the extended curvature.

The production at level 3 describes the formation of the left lower part with high curvature. At level 1 the production describes the decomposition of one extended concave segment into two concave segments with higher curvature joined by a straight segment.

At level 0, the first two production rules describe the decomposition of the two lower concave parts; the third describes the formation of corners at the bottom while the fourth the generation of a strong convexity between two straight segments. This last convexity corresponds to the characteristic upper corner of the A.

Finally, for letter R (figure 7, table VI) we have a description at level 4 which is the same as for A. The production rules at level 3 already allow to discriminate between R and A. The character R has two concavities: one similar to the one in letter A and the second similar to the one in B. The production at level 2 describes the lower convexity. All the other productions add details that could be used to verify an hypothesis but are otherwise irrelevant.

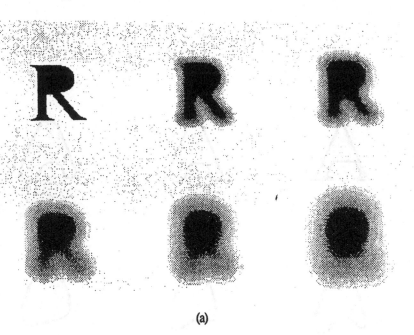

(a)

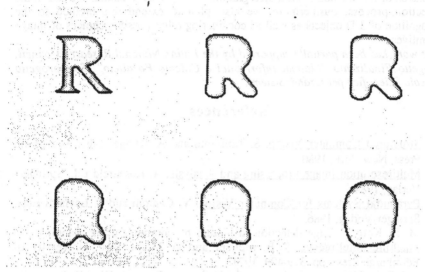

(b)

Figure 7.(a) A six level Gaussian Pyramid of the character R. (b) Contours of the thresholded binary versions.

7. Conclusions

The main limitations of the linguistic approach in the past were essentially due to the noise sensitivity of the obtained descriptions, for this reason such an approach was abandoned in favor of others (based on statistics, knowledge and learning). We overcome such limitation by means of the multiresolution representation which allows many different descriptions of the same object at various definition levels; moreover the use of the coarsest level as a starting point for the matching procedure allows a fast search for the significant details that will enable object recognition so implementing the focus of attention typical of biological systems. Moreover, the hypothesis and testing which may be performed using this method is also an emulation of human behaviour in decision processes. In this respect, this paper is to be considered a concept paper and as such the implementation details have been omitted.

A particular multiprocessor architecture, the pyramid (as widely represented in [3]), well matches both the data structure and the computational nature of the recognition algorithm. The full computation from grammatic description to graph matching may also be fully performed within the pyramidal architecture but the coding of such algorithm requires careful attention both for the information exchange between pyramidal planes and in the neighborhood (local) computations.

Finally, in order to increase the recognition rate and to enable to weigh attributes of features of the objects to be recognized, a possibility to be considered is the extension of this method to attributed grammars [18] so as to include metrics in the detection process. Furthermore, we also plan to extend this method to the recognition of 3-D objects as well as considering other possible choices of image primitives.

This work has been partially supported by the Italian National Research Council, Progetto Finalizzato "Sistemi Informatici e Calcolo Parallelo", Sottoprogetto Calcolo Scientifico per Grandi Sistemi.

References

1. Structured Computer Vision, S. Tanimoto and A, Klinger Eds., Accademic Press, New York, 1980.
2. Multiresolution Image Processing and Analysis, A. Rosenfeld Ed., Springer-Verlag, 1984.
3. Pyramidal Systems for Computer Vision, V. Cantoni and S. Levialdi Eds., Springer-Verlag, 1986.
4. M. D. Kelly, " Edge detection in picture by computer using planning", in Machine Intelligence 6, B. Meltzer and D. Miche Eds., University of Edimburgh Press, pp. 379-409, 1971.
5. P. J. Burt, "Attention Mechanisms for Vision in a Dynamic World", *Proc. 11th Int. Conf. on Pattern Recognition,* 1988, pp. 977-987.
6. C. R. Dyer, "Multiscale Image Understanding," in *Parallel Computer Vision,* L. Uhr, ed., Orlando, FL: Academic Press, 1987, pp. 171-213.
7. A. Califano, R. Kjeldsen, R. M. Bolle, "Data and Model Driven Foveation", Proc. 10th Int. Conf. on Pattern Recognition, 1990, pp. 1-7.
8. C. R. Carlson, R. W. Klopfenstein, and C. H. Anderson, "Spatially Inhomogeneous Scaled Transforms for Vision and Pattern recognition", Optics Letters, vol. 6, pp 386-388, 1981.
9. C. Braccini, and A. Grattarola, "Scale-invariant Image Filtering with Point and Line Symmetry," in Digital Image Analysis, S. Levialdi, Ed., Pitman, London, pp. 183-192, 1984.
10. Eye Movements and Vision, A. L. Yarbus, Plenum Press, New York, 1967,
11. D. H. Ballard, "Generalizing the Hough Transform to detect arbitrary shapes", Pattern Recognition, Vol. 13, No 2, pp. 111-122, 1981.
12. K. S. Fu, "Recent developments in pattern recognition," *IEEE Transactions on Computers,* Vol C-29, No 10, 1980, pp. 845-854.
13. R. A. Kirsch, "Computer Interpretation of English Text and Picture Patterns", IEEE Trans. on EC 14, 1964, pp.363-376.
14. M. F. Dacey, "The Syntax of a Triangle and some other Figures", Pattern Recognition 2, 1970, pp. 11-31.
15. P. J. Burt,"The Pyramid as a Structure for Efficient Computation", in Multiresolution Image Processing and Analysis, A. Rosenfeld Ed., Springer-Verlag, 1984, pp. 6-35.
16. N. Nilsson, Problem Solving Methods in Artificial Intelligence, McGraw Hill, New York, 1971.
17. R. S. Michalsky, "Pattern Recognition as Rule-Guided Inductive Inference", IEEE Trans. on PAMI, Vol. 2, N° 4, 1980.
18. W. H. Tsai, K. S. Fu, "Attributed Grammar: a tool for combining syntactic and statistical approaches to pattern recognition", IEEE Trans. on Systems, Man and Cybernetics, Vol. SMC-10, No 12, 1980, pp. 873-884.

Parallel Manipulations of Octrees and Quadtrees*

Vipin Chaudhary[1], K. Kamath[2], P. Arunachalam[3], and J. K. Aggarwal[3]

[1] Department of Electrical and Computer Engineering
Wayne State University, Detroit, MI 48202, USA
[2] Sun Microsystems, San Jose, CA 95134, USA
[3] Department of Electrical and Computer Engineering
The University of Texas at Austin, Austin, TX 78712, USA

Abstract. Octrees offer a powerful means for representing and manipulating 3-D objects. This paper presents an implementation of octree manipulations using a new approach on a shared memory architecture. Octrees are hierarchical data structures used to model 3-D objects. The manipulation of these data structures involves performing independent computations on each node of the octree. Octrees are much easier to deal with than other forms of representations used to model 3-D objects especially where extensive manipulations are involved. When these operations are distributed among multiple processing elements (PEs) and executed simultaneously, a significant speedup may be achieved. Manipulations such as a complement, a union, an intersection and other operations such as finding the volume and centroid which this paper describes are implemented on the Sequent Balance multiprocessor. In this approach the PEs are allocated dynamically, resulting in a uniform load balancing among them. The experimental results presented illustrate the feasibility of the approach. Although this evaluation has been originally done for shared memory machines, it will provide insight for the evaluation on other architectures.

1 Introduction

Efficient manipulations of 3-D objects are important in various applications such as computer graphics, computer vision and other related areas. These schemes can be categorized as surface descriptions or volumetric descriptions. Chien and Aggarwal [1] summarize advantages and disadvantages of each category. Most representation techniques suffer from severe memory and processing requirements with increasing input requirements [2]. The octree is a well known data structure in the representation of 3-D objects. It is used to determine various geometric properties such as volume and centroid and to manipulate objects by computing their complement, union, and intersection. An octree representation scheme uses efficient tree traversal algorithms to overcome the drawbacks stated earlier. Chen and Huang [3] survey in detail the construction of octrees. Though

---

* This research was supported in part by IBM.

this can be done on a sequential machine, the nature of the algorithm suggests the use of a parallel machine. As these algorithms use three orthogonal views to generate the octree, they may pose problems for objects with cavities as three views are insufficient to generate an exact 3-D description of an object with cavities. Chien and Aggarwal [2] elaborate on an octree generation from more than three views. While a tree structure indicates an increase in the data dependencies, the regularity of the structure presents ways to avoid this problem. Samet [8] presents a detailed study of the complexity of the tree traversal algorithms. Moitra and Iyengar [9] also discuss an idea of the parallelism which can be found in such algorithms.

The octree structure for the representation of 3-D objects is an extension of the quadtree structure for 2-D objects. The octree manipulation is computationally expensive because of the huge volume of data. Hence, it makes sense to parallelize the operations especially in real time systems. However, due to the tree nature of the algorithm the parallelization is not easy and requires more complicated techniques. Another problem with these algorithms is that they are not computationally intensive and require more data communication than inherent computation on a single node. Due to this reason the speedup increase is not linear with the increase in number of PEs.

The rest of the paper is organized as follows. The next section describes the parallel algorithms for generating octrees from three orthogonal views. Section three describes the manipulation of octrees involving the union and intersection of two octrees, evaluating the volume and centroid of an octree, and finally the displaying of the octree as an object. The results of the implementation of the above algorithms on the Sequent Balance multiprocessor are detailed in section four. We conclude in section five with comments on the results of our implementation and possible extensions of this work.

2 Representation of Octrees

The hierarchical representation of an octree represents a binary image in a compressed form. Computations to be performed on these data structures can be considered as simple tree traversal algorithms which can be efficiently implemented in parallel. It is assumed that the objects are specified in a binary format with an image represented by white pixels and background by black pixels. The figures representing the objects have been drawn in the inverse format. Figure 1 illustrates an example of an octree and its three orthogonal views.

2.1 Parallel Method for Octree Construction

The octree of a binary image is constructed by subdividing the image into eight octants recursively until each octant is either fully white (object) or fully black (non-object). Each octant is a node in the tree, and each node can be a terminal node (leaf node) or a non-terminal node (grey node). A leaf node can be white or black and a grey node which is non-terminal, is a subtree which defines a part of the object neither completely white nor black. Each node in an octree contains information regarding the structure of the octree. This information includes the color, the surface pointers, the children, and the node pointers. Three quadtrees corresponding to the top, front, and the side view of the object are first constructed from the scans of the three images. These are then intersected to get the final octree of the object. Hence, the quadtree generation algorithm is used thrice to get the quadtrees of the individual views, and the octree is obtained from this. Table 1 gives the various octants of an image depending on the view direction and the position of the quadrant.

Table 1. Illustration of the various octants of an image depending on the view direction and the position of the quadrant.

View	NE	NW	SE	SW
Front	0, 4	1, 5	2, 6	3, 7
Top	4, 6	5, 7	0, 2	1, 3
Side	4, 5	0, 1	6, 7	2, 3

This part presents the parallel algorithm used for generating an octree. The three quadtrees are traversed in parallel, and a logical AND operation is performed on the node pairs. This intersection table data is maintained in the shared memory. A task queue is set up, and the root pointers of the three quadtrees and the octree pointer are inserted as the task in the task queue. When the algorithm is invoked, a check is performed to see if all the three octnodes (octree nodes) have the same color. If not, eight children corresponding to the eight octants are created and appended to the task queue which are obtained from the children of the various quadtrees. This process continues until all the three quadtree nodes checked are of the same color. An idle PE (Processing Element) picks up the next available task in the task queue and executes it. The entire job is complete when the task queue is empty and the entire image has been created. It should be noted that an empty queue does not imply a completion of the job because the task may not have been appended to the queue. Table 2 gives the intersection table between the quadtrees and the final octree.

2.2 The Parallel Algorithm for Octree Generation

This section presents (more elaborately) the actual algorithm used is the study. An idle PE takes the task from the task queue and performs the following operation:

1. If all the three quadtree nodes are grey, then the octnode is marked grey, eight child nodes are appended to it and eight tasks are created in the task queue. Each task contains a child node of the octree and its corresponding three intersection quadtree nodes referred from the intersection table.

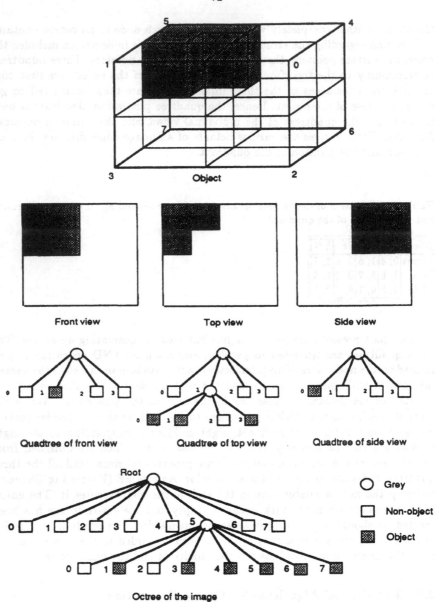

Fig. 1. Example of an octree generation by three orthogonal views of the object

Table 2. Intersection table between the quadtrees corresponding to the final octree.

Quadtree node of front view	Quadtree node of top view	Quadtree node of side view	Corresponding octree node
0	2	1	0
1	3	1	1
2	2	3	2
3	3	3	3
0	0	0	4
1	1	0	5
2	0	2	6
3	1	2	7

2. If all the three quadtree nodes are black, then the corresponding octree node is marked black.

3. If one of the quadtree nodes is marked white, then the corresponding octree node is marked white.

4. If two of the quadtree nodes are marked grey and the third is black or null, then the corresponding octree node is marked grey. Then, eight nodes are added to the octnode, and eight tasks are appended to the task queue. Each entry in the task queue has two pointers to the children of the nodes marked grey, and the third is marked null.

5. If one of the quadtree nodes is grey, one is black, and the third is null, then the corresponding octree node is marked grey. Then, eight nodes are added to the octnode, and eight tasks are appended to the task queue. Each entry in the task queue has a pointer to a child node of the grey node while the other two pointers are marked null.

6. If two of the nodes are marked black and the third is marked null, then the corresponding octree node is marked black.

7. If two of the nodes are marked white and the third is marked black, then the corresponding octree node is marked white.

In algorithm 1,

1. the total object and background pixels are assumed to be a cube of dimension *bsize*.

2. q_ptr_1, q_ptr_2, and q_ptr_3 represent the pointers of the corresponding quadtrees for the octree.

3. q_size_1 represents the size of the first quadtree.

4. T_const(meaning Tree constant) is a constant dependent on the image being a quadtree or an octree. This is 4 for a quadtree and 8 for an octree.

Algorithm 1. <u>Octree Generation</u>.

begin
   while(IMAGE_SIZE$\neq bsize^3$)
      while(TASK_QUEUE$\neq Empty$)
         if($q_ptr_1 = q_ptr_2 = q_ptr_3 = $ GREY)
          O_ptr = O_ptr + T_const
          O_color = GREY
          TASK_QUEUE = TASK_QUEUE + T_const
         else if($q_ptr_1 = q_ptr_2 = q_ptr_3 = $ BLACK)
          O_color = BLACK
          B_SIZE = q_size_1
         else if(q_ptr_1 or q_ptr_2 or $q_ptr_3 = $ WHITE)
          O_color = WHITE
          B_SIZE = q_size_1
         else if(two q_ptrs are GREY, third q_ptr is BLACK or NULL)
          O_ptr = O_ptr + T_const
          O_color = GREY
          TASK_QUEUE = TASK_QUEUE + T_const and $q_ptr_3 = $ NULL
         else if(one q_ptr is BLACK, one q_ptr is GREY, one q_ptr is NULL)
          O_ptr = O_ptr + T_const
          O_color = GREY
          TASK_QUEUE = TASK_QUEUE + T_const
         else if(two q_ptrs are BLACK and the third is NULL)
          O_color = BLACK
          B_SIZE = q_size
         else if(one of the two q_ptrs is WHITE, and the third is NULL)
          O_color = WHITE
          B_SIZE = q_size
      IMAGE_SIZE = IMAGE_SIZE+ B_SIZE
end

Each PE executes the above code concurrently. In the implementation on a shared memory machine, all the shared variables are locked when they are updated to prevent simultaneous accesses by many processors and thus to avoid erroneous values. The entire task queue, tail, and image size are the variables shared by the entire process whereas all other variables are local to a processor. All locked variables are accessible by only one processor; the other processors must wait until it is released. This retards the speedup achieved by parallel processing with the shared memory paradigm.

To get a picture of the object, one needs to store surface information explicitly in the octree nodes. Using this, a 2-D shaded image of the object can be obtained. Such an octree is called a volume/surface (VS) octree. This is usually done using a *multi level boundary scan*. This scheme was first suggested by Chien and Aggarwal [1] and used to detect all the interfaces between the object and the surrounding volume. As no neighbor finding operations are involved, the implementation is easier and faster. This is similar to that of Jackins and Tanimoto [6] and more generalized than that which was suggested by Doctor and Torborg [7].

3 Manipulation of Octrees

A wide variety of information can be obtained from the octrees. Such information as evaluating the volume and centroid of the object, getting 2-D projections of the object from various views and angles, and finding the complement, the union, and the intersection of the object can be obtained.

3.1 Union and Intersection of Octrees

The union and the intersection of octrees also involve the tree traversal concepts. Figure 2 illustrates the union and Fig. 3 illustrates the intersection of two objects (and the corresponding octrees) respectively. It can be seen that the resulting octree is obtained by manipulating the two octrees of the individual objects themselves. The intersection involves traversing the trees in parallel and performing a logical AND operation between them. The logical AND is necessary as it generates a 0 if either object is absent and generates a 1 if both are present. As the union of objects implies that the final image should have a pixel at any point if either of the two objects are present at that location, the equivalent logic operation is used. The union involves performing a logical OR operation between the two trees.

Algorithm 2.<u>Union of octrees</u>

```
begin
    while(IMAGE_SIZE≠ bsize³)
        while(TASK_QUEUE≠ Empty)
            if(o_ptr₁ = o_ptr₂ =BLACK)
                U_color = BLACK
                B_SIZE = min(o_ptr₁.size, o_ptr₂.size)
            else if(o_ptr₁ or o_ptr₂ = WHITE
                U_color = WHITE
                B_SIZE = min(o_ptr₁.size, o_ptr₂.size)
            else
                U_ptr = U_ptr + T_const
                U_color = GREY
                TASK_QUEUE = TASK_QUEUE + T_const
    IMAGE_SIZE = IMAGE_SIZE+ B_SIZE
end
```

In algorithm 2 and 3,

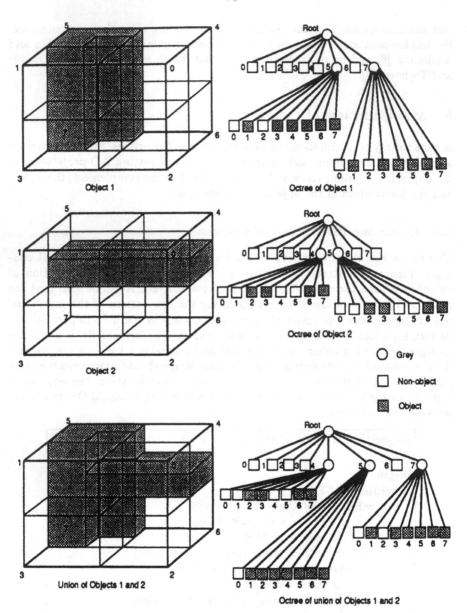

Fig. 2. Union of two objects and the corresponding octrees

1. the total object and background pixels are assumed to be a cube of dimension *bsize*;
2. o_ptr_1 and o_ptr_2 are the pointers of the two octrees whose union is sought; and
3. o_ptr_1.size and o_ptr_2.size represent the sizes of the nodes of the octree.
4. U_ptr and U_color are the pointers to and the colors of the union respectively.

In intersection, the root nodes of both the octrees and the intersection octree are inserted as a task in the task queue. An idle PE takes the task from the task queue and starts executing using the algorithm stored in the local memory. If the node pointed to by q_ptr$_1$ and q_ptr$_2$ is white, the resultant node in the intersection is marked as white. If one of the nodes is black, the resultant node in the intersection is marked as black. If both are grey, or if one is grey and the other is white, the resultant node in the intersection is marked as grey. Now eight child nodes are created and are appended to the queue. Many PEs can take these tasks and process them independantly resulting in an increased speed of operation. The operation of the union is similar. If the node pointed to by q_ptr$_1$ and q_ptr$_2$ is black, the resultant node in the union is marked black. If one of the nodes is white, the resultant node in the union is marked white. If both are grey, or if one is grey and the other is black, the resultant node in the intersection is marked grey. Now eight child nodes are created and are appended to the queue.

Algorithm 3. Intersection of octrees

```
begin
   while(IMAGE_SIZE≠ bsize³)
      while(TASK_QUEUE≠ Empty)
         if(o_ptr₁ = o_ptr₂ =WHITE)
            U_color = WHITE
            B_SIZE = min(o_ptr₁.size, o_ptr₂.size)
         else if(o_ptr₁ or o_ptr₂ = BLACK
            U_color = BLACK
            B_SIZE = min(o_ptr₁.size, o_ptr₂.size)
         else
            U_ptr = U_ptr + T_const
            U_color = GREY
            TASK_QUEUE = TASK_QUEUE + T_const
   IMAGE_SIZE = IMAGE_SIZE+ B_SIZE
end
```

3.2 Displaying the Object

In order to display the octree as an object, an operation needs to be performed which will detect the boundary surfaces of the object. Meagher [5] first suggested

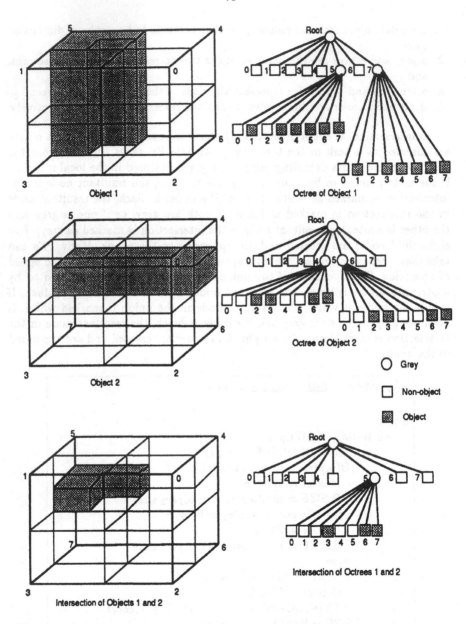

Fig. 3. Intersection of two objects and the corresponding octrees.

attaching the surface normals to the nodes of the octrees. An octree can have the surface information stored explicitly in its nodes. Such an octree is called a volume/surface (VS) octree. The surface information is computed using the *Multi level Boundary Search* (MLBS) method. Here, the octree is traversed, and the surface normals of the surface nodes are computed from the adjacency information. The orientation of the surface normals is coarsely quantized into 26 directions and stored as the surface information in the nodes.

In 3-D space there are 12 interfaces for 12 different combinations of the child node pairs that are adjacent to each other. The orientation of the surface normals is maintained in a table. The root node of the octree is inserted as a task into the task queue. An idle PE picks up the task and refers to the adjacency table to find the nodes adjacent to the node picked from the queue. For each of those adjacency nodes the following operations are performed.

1. If both of the octree nodes are grey, then the four pairs of child nodes adjacent to each other are appended to the task queue.

2. If one of the octnodes is black and the other is white, then the surface information is stored in the black node.

3. If one octnode is grey and the other non-grey, then the child nodes of the grey node and the non-grey node are appended to the task queue.

As the positions of the black and white nodes are known, the surface normals of each block can be computed by averaging the directions of all black and white interfaces of each block. The directions are assumed to be moving from the white to the black (object to surounding). The following is the algorithm for this operation.

Algorithm 4. <u>Multi-level boundary search.</u>

begin
  while(IMAGE_SIZE$\neq bsize^3$)
    while(TASK_QUEUE$\neq Empty$)
      if(o_ptr_1 = GREY and o_ptr_2 = NULL)
        TASK_QUEUE = TASK_QUEUE +20
        TASK$_{o\_ptr_1}$ = TASK$_{o\_ptr_1}$+T_const
        TASK$_{o\_ptr_1}$ = TASK$_{o\_ptr_1}$ + 12 for each adjacent pair
      else if(o_ptr_1 = o_ptr_2 = GREY)
        TASK_QUEUE = TASK_QUEUE +4
        TASK$_{o\_ptr_2}$ = TASK$_{o\_ptr_2}$ + 4 for pairs adjacent to o_ptr_1
        TASK$_{o\_ptr_1}$ = TASK$_{o\_ptr_1}$ + 4 for pairs adjacent to o_ptr_2
      else if(o_ptr_1 =BLACK and o_ptr_2 =WHITE)
        o-ptr$_{BLACK}$ = surface
        B_SIZE = min(o_ptr_1.size , o_ptr_2.size)
      else if(o_ptr_1 =GREY and $o\_ptr_2 \neq$ GREY)
        TASK_QUEUE = TASK_QUEUE +4
        TASK$_{o\_ptr_1}$ = TASK$_{o\_ptr_1}$ + 12 for each adjacent pair
    IMAGE_SIZE = IMAGE_SIZE + B_SIZE
end

---

1. the TASK_QUEUE is the queue made for the elements of the search.
2. the TASK$_{o\_ptr1}$ implies those tasks of the first octree.
3. the o-ptr$_{BLACK}$ are those octree pointers which point to octnodes which are BLACK.
4. surface implies that the octree pointer referred to is on the surface.

Using this method, a 3-D object can be given a 2-D projection. The visibility of each block is determined by the dot product of its surface normal and the viewing direction. Since the surface directions are quantized, a set of surface directions visible are obtained by taking a dot product of the viewing direction and the 26 orientations of surface normals. Once the MLBS is carried out, the surface information is stored in each black terminal node. To get the projection, the root node is inserted in the task queue. An idle PE takes up this task and starts executing it. If the color is grey, eight tasks are appended to the task queue, and the tree is traversed to a lower level. If the octree node pointed to is black, its surface normal is compared with the set of visible orientations. If there is a match, the node is projected onto the screen.

3.3 Other Operations on Octrees

The volume of the object can be calculated easily from the octree by traversing down the tree until every white node has been visited. From the level of the node, the volume of each individual octnode can be computed and summed up to get the total volume of the object. If n is the level of the octnode, then $2^{3 \cdot n}$ is the volume of the octnode.

The centroid of the object is a point which is the average of all the white pixels in that coordinate. To locate the centroid the octree must have a structure which includes the starting points of the three coordinates. A procedure can be constructed which computes the centroid of the object by summing up the products of the centroid and the volume of each node, and then dividing this sum by the total volume. If X_{cent}, Y_{cent} and Z_{cent} are the coordinates of the centroid, the centroids of the individual nodes can be easily obtained by using the formula $X_{cent} = X_{start} + 2^{n-1}$ where X_{start} is the coordinate of the starting address of the node, and similarly for Y_{cent} and Z_{cent}. Thus the final centroid is given by

$$X_{cent} = \frac{1}{Vol} \sum_{i=1}^{n} X_{cent_i} * Vol_i,$$

$$Y_{cent} = \frac{1}{Vol} \sum_{i=1}^{n} Y_{cent_i} * Vol_i,$$

$$Z_{cent} = \frac{1}{Vol} \sum_{i=1}^{n} Z_{cent_i} * Vol_i$$

where n is the total number of octnodes, and Vol is the total volume.

The complement of an image is obtained by changing all the white pixels to black and black pixels to white. This is accomplished in octrees by creating a complimentary tree and inserting it in the task queue. Any idle PE starts on the octree and traverses down. If the node is marked black, the node in the complementary tree is marked white. If the node is marked white, the node in the complementary tree is marked black. However, if the node is marked grey, eight child pointers are appended to the task queue, and eight nodes are created in the c_tree where c_tree is the complementary tree. Idle PEs will pick up tasks from the task queue and execute them as long as they are available in the task queue. The algorithm terminates when the task queue is empty, and the volume of the complementary image equals that of the original image.

4   Results

All the algorithms were implemented on the Sequent Balance multiprocessor. We used images of two different sizes for timing analysis. The results of these implementations are shown in Fig. 4. The display was done using the multi-level boundary search. The speedups obtained for images of different sizes are shown separately. It should be mentioned that the amount of computation on each node is much smaller than the amount of time required to create child processes on the Sequent Balance and inserting them into the task queue. Thus, a large

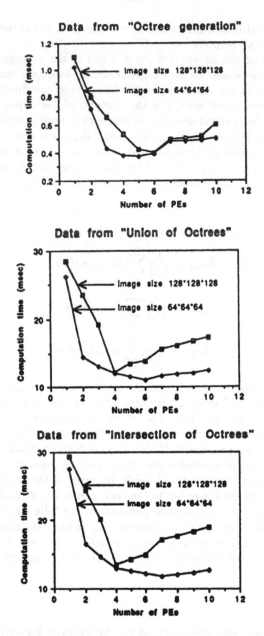

Fig. 4. Results of the various implementations: (a) Octree generation, (b) Union of Octrees, and (c) Intersection of Octrees.

portion of the time is spent in system overhead and not actual processing. This is an inherent drawback of the load balancing paradigm we follow and was found quite visible with lesser image sizes. Examples of intersection and union of some objects are are also shown.

It can be seen that while there is a great variation in the speedups, general trends can be noted. In general, the times taken for smaller images are lesser than those for larger images, provided the images are complicated enough. If the images are simple, then time required for the octree construction would be obviously less. The time necessary for union or intersection of octrees is much more than that of a single octree generation as three octrees need to be constructed (two for the initial objects and one for the final object) with a large increase in the amount of shared data and thus problems with variable locking. It is also seen that the time taken increases after a certain number of PEs. This is mainly due to the increase in the overhead of creating new tasks and the shared data being much larger in volume than the data local to a processor.

We verified the algorithms by displaying the union and intersection of several objects. Figures 5, 6, 7, and 8 verify the correctness of the algorithms.

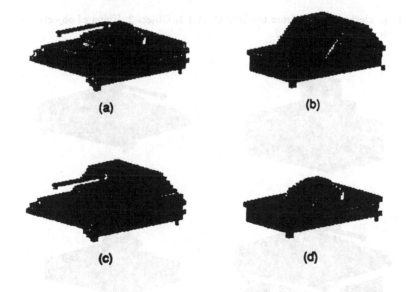

Fig. 5. In clockwise order from top left: Object 1; Object 2; Union of objects 1 and 2; Intersection of objects 1 and 2.

5 Conclusions

It can be inferred from the graphs that low speedups were obtained for smaller image sizes but the speedups are generally low even in a large image size. This

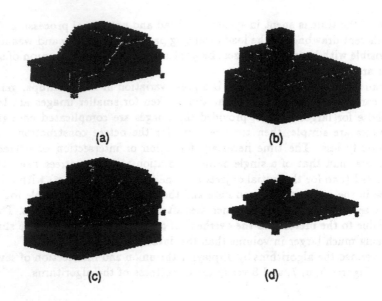

Fig. 6. In clockwise order from top left: Object 1; Object 2; Union of objects 1 and 2; Intersection of objects 1 and 2.

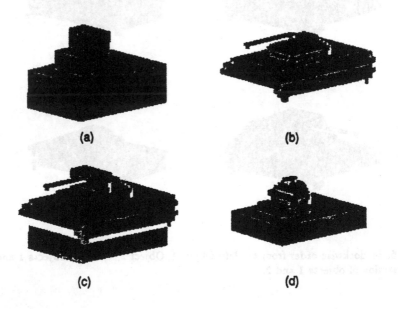

Fig. 7. In clockwise order from top left: Object 1; Object 2; Union of objects 1 and 2; Intersection of objects 1 and 2.

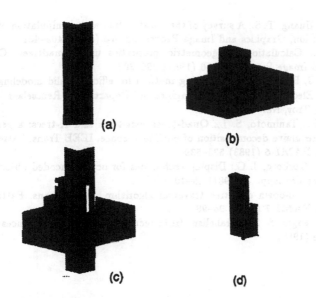

Fig. 8. In clockwise order from top left: Object 1; Object 2; Union of objects 1 and 2; Intersection of objects 1 and 2.

is due to the large overhead involved in the creation of new tasks (50ms for each). While dynamic scheduling may improve the load balancing, it does create other processes which actually increase the load. The bottleneck of the entire computation is the shared task queue whose access is mutually exclusive to the processors. Since the task sizes in all the octree algorithms are extremely small, most of the processors are waiting for access to the globally shared task queue. Thus, there is a tradeoff involved between load balancing and speedup.

It will be worthwhile to compare the speedups obtained by statically partitioning the data or by using a combination of static and dynamic partitioning.

6 Acknowledgements

It is our pleasure to acknowledge the help of Mr. Michael Schulte in debugging a part of the code used for implementation.

References

1. Chien, C. H., Aggarwal, J. K.: Volume/surface octrees for the representation of three-dimensional objects. Computer Vision, Graphics and Image Processing. **36** (1986) 100–113
2. Chien, C. H., Aggarwal, J. K.: Reconstruction and matching of 3D objects using quadtrees/octrees. Proceedings of 3rd Workshop on Computer Vision. (1985) 49–54

3. Chen, H. H., Huang, T. S.: A survey of the construction and manipulation of octrees. Computer Vision, Graphics and Image Processing. **43** (1988) 409–431
4. Schneier, M.: Calculations of geometric properties using quadtrees. Computer Graphics and Image Processing. **16** (1981) 296–302
5. Meagher, D. J. R.: The octree encoding method for efficient solid modeling. Ph. D. dissertation, Electrical and Systems Engineering Department, Rensselaer Polytechnic Institute, Troy, New York 12181.
6. Jackins, C. L., Tanimoto, S. L.: Quad-trees, oct-trees and K-trees: a generalized approach to recursive decomposition of euclidean space. IEEE Trans. Pattern Anal. Mach. Intell. **PAMI-5** (1983) 533–539
7. Doctor, L. J., Torborg, J. G.: Display techniques for octree-encoded objects. IEEE Comput. Graphics Appl. **3** (1981) 29–40
8. Samet, H.: A top-down quadtree traversal algorithm. IEEE Trans. Pattern anal. Mach. Intell. **PAMI-7** (1985) 94–98
9. Moitra, A., Iyengar, S. S.: Parallelism from recursive programs. Advances in Computers. June (1986)

VISION REFLEX OPERATORS

C. Coutelle, C. Fortunel, M. Eccher, B. Zavidovique

Laboratoire Système de Perception, Etablissement Technique
Central de l'Armement, 16 bis Avenue Prieur de la Côte d'Or,
94114 ARCUEIL CEDEX, FRANCE

Abstract. Our goal is to realize a robust, real-time, autonomous and compact perception system for a multisensor robot. It is suggested that such a perception system should rely on a set of basic modules called reflex operators. These operators perform region of interest extraction in preeattentive mode and contribute to object identification in attentive mode. In this paper, we focus on vision reflex operators and their hardware implementation for which we propose an original approach. Vision automata are emulated in real-time using a data-flow machine dedicated to image processing with its associated functional programming langage. Emulation step provides hardware ressource information for subsequent VLSI integration.

1 Introduction

Today, there is an increasing need to build intelligent robots having the capability to adapt to non-controlled or even hostile environments. An essential feature of these robots is their perception system that enables them to acquire data about the world, to process it and interpret it in order to achieve a specific mission. Building such perception systems is a very complex task as they must achieve simultaneously four characteristics [22]:

- Robustness: the perception should not be affected by partial observability (spatial or frequential) or unstructured environments (in the military field: counter measures, decoys...).
- Rapidity: events need to be processed as soon as they happen with delays as short as possible (it is the real-time processing capability).
- Autonomous operation: they need to adapt to changes in the environment or to unknown situations (changing light conditions, presence of obstacles).
- Compactness: the architecture of the system must satisfy physical constraints such as volume constraints or current supply limitations.

The classical approach to building perception systems for mobile robots is a sequential organization of computational modules [7]. In such decomposition, data acquisition, data processing and actuators command are considered individually. Rather than simplifying basic perception problems, this led to ill-posed or unstable problems (in particular the "shape from X" problem class). The inefficiency of this approach and the encouraging results obtained in Active Perception (which has its origin in the observation of animal perception systems) have led to incorporate perception issues more closely in the control aspects of the system [4, 5, 6, 8, 9]. Active Vision views the perception problem as a control problem rather than an algorithmic one: data acquisition, data processing and actuator commands are now tightly coupled tasks also called agents, running independently of each other and of an overall system controller. One speaks more in terms of multiple and concurrent behaviors all participating to the realization of a mission, than in terms of a particular technique needed to be used to accomplish the mission. We believe that the building of a perception system with the qualities formerly described must rely on primitive modules which implement basic visual behaviors (automatic selection of points of fixation, focusing on moving objects, etc.).

2 Reflex Vision Operator

2.1 Concept of Reflex Vision Operator and Reflex Perception

One of the inability of current systems to perform complex tasks is due to the lack, by their perception system, of the ability to extract robust primitives and to handle events in real-time. Many observations of physiological mechanisms in animal and human perception systems [12, 21, 19, 18] suggest several solutions on which features one should extract and how to use them:

- retina signals are functionaly segregated: a lot of primitives are systematically computed and available. Theses primitives are associated with global and simple properties such as linearity, contour closure, regions size, contrast, motion detection and so on. It should be noted that these primitives, in the absence of context, do not permit any identification of observed objects. The cortex selects the primitive which best fit the current perceptual activity and provide the appropriate information. It is surprising to note that today, there does not exist complete operators capable of extracting such information.

- human vision is a sequence of at least two stages: preattentive vision which extracts points of interest in the whole image (attention focalisation) without any identification, and attentive vision which attempts to identify objects pointed to during the preattentive stage (via a local and precise analysis).

By analogy with the human visual system, we have decided to implement primitive visual operators that, given a specific operative context, will provide high level semantic information (for instance, vertical edges correspond to wall corner in a city context) from which complex visual behaviors can be built. Broadly speaking, such operators detect first order or second order changes of primitives computed from the raw sensor signal (video, acoustic, radar, infrared), and localize in time and space the detected situation (appearance of a red object, of a fast moving object, etc.). Operators are able to take local decisions based on processing results in the sense that they can directly control actuators. This is possible because the operator associates a semantic information with its output, as indicated by a cognitive module. This allows the definition of predefined reactions (behaviors) using tight control loops involving the sensor, an operator and an actuator. It is clear that to be tested and used, such operators should work in real-time. We call such operators, "reflex operators". As in the human perceptual mode, they can be used in two modes:

- a preattentive mode where they are used to perform a global analysis and extract the areas where attention should be brought. If necessary, an action can directly be taken upon assessment of the situation. Whenever an occurence of the situation appears, the reflex operator interrupts any decision module interested by the situation assessment.

- an attentive mode where the operator attempts to identify objects based on a specific request. The reflex operator answers to some central decision system questions about the presence or absence, within an area of interest, of an object by analysis of its extracted characteristics: is there a red region larger than ten pixels, or is there a new peak in the intensity histogram? When the question has been answered, the operator waits for a new question or goes back into the preattentive mode.

2.2 Third Generation Robotics

To validate the concept of vision reflex operators, we have built a perception robot PERCEPT (see figure 1).

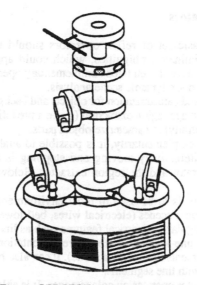

Fig. 1. PERCEPT multisensor robot.

It is designed to support the integration of multiple heterogeneous sensors of acoustic or vision type, that we hope to make as rough as possible in the way they acquire data: we favor an approach where multiple simple and inexpensive sensors cooperate to solve a problem rather than a few sophisticated sensors, which brings the redundancy necessary to obtaining a robust and reliable solution. For example, let us consider the problem of driving a car. The distance between the car in front of you is a vital information, but you do not need to know its precise value to avoid bumping into it. We are suggesting to use rough sensors to measure the distance, and associate with the sensor, a reflex operator that will detect the situation (distance is decreasing fast) that may indicate a collision, so the appropriate action can be taken (brake). Thus it seems feasible to drive a car using only rough information (is there a car in front of you, are stop lights on...). As one can easily foresee, having a series of operators to assess a situation, the perception problem becomes one of control: when do I use a particular piece of information? Which information should I seek? Which one should I pay attention to? The simplicity of reflex operators makes them very reliable because they are better understood and characterized; it is easier to acquire knowledge about their perfomance limitations and the quality of the results they provide. Additionaly they are easier to build and work in real-time.

The robot PERCEPT to which we plan to incorporate our reflex operators supports the concepts previously mentionned. It holds several heterougenous sensors: two B&W cameras, one colour camera, a battery of ultrasonic devices and a ramp of microphones (from bottom to top). Each sensor is mounted on its own set of actuators. All information transits through a central control unit due to hardware architectural constraints. This control unit ensures tight feedback between each sensor and its associated actuator.

3 Reflex Operators Description

3.1 Selected Operators

We think that the basic set of reflex operators should make a coherent vision subsystem able to eliminate ambiguities, which could appear when using only a single operator. So we have chosen three complementary operators dedicated to:
- motion detection for dynamic scene analysis.
- edge detection and caracterization for objects and background separation.
- homogeneous image region extraction (given a predefined homogeneity criteria such as color uniformity) to caracterize object parts.

Because of operator complementarity, it is possible to analyze complex scenes; for instance, an object alternatively moving and stopping is easily tracked using the motion detection operator and the region extractor. Below are described the three chosen operators:
- the line detector is to be used in most environments such as urban scenes, indoor scenes, outdoor scenes (electrical wires, bell-towers, roadside...) which are characterized by regular geometrical features such as line segments. This operator extracts the list of line segments with a given orientation (for instance vertical or horizontal line segments in an indoor scene). It can also be used to identify objects roughly modeled with line segments.
- the region detector operates on color images. It is able to detect regions having a predefined color feature or a given range of colors in the image, and to characterize these regions geometrically (area, center of gravity, perimeter, second order moments...).
- the movement detector is dedicated to extracting areas associated with objects moving relative to the background for a moving sensor. A list of displacement vectors is provided (movement direction and speed) for each detected object.

At this point, we have simulated all selected algorithms and started building the line detector we describe next.

3.2 General Organization

Each operator implements completely a perception function and extracts from sensor data a list of symbols, directly usable by a cognitive module. This provides for a very modular architecture where one can add new operators at wish. More importantly, one is led to build homogeneous control structures independently of the fact that we have a couple of operators or several dozens. All operators furnish the same type of results in symbolic form (for example: whether something is present or not, an answer to a localized binary question or a system interrupt when a situation of interest has been detected). This allows for the integration of operators independently of available technology; they can be highly integrated processors or even use recent technologies such as optical computing. It is an approach that is not limited to vision sensors but in fact applies to any perception module.

As a result, all reflex operators we have chosen possess the same architecture, as shown below (figure 2).

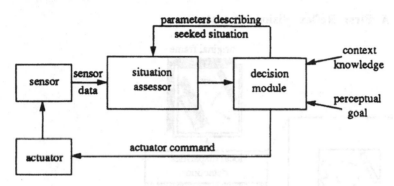

Fig. 2. Reflex operator synopsis.

The situation assessor extracts, either systematically or in response to a request, sets of characteristics in the scene (for example: ten parallel lines) as described by a set of parameters. The decision module, as a function of some knowledge about the scene (context) and a perceptual goal (find building windows) determines first how to program the situation assessor (vertical and horizontal segments, 50 cm aparts) and second, may send a command to its associated actuator to get closer to the fixed perceptual goal. The decision module is implemented as a symbolic controller (rule based system type) that is easy to program. The situation assessor module, which requires most computing ressources as it analyzes sensor data, supports a general mechanism described in figure 3.

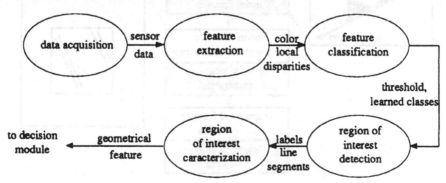

Fig. 3. Situation assessor structure.

Data input is, in the most complex case, a multispectral image sequence. The first processing step consists in extracting scalar or vectorial characteristics such as displacement vectors, local disparities or textural features. Then a classification module decides whether or not a feature of interest (edge point, right color...) is present. This detection is followed by a grouping operation that builds higher level features: connected component labelling, edge linking... The last step associates a rough geometrical description of regions of interest: perimeter, center of gravity and area computation.

3.3 A First Reflex Vision Operator

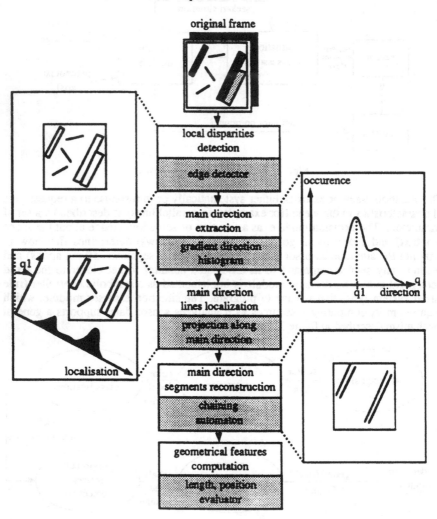

Fig. 4. Line segment extraction reflex operator.

The first reflex vision operator that we have decided to implement is the line extraction operator because it appears to be the simplest of all and could be used in identification tasks when objects are modeled with line segments (horizon, flying deck visual servo-control...). The purpose of this reflex operator is to extract line segments associated with main image directions. Input data are grey level images. The feature extraction step consists in detecting local disparities (any edge operator will suffice). In a preattentive mode, the operator selects the main direction by finding the peaks in the histogram of significant contour points. In an attentive mode, the direction is given by the decision module. All edge points having the right direction are kept to create a binary image of feature points. Line segments are extracted by projecting selected feature points along the right direction. Peaks in this projection correspond to lines in the image where line segments may be present. The last step consists in extracting features of line segments (length, position) as they are reconstructed by a chaining automaton. The various stages of the operator are illustrated below (figure 4).

4 Hardware Implementation

4.1 Discrete Approach

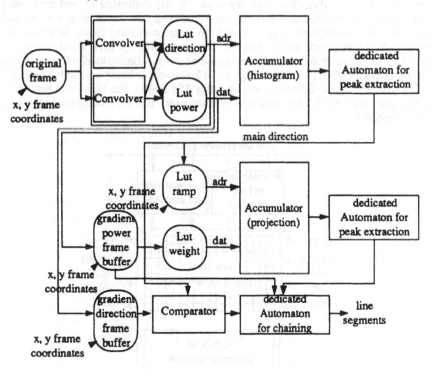

Fig. 5. First reflex operator block diagram.

Having simulated the algorithm and decided it was a useful and robust operator, the question came of determining a hardwired implementation for it, keeping in mind that the solution must satisfy constraints of real-time processing and compactness. To start, we looked at using discrete components. The block diagram solution that resulted is shown below (figure 5).

An appropriate space requirement estimation showed that four Double Europe VME boards would be necessary (about 200 chips). This clearly does not fit with our compactness objective as the building of just three or four operators would fill an entire VME chassis. The solution would require too much power comsumption and volume to allow for an integration into any robotics system. As our goal is to build many of such operators (at least on the order of ten), one needs to look at other technologies to build them.

4.2 The Functional Approach

If it is clear that a classical electronic approach cannot satisfy our building requirements (power comsumption, volume), one needs to use a solution where one partially or totally integrates each processing module. Having at our disposal a multiprocessor functional calculator [17, 14], we decided to evaluate the implementation of our algorithm on this system called the "Functional Calculator". This machine is a data-flow type machine [2, 10] dedicated to real-time image processing. Its architecture (see figure 6) allows low level image processing (convolutions, histograms...) as well as high level processing (peak detection in an histogram, equivalence table updating, list processing...). Low level processing, which is very regular and simple, is executed on a three dimensional network of data-flow processor (DFP) [13]. The complete DFP network is being assembled with 1024 DFPs (the Functional Computer having today 256 DFPs). For high level processing, which involves irregular data structures, 9 transputer T800 are used (three in the actual configuration).

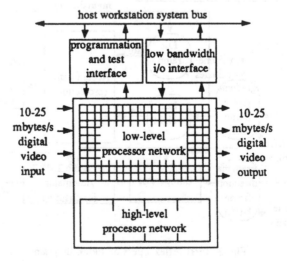

Fig. 6. Data-flow Functional Computer architecture.

Because there is a natural duality between data-flow machines and functional programming, algorithms are programmed with a functional language [3]. So before implementing an algorithm onto the Functional Calculator, a functional decomposition of the algorithm has to be derived which can be expressed as a data-flow graph (figure 7).

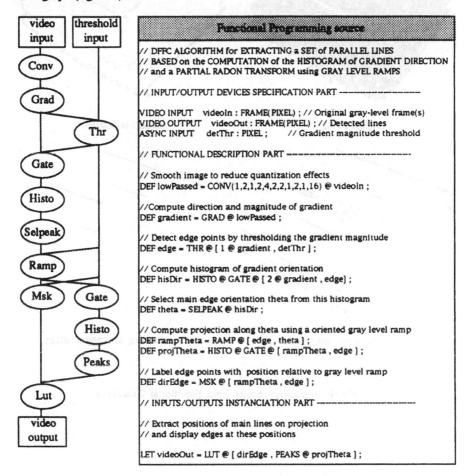

Fig. 7. Functional programming source and associated data-flow graph
for the line segment extraction operator.

This graph shows all data dependencies and intrinsic parallelism of the algorithm. The corresponding functional program is shown below. All operators specified (CONV, GATE...) are macros functions available on the machine. The corresponding functional program is shown above.

This machine was then used to emulate the algorithm in real time. Results of the emulation are shown in figure 8 [11]. The emulation allows us to verify that the algorithm worked as planned on real images, but also that a hardware solution existed for implementing the algorithm.

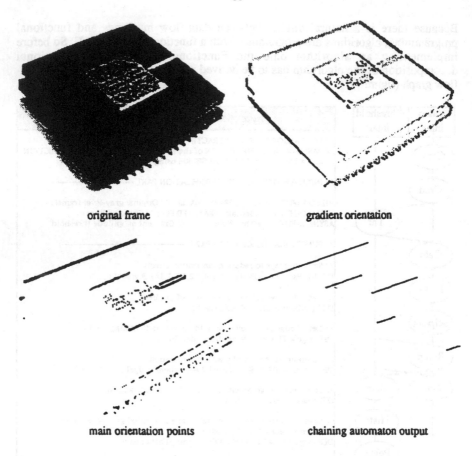

original frame

gradient orientation

main orientation points

chaining automaton output

Fig. 8. Emulation results from Functional Calculator.

It is clear that this solution, as implemented on the emulator, does not yet meet our constraints, because of all the input/output modules associated with the machine. An easy way to reduce the layout surface of the operator is to realize DFPs based boards by eliminating all the processors used exclusively for data routing. But this solution would also use four VME cards (120 DFPs and 2 transputers are necessary for implementing our algorithm), so it does not offer any advantage over the discrete component approach (this is primarily due to the fact that 8 bits processors are used to implement 16 bit arithmetic). Another solution is to collect emulation ressources that have been used and to extract the low level data-flow graph of the algorithm. This can lead, with common routing tools, to the design of a final vision automaton which can be integrated onto a circuit, using for example standard cell libraries. This process integrates in a straight forward fashion all the operations defined by the graph without modifying it (for instance if you need a 512 bytes long FIFO, two 256 bytes long FIFOs will be used as they were on the emulator). We estimate that using Xilinx devices, a few circuits from the 4000 family would suffice. The last approach available to us, is a true synthesis operation where the low level data-flow graph is transformed so that the best compromise between processing time and circuit area is found. In this case, multiple adding operations, as specified in the graph, may in fact be implemented with one simple adder (operator assignment in high level synthesis: [15, 16, 20]). This approach, although we do not yet have a clear estimate of its layout efficiency, is the only way that can, at the same time, ensure a real time version of the algorithm and use the smallest circuit area. For this reason, this is the approach we have selected to build our first reflex operator.

5 Conclusion

We have described in this paper a coherent and modular vision subsystem based on an efficient use of an extensible vision reflex operators set. These complementary operators provide high semantic level information but yet remain simple enough so they can be built efficiently into VLSI circuits and be used in real-time. Such integration is even more justified taken into account the fact that each operator is complementary to the others (we need a lot of them), functions independently and need not share ressources with other operators (they can be highly integrated onto single chips if possible).

We are now engaged in a VLSI conception phase of the first operator. Having built the situation assessor modules for at least two operators, our next step will be to develop control strategies to efficiently use such operators.

References

1. J. Aloimonos, I. Weiss, A. Bandyopadhyay: Active vision. InternationalJournal of Computer Vision, pp. 333-356, 1988.

2. K.P. Arvind, D.E. Culler: Dataflow architectures. Annual Reviews in Computer Science, vol. 1, Palto Alto CA, Annual Reviews Inc 1986, pp.225-253.

3. J. Backus: Can programming be liberated from Von Neumann style? A functional style and its algebra of programs. Comm. of ACM, vol. 21, no 8, August 1978.

4. R. Bajcsy: Active perception. Proceedings of the IEEE, vol. 76, no 8, pp. 996-1005, August 1988.

5. D. H. Ballard: Behavioural constraints on animate vision. Image and Vision Computing, vol 7, no 1, pp. 3-9, February 1989.

6. D. H. Ballard: Animate vision. Artificial Intelligence, vol 48, pp. 57-86, 1991.

7. R.A. Brooks: A robust layered control system for a mobile robot. IEEE Journal of Robotics and Automation, vol. RA-2, no. 1, pp. 14-23, March 1986.

8. C. Brown: Gaze controls cooperating through prediction. Image and Vision Computing, pp. 10-17, vol. 8, no. 1, February 1990.

9. P. J. Burt: Attention mechanisms for vision in a dynamic world. 9th International Conference on Pattern Recognition, pp. 977-987, Rome, Italy, November 1988.

10. J.B. Dennis: Data-flow supercomputers. Computer, vol. 13, pp. 48-56, November 1980.

11. E. Merlet: Développement d'une bibliothèque de programmes de traitement d'images en temps réel pour un co-processeur basé sur un réseau de transputeurs. ETCA/CREA/SP Internal Report.

12. U. Neisser: Cognitive psychology. Appleton Century Crofts, New York, 1967.

13. G. Quenot, B. Zavidovique: A data-flow processor for low-level real-time image processing. IEEE Custom Integrated Circuit Conference-91, San Diego CA, 13-16 May 1991.

14. G. Quenot, J. Serot, B. Zavidovique: Functional programming on a data-flow architecture. applications in real-time image processing. Journal of Machine Vision Applications, to be published.

15. A. Safir, B. Zavidovique: On the synthesis of specific image processing automata form emulation results. EURO ASIC 1989.

16. A. Safir, B. Zavidovique: Towards a global solution to high level synthesis problems. IFIP 1990.

17. J. Serot, G. Quenot: Real time image processing using functional programming on a data-flow architecture. Computer Architecture for Machine Perception, pp.33-44, Paris, France, December 1991.

18. S. J. Thorpe: Traitement d'image par le système visuel de l'homme. Traitement d'Images: du Pixel à l'Interprétation, 2ème atelier scientifique, pp. IX-1-IX-12, Aussois, France, April 1988.

19. A. Treisman: L'identification des objets visuels. Pour la Science, pp. 50-60, January 1987.

20. F. Verdier, A. Safir, B. Zavidovique A high level synthesis algorithm including constraints. Eighteenth EUROMICRO Conference 92, Paris, France, September 1992, to be published.

21. N. Weisstein, C. S. Harris: Visual detection of line segments: an object-superiority effect. Science, vol. 186, pp. 752-755, November 1974.

22. B. Zavidovique, Image Processing Algorithmy Team: Des systèmes R.A.R.E. (Rapides, Autonomes, Robutes, Embarquables). Conference Invitee, Journées Sciences et Défense, Paris-La Villette, France, May 1990.

NOTE ON TWO-DIMENSIONAL PATTERN MATCHING
BY OPTIMAL PARALLEL ALGORITHMS

Maxime Crochemore[1] and Wojciech Rytter[2]

Abstract

We discuss techniques for constructing parallel image identification algorithms : cutting images into small factors, and compressing images by a parallel reduction of large number of such independent factors into smaller objects. A version of Kedem Landau Palem algorithm for parallel one-dimensional and two-dimensional rectangular image recognition on a CRCW PRAM is presented. The crucial part in KLP algorithm is a suffix-prefix matching subprocedure. In our algorithm such a subprocedure is omitted. A novel algorithm for pattern-matching is proposed, more directly designed for two-dimensional objects. It does not use the multi-text/multi-pattern approach as in KLP algorithm. The importance of five types of factors in strings and images is emphasized. A new useful type of two-dimensional factors is introduced : thin factors.

Keywords : analysis of algorithms, parallel algorithms, pattern matching, string-matching, image processing.

[1] LITP, Institut Blaise Pascal, Université Paris 7, 2 place Jussieu, F-75251 PARIS cedex 05.
mac@litp.ibp.fr
Work by this author is supported by PRC "Mathématiques-Informatique"
& NATO Grant CRG 900293.

[2] Institute of Informatics, Warsaw University, Ul. Banacha 2, 00913 Warsaw 59, Poland.
rytter@mimuw.edu.pl
Work by this author is supported by the Polish Grant 2-11-90-91-01.

1. Introduction

We investigate the parallel complexity of the two-dimensional pattern matching problem. Our model of computation is the CRCW PRAM (see [Vi 85] or [GR 88] for instance). A parallel algorithm is said to be optimal if its *total work*, product of time by number of processors, is linear. Our aim is the construction of an optimal algorithm for the two-dimensional pattern matching problem.

Two fast parallel algorithms for 2D pattern matching have been given independently in [KLP 89] and [CR 90] (full version in [CR 91]). We refer to them as KLP algorithm and dictionary algorithm, respectively. The first of them is optimal and the second is optimal within factor $\log(m)$. The KLP algorithm uses $\log(n)$ processors less, but the dictionary algorithm is much simpler and it is related to a series of algorithms for several other problems with similar complexity. The authors in [CR 91] concentrated rather on a broad range of simple applications, than on getting optimality result. In fact, the dictionary approach to matching problems is an old idea. It comes from [KMR 72]. And it essentially appears implicitly in [AILSV 88] and [AL 91]. In [CR 91] we stated that for fixed-size alphabets $n/\log(n)$ processors suffice for the algorithm. The proof was essentially omitted. Later, we were asked by some readers for a full proof of the assertion. This note answers such request. Moreover, it presents also a simplification and an alternative to the KLP algorithm. The crucial part in the KLP algorithm is the suffix-prefix matching subprocedure. In our present algorithm we use a more direct solution that makes the subprocedure useless. A novel algorithm for 2D pattern matching is proposed, more directly designed for two-dimensional objects. It does not use the multi-text/multi-pattern approach as in KLP algorithm and other algorithms (as in [Ba 78] or [Bi 77]). Four types of factors (characteristic pieces) are used : basic, regular basic, small and regular small factors. In this note, we also introduce a fifth useful type of two-dimensional factors : thin factors. Their applicability is demonstrated for 2D pattern matching.

2. Five types of factors

The basic parts of KLP algorithm and the dictionary algorithm from [CR 91] are constructions of dictionaries which enable to check in constant time whether some factors of text or pattern are the same. Factors are rectangular parts of two-dimensional image, but can also be strings (one-dimensional structure). In the two-dimensional case, if the pattern is of shape m by m', then we assume that m' exactly divides m, and the two-dimensional factors of shape r.m/m' by r are considered. We say that the longest side of the factor is its length. In fact, we can assume w.l.o.g. that m=m', that is, we consider 2D factors that are squares. Factors whose length is a power of two are called *basic factors*. One can also assume that the pattern has a size that is a power of two. If not, then in one-dimensional case we can search for two subpatterns which are the prefix and the suffix of the pattern whose size is the largest possible power of two. Similarly, in the two-dimensional case we can take four (possibly overlapping) subpatterns whose common size is a power of two. The one-dimensional basic factor f of the pattern is called a *regular basic factor* iff it starts in the pattern at a position divisible by the size of f. Analogously, in the 2D case we say that the 2D factor of shape s by s' is regular iff it starts (left upper corner) at a position such that horizontal and vertical coordinates are multiples of s and s', respectively.

The size of a 2D image is its area. The basic property of regular factors is given below.

Fact 1 : a pattern of size N has O(N) regular basic factors.

This fact implies easily our key lemma, which is essentially already proved in [CR 91] :

Lemma 1 (key lemma)
  Assume that we have t patterns, each of size M. Then we can identify all of them in a one-dimensional or a two-dimensional image of size N in time O(log M) with total work N.log(M) + t.M.

Proof. If we look closer at the pattern matching algorithms in [CR 91] then it is easy to see that the total work of the algorithm is proportional to the number of considered factors. But, in the pattern, only regular factors need to be processed if their size is a power of two (which can be assumed). Hence the work spent on the pattern is O(M). Similarly for t patterns of size M, it is O(t.M). This completes the proof. ◆

  Let k=log(m). We can assume w.l.o.g. that n, m are divisible by k. The factors of length k of the string are called *small factors*. The small factors that start at positions divisible by their size are called *regular small factor*. Let x be a string of length n and let small(x, i) be the small factor of x starting at position i≤m.. Assume, for technical reasons, that the text has k-1 special endmarkers at the end. Define the string $\underline{x}$ such that

$$\underline{x}_i = name(small(x, i)),$$

where name(f) is the name of the regular small factor equal to f (if there is no such factor than it is some special symbol). The names should be consistent :

if small(x, i) = small(x, j) then $\underline{x}_i = \underline{x}_j$.

The string $\underline{x}$ is called here the dictionary of small factors. A simple way to compute such dictionary is to use the parallel version of the Karp-Miller-Rosenberg algorithm. This gives n.log(log(n)) total work, very close to optimal. the optimality can be achieved easily if the alphabet is constant using a kind of the "four Russians trick".

Preprocessing for constant size alphabet

Assume for simplicity that the alphabet is binary and we take k=log(m)/4, which is of the same use later as log(m). There are potentially only $m^{1/2}$ binary strings of length 2.k. For each of them we can precompute names of all small factors of length k, these names could be the integers corresponding to binary representations of factors. We have enough processors for each of $m^{1/2}$ binary strings of length 2.k. The total number of processors is n/log(m), so we have about $m^{1/2}$ processors for each small segment of logarithmic size. Next, take independently and at the same time each segment of size 2.k starting at a position divisible by k. One processor can encode it as binary number in log(m) time; look at the precomputed table for names of its small factors and write down k consecutive entries of string $\underline{x}$ in time k.

However the breakthrough was done by Kedem, Landau and Palem. They showed that the assumption on fixed-size alphabet can be dropped. They observed that only names corresponding to regular small factors have to be considered in the naming procedure for all factors (two factors whose names are not equal to any regular small factor can have the same name, even if they are unequal). The total length of all small regular factors is linear and each of them has logarithmic size.The Aho-Corasick pattern machine for all regular basic factors can be constructed by an optimal algorithm, due to the linearity of their total length. Such approach does not work if we want to make pattern matching machine for all small factors.

However if the alphabet has a constant size then a rough approach is possible, the introduction of regular small factors is not needed. In this case the lemma below follows by a very simple application of the "four Russians" described above. The "four Russian" trick of encoding small segments by numbers is used in [Ga 85] and [ML 85] for string problems.

Lemma 2
  The dictionary of small factors can be computed in log n time with linear work.

The fifth type of factors is introduced in this paper : *thin factors*. It is a natural generalization of small factors to the 2D case. Thin factors are m by log(m) subarrays of the text and pattern arrays. They arise if we cut the 2D pattern by m/log(m) cut-lines, the distance between consecutive lines is log m, see Figure 3. The construction of the dictionary for such factors is discussed in the section of 2D pattern-matching.

We start with the section on one-dimensional string-matching. The algorithm for two dimensions is conceptually a natural extension of the one-dimensional case.

3. One-dimensional string-matching

The basic parts of one- and two-dimensional pattern-matching algorithms presented in this paper are similar : computation of the dictionary of small factors, compression of strings by encoding disjoint log(n) sized blocks by their names, and an application of the algorithm from Lemma 1. Two auxiliary functions are needed : shift and compress (compr, in short).
Let x be the pattern of size m, y be the text of size n and k=log m. We assume that n, m are multiples of k. For $0 \leq r \leq k-1$ denote by shift(x, r) the string x[1+r.. m-k+r], see Figure 1.

For a string z denote compr(z) = $z_1 z_k z_{2k} \ldots z_{(h-1)k}$, where h=|z|/k. The string compr(z) contains the same information as z and is shorter by a logarithmic factor. Essentially each letter of the new string encodes a logarithmic-size block of z.

Intuitively speaking compress(z) is a concatenation of names of consecutive small factors, which compose a given string. The *compression ratio* is the reduction of information, in this case small factor (string of length log m) is replaced by one symbol. The compression ratio is log(m). One-dimensional objects of size log m are reduced to zero-dimensional. We generalize the same idea for more dimensions in the next section.

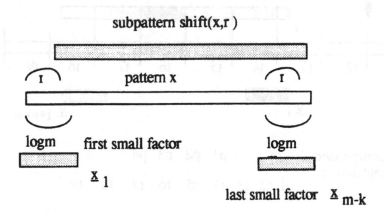

Figure 1. Illustration of the operation shift. To identify the match it is enough to identify the part shift(x, r) and the first and last full small factor of the pattern x (shadowed factors). The identification of shift(x, r) is done by searching for its compressed version.

The correctness of the constructed algorithm is based on the following obvious observation :

Fact 1 : An occurrence of x starts at position i in the text y iff the following conditions are satisfied

(*) compr(shift(x, k-(i mod k))) starts in compr(y) at (i div k) & $x_1 = y_i$ & $x_{m-k} = y_{i+m-k}$.

The conditions (*) are illustrated in Figure 2 (see also Figure 1).

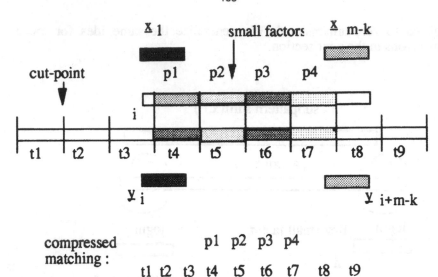

Figure 2. Let $p_1p_2p_3p_4 = compr(shift(x, k-(i \bmod k)))$, and $compr(y)=t_1t_2t_3t_4t_5t_6t_7t_8t_9$. The pattern x starts at position i in y iff $p_1p_2p_3p_4$ starts at the 4th position in $t_1t_2t_3t_4t_5t_6t_7t_8t_9$, $x_1 = y_i$,

and $x_{m-k} = y_{i+m-k}$.

The structure of the algorithm based on the equivalence (*) is given below.

Algorithm 1 {one-dimensional pattern matching}

begin
    compute the dictionary of small factors together for x and y;
    construct tables x and y ;
    identify all patterns $compr(shift(x, 0)),\ldots, compr(shift(x, k-1)$
    in the text compr(y), by the algorithm from Lemma 1;
  for each position i **do in parallel**
{in constant time by one processor for each i due to information already computed}
    if conditions (*) are satisfied for i **then** report the match at i
end algorithm.

Theorem 1
Algorithm1 solves the string-matching problem on a CRCW PRAM in log(m) time with O(n/log(m)) processors (linear total work).

Proof. The compressed text has size m/log(m) and we have log(m) shifted compressed patterns, each of size m/log(m). Hence, according to Lemma 1, the total work, when applying algorithm from this lemma, is of order m/log(m).log(m) + log(m).m/log(m) which is O(m). This completes the proof. ◆

4. Recognizing images: two-dimensional counterpart of Algorithm 1

We give a novel approach with greater utilization of the two-dimensionality of the image. It is based on the fact that the key lemma (Lemma 1) works in the same way for strings as for multidimensional images. In [CR 91] the algorithm for 2D patterns is not multiple application of a 1D-pattern algorithm. It takes fully into account the 2D structure of images.

Let Y be the n by n host array and let X be the m by m pattern array. The general case of rectangular (non-square) arrays can be handle in the same way. However we should assume that the shorter side is at least log m long, pattern is not thin. Otherwise Algorithm1 can be differently used. We say that a subarray occurs at position (i, j) in a given table iff its left upper corner is placed at position (i, j).

There are two possible algorithms. The first one is to parallelize the Baker and Bird algorithms. Such parallelization uses the multi-text/multi-pattern approach. The two-dimensional matching is reduced to a one-dimensional one by multi-text/multi-pattern matching.
This means that Algorithm1 should have easy extension to multi-dimensional case. We show that this is true. We cut two-dimensional patterns by cut-lines in the same way as strings are cut. These lines cut the pattern into small 2D pieces : thin factors. Similarly as in Algorithm1 our idea is now to cut the pattern image into small pieces and make compression, this time the pieces are two dimensional.
Assume for simplicity of presentation that we deal with images whose sides have lengths divisible by m.
Formally define the cut-lines as columns log(m), 2.log(m),..., n-log(m) of the text arrays, see Figure 3. There are n/log(m) cut lines.

Let X be the pattern image of length m, Y be the text of length n and k=log m. We assume that n, m are multiples of k. Denote by $X_{[j]}$ the j-th column of X. For $0 \leq r \leq k-1$ denote by SHIFT(X, r) the rectangle composed of columns $X_{[1+r]}.. \ X_{[m-k+r]}$.

For a rectangle Z denote $COMPR(Z) = Z_{[1]}Z_{[k]}Z_{[2k]}...Z_{[(h-1)k]}$, where h=|z|/k. The rectangle COMPR(z) contains the same information as z and is shorter by a logarithmic factor. Essentially each column of the new rectangle encodes a thin factor.

Intuitively speaking COMPR(z) is a concatenation of compressed consecutive thin factors, which compose a given pattern image. Each thin factor is reduced to a single column. The *compression ratio* is in this case log(m). Two-dimensional objects are reduced to one-dimensional. We use the same idea as in the former section. The correctness of the constructed algorithm is based on the following obvious observation, see Figure 3 :

Fact 2 : An occurrence of X occurs at (i, j) in the image Y iff the following conditions are satisfied

(**)
    COMPR(SHIFT(X, k-(j mod k))) occurs in COMPR(Y) at (i, j div k);
    $X_{[1]}$ occurs at i-th place of j-th column of Y;
    X_{m-k} occurs at i-th place of (j+m-k)-th column of Y.

The structure of the algorithm parallels that of Algorithm1.

Algorithm 2 {two-dimensional pattern matching}
    begin
        compute dictionary of small factors together for rows of X and Y;
        construct tables $\underline{X}$ and $\underline{Y}$;
        identify patterns COMPR(SHIFT(X, 0),.., COMPR(SHIFT(X, k-1)
        in the image COMPR(Y) by the algorithm from Lemma 1;
        for each j do in parallel
            find all occurrences of the first and last column of $\underline{X}$
            in the j-th column of $\underline{Y}$;
    {of the first and last thin factor of image pattern, one-text/one-pattern algorithm }
        for each position (i, j) of Y do in parallel
    {in constant time by one processor for each (i, j) due to information already computed}
        if condition (**) is satisfied for (i,j) **then** report the match at (i,j);
    end algorithm.

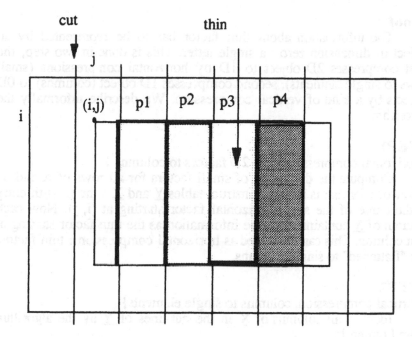

Figure 3. Partitioning of the pattern and the text arrays by cut-lines. We search for the "essential" part of pattern image (composed of dashed thin factors), and the first and last thin factors of X.

5. An alternative recognition algorithm

Yet another 2D pattern-matching algorithm can be constructed by giving consistent names to some thin factors, see Figure 3. A notion of dictionary of thin factors is introduced and discussed. Also we introduce the idea of *vertical compression* and *horizontal compression*.

The *dictionary of thin factors* for Y and X is the problem to identify all thin factors of the pattern array X occurring in Y at positions lying on cut-lines.

Lemma 3

The dictionary of thin factors for Y and X, can be computed in log(m) time with linear total work.

Proof
 The information about thin factor has to be represented by an object of dimension zero : a single letter. This is done in two step, the first compresses 2D object to 1D by horizontal compressions (small rows to single elements), second compresses 1D object (columns) to 0D objects by a kind of vertical compression. We describe informally the algorithm.

STEP1
{horizontal compression, thin 2D factors to columns }
 Compute the dictionary of small factors for all rows of X and Y. The work for that is linear. Construct tables $\underline{Y}$ and $\underline{X}$; the (i, j)-th entry is the name of the small horizontal factor starting at (i, j). Now each column of $\underline{X}$ contains the same information as the thin factor starting at that column. This can be treated as horizontal compression : thin factors are "flattened" to single columns.

STEP2
{vertical compression, columns to single elements}
 Identify all columns of $\underline{X}$ in the cut-lines of $\underline{Y}$ by the algorithm from Lemma 1;

 In this way we identified all columns of $\underline{X}$ occurring on cut-lines of $\underline{Y}$, each such columns gets a single element name (zero dimensional object). The columns of $\underline{X}$ correspond to thin factors, each of them is given a single name identifying itself. Hence it gives the required data structure.

 The total work is linear, since in step2 we deal only with n/log(m) columns of Y. This completes the proof of the lemma. ◆

 Once the dictionary of thin factors is computed we can construct two arrays : the string flat(X) of size m and the n by (n/k-1) array C(Y), where
 flat(X)[i] = name of thin factor occurring at position (1, i) in X ;
 C(Y)[i, j] = name of thin factor occurring at the j-th cut-point in the i-th row;
 (i, j.log(m)) is the position of the j-th cut-point in the i-th row).

 The table C(Y) is a collection of compressed rows of $\underline{Y}$ and flat(X) is a flat representation of the array $\underline{X}$. One can redefine conditions (**) in terms introduced above, getting equivalent conditions denoted here by (***).

Algorithm 3 {alternative two-dimensional pattern matching}
  begin
    compute dictionary of small factors for all rows of X, Y and construct tables $\underline{X}$, $\underline{Y}$;
    compute dictionary of thin factors and construct string X'=flat(X) and array C(Y);
  for each row i of C(Y) **do in parallel**
    identify all patterns compr(SHIFT(X', 0), .., compr(SHIFT(X', k-1)
    in the i-th row of C(Y) by the algorithm from Lemma 1;
  find occurrences of the first and the last column of $\underline{X}$ in all columns of $\underline{Y}$;
  for each position (i, j) **do in parallel**
    {in constant time by one processor for each (i, j) due to information already computed}
    if conditions (***) are satisfied for (i, j) **then** report the match at (i, j);
  end algorithm.

Theorem 3
  Under the CRCW PRAM model, algorithms 2 and 3 are optimal parallel log(n)-time algorithms for two-dimensional pattern matching.

Concluding remark

The Kedem-Landau-Palem algorithm is a rough algorithm. It does not use any special mathematics on strings or images. Nevertheless, it is very efficient and easy to understand. The idea of cutting an object into small pieces and compressing them is a useful one. In this paper, we introduced a new notion of thin factors. We consider that the notion of small two-dimensional pieces of the image (thin factors) is a natural notion when dealing with two- dimensional objects, and could be helpful in other applications. Our approach to two-dimensional pattern matching is here more oriented to the two-dimensionality of the problem. It also shows applicability of cutting/compression technique.

Acknowledgment

We are grateful to Gadi Landau for pointing out non-triviality of the 2D pattern matching for constant size alphabets, and activating our work on optimal parallel computation for the problem.

References

[AILSV 88] A. Apostolico, C. Iliopoulos, G. Landau, B. Schieber, U. Vishkin, Parallel construction of a suffix tree with applications, *Algorithmica* 3 (1988) 347-365.

[AL 91] A. Amir, G.M. Landau, Fast parallel and serial multidimensional approximate pattern matching, *Theoret. Comput. Sci.*81 (1991) 97-15.

[Ba 78] T. Baker, A technique for extending rapid exact string matching to arrays of more than one dimension, *SIAM J. Comp.* 7 (1978) 533-541.

[Bi 77] R.S. Bird, Two-dimensional pattern-matching, *IPL* 6 (1977) 168-170.

[CR 90] M. Crochemore, W. Rytter, Parallel computations on strings and arrays, in: (*STACS'90*, Choffrut and Lengauer eds, LNCS 415, Springer-Verlag, 1990) 109-125.

[CR 91] M. Crochemore, W. Rytter, Usefulness of the Karp-Miller-Rosenberg algorithm in parallel computations on strings and arrays, *Theoret. Comput. Sci.* 88 (1991) 59-82.

[[GR 88] A. Gibbons, W. Rytter, *Efficient parallel algorithms*, Cambridge University Press (1988).

[KMR 72] R. Karp, R. Miller, A. Rosenberg, Rapid identification of repeated patterns in strings, arrays and trees, in proceedings of *ACM Symposium on Theory Of Computation*, 4 (1972) 125-136.

[KLP 89] Z. Kedem, G.M. Landau, K. Palem, Optimal parallel suffix-prefix matching algorithm and its applications, *SPAA'89*.

[Vi 85] U. Vishkin, Optimal parallel pattern matching in strings, *Information and Control*, 67 (1985) 91-113.

An Efficient Line Drawing Algorithm
For Parallel Machines

Phil Graham, S. Sitharama Iyengar, and Si-Qing Zheng

Department of Computer Science
Louisiana State University
Baton Rouge, LA 70803

Abstract. Fractals have recently been used to draw self-similar objects such as trees, coastlines, and mountains in computer graphics. This paper shows that there are advantages in using fractals to draw lines in an MIMD environment as well. In the course of developing our parallel algorithm, methods which reduce the space requirements and increase the speed of the sequential algorithm upon which it is based are also discussed. While the only algorithm presented here deals with lines, it is possible that there are advantages in developing comparable algorithms which draw other self-similar geometric shapes, such as circles.

Keywords and Phrases. Fractal, algorithm, line drawing, MIMD environment, integer logic, rubber band lines, binary tree.

1 Introduction

Over the years, much attention has been devoted to developing efficient line drawing algorithms in computer graphics. In addition to drawing lines and modeling geometric shapes, such as squares, line drawing algorithms are used to approximate other shapes such as circles. They are also used in ray tracing and other applications involving lines.

Truly, line drawing algorithms are a fundamental topic in the area of graphics algorithms. Various algorithms have been presented which eliminate or greatly reduce the need to perform multiplications and divisions. For instance, the traditional Bresenham algorithm [1] performs integer logic on the error term with respect to one coordinate as values of the other coordinate are incremented. The algorithm in [4] uses integer logic to generate line points by instead using a double step increment with respect to one of the coordinate axes. Bresenham's run length slice algorithm [2, 3] calculates a slice of movements with respect to a particular coordinate axis in each iteration of a loop. There are also line drawing algorithms which are derivations or variations of these algorithms [5].

Although the algorithms just described greatly increase the speed, lower response times are desired in real-time or interactive graphics and visualization in large-scale scientific computing. This is achieved by running any of the above algorithms on multiple processors. When line drawing algorithms are run in a multiprocessing environment, there are two approaches. In the first approach, each processor simultaneously draws a different line. If the number of lines to be drawn is large, this approach is highly efficient. No additional multiplications/divisions are needed during startup, all processors are active, and the throughput is high. In the second approach, each individual line is divided up, with each processor drawing a portion of it. Here, at least two multiplications/divisions are needed during startup, assuming no communication is performed and the work is divided equally. This is because it is necessary for each processor to determine its starting y coordinate at an arbitrary location of x (a division is required to determine the slope and a multiplication is required to determine $y = slope \cdot x$). In spite of the additional work during startup, this second approach can be faster than the first when the number of lines to be drawn is small, as is in the case of rubber band lines.

It has been shown in [6] that when Bresenham's line drawing algorithm is modified to be used for the second approach, four multiplications/divisions are needed during startup if the number of processors, P, is a power of two. If P is not a power of two, five multiplications/divisions are needed during startup. Because of these time consuming operations, the speedup only approaches the perfect value of P for longer lines. Presumably, the other algorithms also require a relatively large number of multiplications/divisions if no communication is performed and the work is evenly divided. This is because each processor must determine its starting x coordinate value, starting y coordinate value, the number of pixels it will set, and the value of any variables used to perform integer logic at the starting point. Some algorithms such as the one in [3] and the repeated pattern bidirectional algorithm in [5] also require multiplications/divisions during startup when they are performed sequentially. Therefore, it appears that a different algorithm must be used in order to lower the number of multiplications/divisions performed.

Recently, a method called the *recursive bisection* (RB) algorithm was proposed which uses a fractal approach to draw approximations to lines [7]. In other words, the pixels set by the algorithm are not always the ones that are closest to the true line. However, as noted in [7], the RB algorithm gives good line approximations on low resolution devices. On high resuolution devices, the difference between the line drawn by the RB algorithm and a line drawn by setting the pixel closest to the true line is imperceptible. Moreover, [7] shows that the RB algorithm is faster than the traditional Bresenham line drawing algorithm in cases such as when the lines drawn are at or near horizontal, diagonal, or vertical.

This paper shows that when each processor draws a portion of an individual line, the RB algorithm can be more suitable for use in parallel machines than the traditional Bresenham algorithm. It is also likely that our algorithm can be more efficient than any of the other line drawing algorithms previously mentioned. In fact, there does not exist an algorithm that has fewer steps which require time proportional to multiplication/division since there are only two such steps in the algorithm presented here. Furthermore, the amount of logic used in our algorithm is very small.

The remainder of the paper is organized as follows. Our algorithm is described, and its proofs of correctness are given in the next section. Section 3 describes the time complexity in detail and compares the results with the parallel Bresenham algorithm. Section 4 concludes the paper and discusses future work. The detailed algorithm is given in the appendix.

2 Algorithm Description

In the subsections which follow, we will describe the sequential RB method briefly and our parallel algorithm in detail. In these algorithms and in the remainder of this paper, the following notation is used when drawing a line from point (x_s, y_s) to point (x_f, y_f):

$$u = x_f - x_s,$$
$$v = y_f - y_s, \text{ and}$$
$$w = u - v.$$

As in [7], only lines such that $w > v > 0$ (i.e., lines in the first hexadecimant) will be considered in order to simplify descriptions of the algorithms. For the remaining cases, the algorithms are similar.

(i) Initially we have $S^{10}D^4$.

(ii) After one application of the replacement rule, we have $S^5D^2S^5D^2$.

(iii) After applying the replacement rule to each SD pair in (ii), we have $S^2D^1S^3D^1S^2D^1S^3D^1$.

Figure 1. The lines that result from various applications of the replacement rule when $w = 10$ and $v = 4$.

2.1 The Sequential RB Algorithm

Before describing our proposed algorithm, we will first give an overview of the sequential algorithm upon which it is based, the recursive bisection algorithm [7]. The sequential recursive bisection line drawing algorithm uses the symbols S and D to represent sideways and diagonal movements when drawing the line from point (x_s, y_s) to point (x_f, y_f). In short, it defines lines fractally with the recursive replacement rule:

$$S^wD^v \rightarrow S^{wl}D^{vl}S^{wr}D^{vr}$$

where $wl = w$ div 2, $vl = v$ div 2, $wr = w - wl$, and $vr = v - vl$. Of course, "div" is defined to be integer division, as it is in the Pascal language. The replacement rule is no longer applied when either w or v equals 0 or 1. Figure 1 gives an example of the lines that result from each application of the replacement rule when $w = 10$ and $v = 4$.

As shown in [7], the solution generated by the RB method for any line parameter can be characterized as a binary tree (Figure 2). We will call this tree the SD-partition tree (or simply the SD-tree). Any subtrees of the SD-tree will be called an SD-subtree. The nodes at a given level k of the SD-tree (with the root at level 0) will be numbered 0, 1, 2, ... , $2^k - 1$ from left to right. The definitions below are used in the algorithms and proofs which follow.

Definition 1: $n_{k, i}$ denotes node number i in level k of the SD-tree.
Definition 2: $v_{k, i}$ ($w_{k, i}$) is the value of v (w) at $n_{k, i}$.
Definition 3: $x\_start_{k, i}$ ($y\_start_{k, i}$) is the x (y) coordinate value from which the leftmost operator in the SD-subtree rooted at $n_{k, i}$ originates.
Definition 4: $D\_ops_{k, i}$ ($S\_ops_{k, i}$) is the number of D (S) operators that occur to the left of the SD-subtree rooted at $n_{k, i}$.

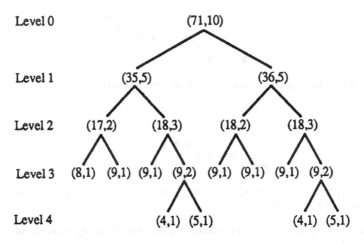

Figure 2. An example of an SD-tree for $w = 71$ and $v = 10$. In the diagram, the numbers in parentheses show the values of w and v at each node. The resulting run-length sequence is:
$$S^8D^1S^9D^1S^9D^1S^4D^1S^5D^1S^9D^1S^9D^1S^9D^1S^4D^1S^5D^1.$$

2.2 The Parallel RB Algorithm

An overview of our parallel algorithm is given in this section. Before this is done, it is necessary to discuss the model of computation and any assumptions made. Our algorithm runs in an MIMD environment in which each processor has local data and access to shared memory. The raster memory is also shared, and each processor sets the appropriate bits in raster memory when drawing its portion of the line. In addition, the processors are dedicated and assigned to the task of line drawing so there is no need to "fork" or otherwise activate them. It is also assumed that each pixel position has a distinct memory address. Therefore, synchronization should require very little time because the algorithms are designed to avoid address contention. If these assumptions are not true, then the algorithms must be modified so that proper coordination is provided or the possibility of contention is eliminated. This paper does not include such modifications. These assumptions are the same as those in [6]. The only additional assumption we make is that the number of processors is a power of two.

Conceptually, our method of parallelizing the the sequential recursive bisection algorithm assigns processors to all SD-subtrees rooted at a given level. Next, the values of v and w at the root of the SD-subtree to which the processor has been assigned and the starting point of the line segment must be determined. Once all of the above information is known, each processor can set the pixel values of its SD-subtree by running the sequential RB algorithm. There are two cases: either P, the number of processors, is less than or equal to $\min(|v|, |w|)$, or it isn't. In the following discussions, each of the P processors are referred to as P_i because each processor is assigned a unique number, i where $0 \leq i < P$. The $\log_2 P$-bit binary

representation of i will be denoted by $b^i = b^i_{\log_2 P - 1} b^i_{\log_2 P - 2} b^i_{\log_2 P - 3} \cdots b^i_0$ and the binary representations of other variables will be denoted in a similar fashion.

We now summarize the algorithm performed by each of the processors, assuming $w > v > 0$.

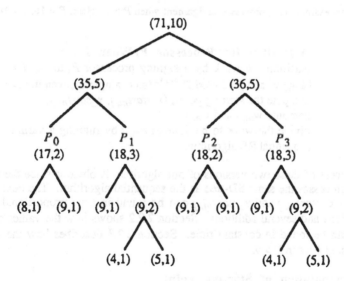

Figure 3. An example of a processor assignment when $P \leq v$. Here, $P = 4$, $v = 10$, and $w = 71$.

Algorithm for Processor P_i When $P \leq v$

step 1:   partition the work by assigning processor P_i to $n_{m,j}$ where $m = \log_2 P$ and $j = i$ (an example is given in Figure 3).

step 2:   compute the starting point, $(x\_start_{m,j}, y\_start_{m,j})$.

step 3:   compute $w_{m,j}$ and $v_{m,j}$.

step 4:   run the sequential RB algorithm on the SD-subtree rooted at $n_{m,j}$.

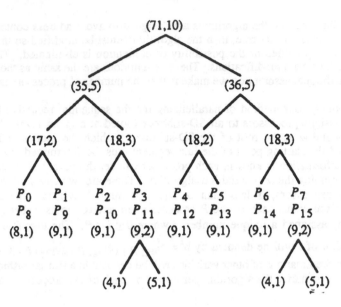

Figure 4. An example of a processor assignment when $P > v$. Here, $P = 16$, $v = 10$, and $w = 71$.

Algorithm for Processor P_i When $P > v$

step 1: partition the work by assigning processor P_i to $n_{m, j}$ where $m = \lfloor \log_2 v \rfloor$ and $j = i \bmod 2^{\lfloor \log_2 v \rfloor}$ (an example is given in Figure 4).

step 2: compute the starting point, $(x\_start_{m, j}, y\_start_{m, j})$.

step 3: compute $w_{m, j}$ and $v_{m, j}$.

step 4: divide the work to be done at $n_{m, j}$ by running a variation of the sequential RB algorithm.

The correctness of these two versions of our algorithm is obvious since the parallel algorithm processes the same SD-tree as the sequential algorithm. The next section shows how $x\_start_{m, j}$ and $y\_start_{m, j}$ can be found in time proportional to two multiplications and several additions. Section 2.2.2 shows how the values of $w_{m, j}$ and $v_{m, j}$ can be found in constant time. Section 2.2.3 describes how the work is divided in step 4 when $P > v$.

2.2.1 Computation of Starting Point

In this section, we will explain how $x\_start_{m, j}$ and $y\_start_{m, j}$ can be found efficiently. As shown above, the values of m and j for a given processor depend upon whether $P \leq v$ or $P > v$. For either case, the starting x and y coordinate values of the point from which P_i processes its first operator are determined to be:

$$x\_start_{m, j} = x_s + S\_ops_{m, j} + D\_ops_{m, j}$$
$$y\_start_{m, j} = y_s + D\_ops_{m, j}.$$

The term $D\_ops_{m,\,j}$ is added to x_s when finding $x\_start_{m,\,j}$ because $u = v + w$; when a diagonal operator is processed, a movement is made one unit in both the x and y directions. The problem now is to be able to find $S\_ops_{m,\,j}$ and $D\_ops_{m,\,j}$ efficiently. In order to determine the number of S operators that have been processed before the SD-subtree rooted at $n_{m,\,j}$ the least significant m digits of b^j must be examined because $n_{m,\,j}$ is the jth node at level m. For instance, if $b^j_{m-1} = 1$ then $n_{m,\,j}$ is in the right SD-subtree of the root node, and at least w div 2 operators have been processed. If $b^j_{m-2} = 1$, then $n_{m,\,j}$ is in the right SD-subtree of a node at level 1, and at least w div 2^2 additional operators have been processed, etc. Some of these terms must be incremented because the value of w at a node in a given level k of the SD-tree may be $(w$ div $2^k) + 1$, as noted in the corollary of Theorem 8 in [7]. Therefore, the number of S operators processed before the SD-subtree rooted at $n_{m,\,j}$ is shown by the following relationship:

$$S\_ops_{m,\,j} = \sum_{y=0}^{m-1} (t_y + inc_y)$$

where

$$t_y = \begin{cases} w \text{ div } 2^{y+1} & \text{if } b^j_{m-y-1} = 1; \\ 0 & \text{if } b^j_{m-y-1} = 0, \end{cases}$$

and inc_y is either 0 or 1, depending on whether term t_y should be incremented. Thus, it intuitively appears that if the wordsize of a computer is n, then finding the number of S (or D) operators that have been processed before a given SD-subtree rooted at level m may require at least $\Omega(nm - n)$ time because up to $m - 1$ additions may be necessary.

However, we will now show that finding the number of S operators processed before a given SD-subtree can be done in $O(n\log n \ \log\log n)$ time since the problem is reducible to multiplication [8]. Again examining $n_{m,\,j}$, it is apparent that if $n_{m,\,j}$ is in the right SD-subtree of a node at level y (which occurs when $b^j_{m-y-1} = 1$) and the value of w at node number $b^j_{m-1}b^j_{m-2}...b^j_{m-y}0$ in level $y + 1$ is $[(w$ div $2^{y+1}) + 1]$ (by Theorem 2, this occurs when the value of the complement of $0b^j_{m-y}...b^j_{m-2}b^j_{m-1}$, denoted as $\overline{0b^j_{m-y}...b^j_{m-2}b^j_{m-1}}$, is less than $b^w_y b^w_{y-1}...b^w_0$), then term t_y should be incremented. That is, increments are added to terms when the conditions shown in Table 1 are satisfied. Equivalent forms are given in Table 2. Naturally, all the increment values can be found in $O(m)$ time by starting with the increment associated with t_1 and saving the results of the current comparison in the following manner. Once the value of inc_1 is found, it will be determined whether t_2 should be incremented by finding the value of $b_2^w b_{m-3}^j$ and the result of the comparison $\overline{b_{m-2}^j b_{m-1}^j} < b_1^w b_0^w$. However, it isn't necessary to examine all the digits when comparing since the results of the previous comparison have been saved. In general, when finding the result of the comparison for determining whether t_y should be incremented, only digits $\overline{b_{m-y}^j}$ and b_{y-1}^w need be inspected if the result of the comparison for inc_{y-1} is known. Once this information is obtained, the result of the comparison for determining whether t_y should be incremented is as shown in the truth tables in Figure 5. After determining the value of inc_2, the values of inc_3, inc_4, ... , inc_{m-1} are found in a similar manner.

Table 1. Conditions when terms are incremented when determining the number of S operators processed before the SD-subtree rooted at the jth node in level m of the SD-tree.

Term	Condition When Term is Incremented
$t_0 = w \text{ div } 2^1$	$(b_{m-1}^j = 1)$ and $(\overline{0} < b_0^w)$
$t_1 = w \text{ div } 2^2$	$(b_{m-2}^j = 1)$ and $(\overline{0 b_{m-1}^j} < b_1^w b_0^w)$
$t_2 = w \text{ div } 2^3$	$(b_{m-3}^j = 1)$ and $(\overline{0 b_{m-2}^j b_{m-1}^j} < b_2^w b_1^w b_0^w)$
$t_3 = w \text{ div } 2^4$	$(b_{m-4}^j = 1)$ and $(\overline{0 b_{m-3}^j b_{m-2}^j b_{m-1}^j} < b_3^w b_2^w b_1^w b_0^w)$
...	...
$t_{m-1} = w \text{ div } 2^m$	$(b_0^j = 1)$ and $(\overline{0 b_1^j b_2^j ... b_{m-1}^j} < b_{m-1}^w b_{m-2}^w ... b_0^w)$

Table 2. Forms equivalent to Table 1.

Term	Condition When Term is Incremented
$t_0 = w \text{ div } 2^1$	never
$t_1 = w \text{ div } 2^2$	$(b_1^w b_{m-2}^j = 1)$ and $(\overline{b_{m-1}^j} < b_0^w)$
$t_2 = w \text{ div } 2^3$	$(b_2^w b_{m-3}^j = 1)$ and $(\overline{b_{m-2}^j b_{m-1}^j} < b_1^w b_0^w)$
$t_3 = w \text{ div } 2^4$	$(b_3^w b_{m-4}^j = 1)$ and $(\overline{b_{m-3}^j b_{m-2}^j b_{m-1}^j} < b_2^w b_1^w b_0^w)$
...	...
$t_{m-1} = w \text{ div } 2^m$	$(b_{m-1}^w b_0^j = 1)$ and $(\overline{b_1^j b_2^j ... b_{m-1}^j} < b_{m-2}^w b_{m-3}^w ... b_0^w)$

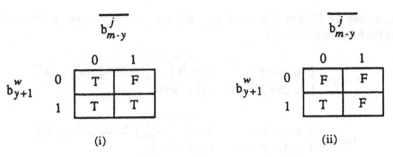

Figure 5. Result of comparison used to determine whether t_y should be incremented when (i) the result of the comparison for t_{y-1} is true and (ii) the result of the comparison for t_{y-1} is false.

Thus, finding the value of $S\_ops_{m,\,j}$ is similar to finding the result of $(jw$ div $2^m)$. For example, if $n = 8$ and $2^m = 16$, $(jw$ div $2^m)$ is:

$$
\begin{array}{c}
b_7^w b_0^j \ \ b_6^w b_0^j \ \ b_5^w b_0^j \ \ b_4^w b_0^j \ \ b_3^w b_0^j \ \ b_2^w b_0^j \ \ b_1^w b_0^j \ \ b_0^w b_0^j \\
b_7^w b_1^j \ \ b_6^w b_1^j \ \ b_5^w b_1^j \ \ b_4^w b_1^j \ \ b_3^w b_1^j \ \ b_2^w b_1^j \ \ b_1^w b_1^j \ \ b_0^w b_1^j \\
b_7^w b_2^j \ \ b_6^w b_2^j \ \ b_5^w b_2^j \ \ b_4^w b_2^j \ \ b_3^w b_2^j \ \ b_2^w b_2^j \ \ b_1^w b_2^j \ \ b_0^w b_2^j \\
+ \ \ b_7^w b_3^j \ \ b_6^w b_3^j \ \ b_5^w b_3^j \ \ b_4^w b_3^j \ \ b_3^w b_3^j \ \ b_2^w b_3^j \ \ b_1^w b_3^j \ \ b_0^w b_3^j \\
\hline
\qquad\qquad jw \text{ div } 2^m \qquad\qquad\Big| \qquad jw \bmod 2^m
\end{array}
$$

and $S\_ops_{m,\,j}$ equals:

$$
\begin{array}{c}
b_7^w b_0^j \ \ b_6^w b_0^j \ \ b_5^w b_0^j \ \ b_4^w b_0^j \ \ 2[inc_3] \\
b_7^w b_1^j \ \ b_6^w b_1^j \ \ b_5^w b_1^j \ \ b_4^w b_1^j \ \ b_3^w b_1^j \ \ 2[inc_2] \\
b_7^w b_2^j \ \ b_6^w b_2^j \ \ b_5^w b_2^j \ \ b_4^w b_2^j \ \ b_3^w b_2^j \ \ b_2^w b_2^j \ \ 2[inc_1] \\
+ \ \ b_7^w b_3^j \ \ b_6^w b_3^j \ \ b_5^w b_3^j \ \ b_4^w b_3^j \ \ b_3^w b_3^j \ \ b_2^w b_3^j \ \ b_1^w b_3^j \ \ 2[inc_0] \\
\hline
\text{number of S operators processed} \qquad\Big|
\end{array}
$$

Since the values of v are distributed in the tree structure in the same way as the values of w, $D\_ops_{m,\,j}$ is found by replacing the values of w in the above arguments with those of v. This completes the reduction of $S\_ops_{m,\,j}$ and $D\_ops_{m,\,j}$ to multiplication. Therefore, $x\_start_{m,\,j}$ and $y\_start_{m,\,j}$ can be found in time proportional to two multiplications and several additions.

2.2.2 Computation of $v_{m,\,j}$ and $w_{m,\,j}$

Of course, $w_{m,\,j}$ and $v_{m,\,j}$ must be determined in order for processor P_i to run the sequential algorithm on the SD-subtree rooted at $n_{m,\,j}$. These values could be found by traversing the SD-tree. However, $v_{m,j}$ and $w_{m,j}$ can be obtained in constant time using the following formulas:

$$w_{m,j} = \begin{cases} w \text{ div } 2^m & \text{if } \overline{b_0^j b_1^j ... b_{m-1}^j} \geq b_{m-1}^w b_{m-2}^w ... b_0^w; \\ w \text{ div } 2^m + 1 & \text{otherwise.} \end{cases}$$

$$v_{m,j} = \begin{cases} v \text{ div } 2^m & \text{if } \overline{b_0^j b_1^j ... b_{m-1}^j} \geq b_{m-1}^v b_{m-2}^v ... b_0^v; \\ v \text{ div } 2^m + 1 & \text{otherwise.} \end{cases}$$

In these formulas $\overline{b_0^j b_1^j b_2^j ... b_{m-1}^j}$ is the binary number obtained by complementing all the bits of $b_0^j b_1^j b_2^j ... b_{m-1}^j$. The above formula for $w_{m,j}$ holds because it is known from Theorem 8 and its corollary of [7] that all values of w at any level k are either (w div 2^k) or [(w div 2^k) + 1]. In order to determine whether $w_{m,j}$ equals (w div 2^m) or [(w div 2^m) + 1], $n_{m,j}$ is renumbered according to the order in which it is incremented (i.e., the value of w at $n_{m,j}$ becomes [(w div 2^m) + 1]) if w should be 2^m and then increased to 2^{m+1}. Since $n_{m,j}$ is the jth node at level m it is renumbered as $\overline{b_0^j b_1^j b_2^j ... b_{m-1}^j}$. $\overline{b_0^j b_1^j b_2^j ... b_{m-1}^j}$ is then compared with the number of nodes that are incremented, which equals $b_{m-1}^w b_{m-2}^w b_{m-3}^w ... b_0^w$, to determine the value of $w_{m,j}$. $v_{m,j}$ is found in a similar manner since the values of v are distributed the same way in the SD-tree as the values of w. Formal proofs are given as Theorems 1 and 2 which appear at the end of this section.

Lemma 1: If $m \geq 1$ and $2^m \leq v < 2^{m+1}$, then

$$v_{m,j} = \begin{cases} 1 & \text{if } \overline{b_0^j b_1^j ... b_{m-1}^j} \geq b_{m-1}^v b_{m-2}^v ... b_0^v; \\ 2 & \text{otherwise.} \end{cases}$$

Proof: The lemma will be proven by induction on m.

Basis: Assume $m = 1$. Then the lemma can only be applied to two nodes, $n_{1,0}$ and $n_{1,1}$. Also, there are two values of v for which the lemma is applicable, when it equals 2 or 3. According to the lemma, $v_{1,0}$ equals one when $v = 2$ since the comparison value of node number zero, $\overline{b_0^j} = \overline{0} = 1$, is greater than b_0^v. $v_{1,1}$ should also equal one since the comparison value of node number one, $\overline{b_0^j} = \overline{1} = 0$, equals b_0^v. When $v = 3$, $v_{1,0}$ should equal one since its comparison value, $\overline{b_0^j} = \overline{0} = 1$, equals b_0^v. $v_{1,1}$ should equal two because its comparison value, $\overline{b_0^j} = \overline{1} = 0$, is less than b_0^v. In all of the above cases, the tests of the comparison values with v give the correct results.

123

Induction: Assume that the lemma holds when $m \leq k$. It will now be shown that the lemma also holds when $m = k + 1$. There are 2 cases: either v is even, or it is odd. For each case, tests are formed for the nodes at level $k + 1$ by finding the value of v at a node in level one and applying the induction hypothesis (the choice of whether to find $v_{1,\,0}$ or $v_{1,\,1}$ depends on whether $n_{m,\,j}$ is in the left or right SD-subtree of the root node). The induction hypothesis can be applied because the nodes are numbered from 0 to $2^k - 1$ when only the least significant k bits of the node labels are considered. Figure 6 gives an example where $k + 1 = 3$. Of course, the other conditions for the use of the lemma are also satisfied. After an assumption is made that a test holds for an SD-subtree, it is shown that the lemma must hold for the nodes in an SD-tree of height $k + 1$ because the assumption leads to a test which is identical to that in the lemma.

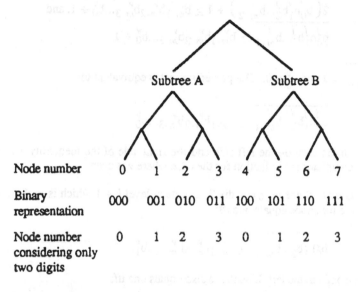

Figure 6. An example of the numberings of the nodes of an SD-tree having height 3 when only the 2 least significant binary digits are considered. The values of the nodes in both SD-subtree A and B increase from left to right, going from 0 to $2^k - 1$.

case 1: v is even. Since $v_{1,\,1} = v - (v\ \mathrm{div}\ 2)$, which equals $(v\ \mathrm{div}\ 2)$ when v is even, a node at level $k + 1$ which is in the right SD-subtree of the root node equals one iff:

$$\overline{b_0^j b_1^j b_2^j ... b_{m-2}^j} \geq b_{m-1}^v b_{m-2}^v b_{m-3}^v ... b_1^v.$$

Therefore, a node in the right SD-subtree of the root node also equals one iff:

$$\overline{b_0^j b_1^j b_2^j ... b_{m-2}^j} \geq b_{m-1}^v b_{m-2}^v b_{m-3}^v ... b_1^v,$$

$$2\left(\overline{b_0^j b_1^j b_2^j \ldots b_{m-2}^j}\right) \geq 2\left(b_{m-1}^v b_{m-2}^v b_{m-3}^v \ldots b_1^v\right), \text{ and}$$

$$b_0^j b_1^j b_2^j \ldots b_{m-1}^j \geq b_{m-1}^v b_{m-2}^v b_{m-3}^v \ldots b_0^v.$$

because $b_{m-1}^j = 0$ and $b_0^v = 0$. Since $v_{1,0} = v$ div 2, a node at level $k+1$ which is in the left SD-subtree of the root node equals one iff:

$$b_0^j b_1^j b_2^j \ldots b_{m-2}^j \geq b_{m-1}^v b_{m-2}^v b_{m-3}^v \ldots b_1^v,$$

$$2\left(\overline{b_0^j b_1^j b_2^j \ldots b_{m-2}^j}\right) + 1 \geq 2\left(b_{m-1}^v b_{m-2}^v b_{m-3}^v \ldots b_1^v\right) + 1,$$

$$2\left(\overline{b_0^j b_1^j b_2^j \ldots b_{m-2}^j}\right) + 1 \geq b_{m-1}^v b_{m-2}^v b_{m-3}^v \ldots b_0^v + 1, \text{ and}$$

$$b_0^j b_1^j b_2^j \ldots b_{m-1}^j \geq b_{m-1}^v b_{m-2}^v b_{m-3}^v \ldots b_0^v + 1$$

because $b_{m-1}^j = 1$ and $b_0^v = 0$. The preceding test is equivalent to:

$$b_0^j b_1^j b_2^j \ldots b_{m-1}^j \geq b_{m-1}^v b_{m-2}^v b_{m-3}^v \ldots b_0^v$$

because both the term on the left side and the right side of the inequality are always odd. This completes the induction for the case where v is even.

case 2: v is odd. Since $v_{1,0} = v$ div 2, a node at level $k+1$ which is in the left SD-subtree of the root node equals one iff:

$$b_0^j b_1^j b_2^j \ldots b_{m-2}^j \geq b_{m-1}^v b_{m-2}^v b_{m-3}^v \ldots b_1^v.$$

Therefore, a node in the left SD-subtree also equals one iff:

$$b_0^j b_1^j b_2^j \ldots b_{m-2}^j \geq b_{m-1}^v b_{m-2}^v b_{m-3}^v \ldots b_1^v,$$

$$2\left(\overline{b_0^j b_1^j b_2^j \ldots b_{m-2}^j}\right) + 1 \geq 2\left(b_{m-1}^v b_{m-2}^v b_{m-3}^v \ldots b_1^v\right) + 1,$$

$$b_0^j b_1^j b_2^j \ldots b_{m-1}^j \geq 2\left(b_{m-1}^v b_{m-2}^v b_{m-3}^v \ldots b_1^v\right) + 1, \text{ and}$$

$$b_0^j b_1^j b_2^j \ldots b_{m-1}^j \geq b_{m-1}^v b_{m-2}^v b_{m-3}^v \ldots b_0^v.$$

because $b_{m-1}^j = 1$ and $b_0^v = 1$. Since $v_{1,1} = v - (v \text{ div } 2)$, which equals $[(v \text{ div } 2) + 1]$ when v is odd, there are now two subcases if a node is in the right SD-subtree. In the first subcase, $v = 2^{k+1} - 1$. Since $v_{1,1} = 2^k$, the induction hypothesis cannot be applied to the SD-subtree rooted at $n_{1,1}$. However, the lemma obviously holds for the nodes in the subtree because the value of v at each node of interest equals two and all the comparison values are smaller than $b_{m-1}^v b_{m-2}^v b_{m-3}^v ... b_0^v$, which equals $111...1_2$. In the second subcase, the induction hypothesis can always be applied to the SD-subtree because $2^{k-1} \le v_{1,1} < 2^k$. Therefore, the value of v at a node in the right SD-subtree will equal one iff:

$$\overline{b_0^j b_1^j b_2^j ... b_{m-2}^j} \ge \left(b_{m-1}^v b_{m-2}^v b_{m-3}^v ... b_1^v \right) + 1,$$

$$2\left(\overline{b_0^j b_1^j b_2^j ... b_{m-2}^j} \right) \ge 2\left[\left(b_{m-1}^v b_{m-2}^v b_{m-3}^v ... b_1^v \right) + 1 \right],$$

$$2\left(\overline{b_0^j b_1^j b_2^j ... b_{m-2}^j} \right) \ge b_{m-1}^v b_{m-2}^v b_{m-3}^v ... b_0^v + 1, \text{ and}$$

$$\overline{b_0^j b_1^j b_2^j ... b_{m-1}^j} \ge b_{m-1}^v b_{m-2}^v b_{m-3}^v ... b_0^v + 1$$

because $\overline{b_{m-1}^j} = 0$ and $b_0^v = 1$. The preceding test is equivalent to:

$$\overline{b_0^j b_1^j b_2^j ... b_{m-1}^j} \ge b_{m-1}^v b_{m-2}^v b_{m-3}^v ... b_0^v$$

because both the term on the left side and the right side of the inequality are always even. This completes the induction and the proof of the lemma. $\square$

Theorem 1: If $1 \le m \le \lfloor \log_2 v \rfloor$, then

$$v_{m,j} = \begin{cases} v \text{ div } 2^m & \text{if } \overline{b_0^j b_1^j ... b_{m-1}^j} \ge b_{m-1}^v b_{m-2}^v ... b_0^v; \\ v \text{ div } 2^m + 1 & \text{otherwise.} \end{cases}$$

Proof: The theorem is easily proven by contradiction once several observations concerning the distribution of v values in the SD-tree have been made. The first observation is that for any integer q, where $q = (v \bmod 2^m) + 2^m$, Lemma 1 is applicable for any node at level m. We also note that for each multiple of 2^m by which q is increased, the value of v at each node in level m is increased by one (hence, the v div 2^m terms in the theorem). Therefore, if the theorem does not hold at $n_{m,j}$ when $v = r$, then there is a contradiction of Lemma 1 at $n_{m,j}$ when $v = (r \bmod 2^m) + 2^m$. $\square$

Theorem 2: If $1 \le m \le \lfloor \log_2 v \rfloor$, then

$$w_{m,j} = \begin{cases} w \text{ div } 2^m & \text{if } \overline{b_0^j b_1^j ... b_{m-1}^j} \ge b_{m-1}^w b_{m-2}^w ... b_0^w; \\ w \text{ div } 2^m + 1 & \text{otherwise.} \end{cases}$$

Proof: The proof is similar to that for Theorem 1. For brevity, we omit it. $\square$

Theorems 1 and 2 ensure the correctness of our algorithm.

2.2.3 Dividing the Work When $P > v$

If $P \leq v$, the only work that remains is for each processor to run the sequential RB algorithm on the SD-subtree to which it is assigned. However, the work must be divided further when $P > v$ since more than one processor is assigned to each SD-subtree. Stated another way, the responsibility for setting the pixels having x coordinates from $x\_start_{m, j} + 1$ to $x\_start_{m, j+1}$ must be divided equally among the processors assigned to the SD-subtree rooted at $n_{m, j}$. This is done by having each processor P_i set pixels having x coordinates $x\_start_{m, j} + 1 + pos_i, x\_start_{m, j} + 1 + pos_i + num, x\_start_{m, j} + 1 + pos_i + 2(num), \ldots$ where num is the number of processors assigned to each SD-subtree and pos_i is a unique number in the range from 0 to $num - 1$ which is determined by the value of i. The values of num and pos_i are found as follows:

$$num = 2^{\log_2 P - m}$$
$$pos_i = i \text{ div } 2^m$$

For the detailed algorithms, refer to the appendix. We add that the code for the case where $P > v$ is not optimized so that the steps of the algorithm can be shown more clearly.

3 A Comparison of Performance

Up to this point, it has been assumed that the parallel RB algorithm divides the work into approximately equal loads even though no proof has been given. However, its truth is evident when the values of v (and w) at the nodes in the SD-tree are examined. As noted earlier, the values of v (and w) at level m of the SD-tree differ by at most one. Since the processors are also equally divided among the nodes at level m, the workloads for each processor are all approximately the same size. Therefore, the time complexities of each algorithm are as follows:

Parallel RB ($P \leq v$):	$c_1 + c_2 (u)/P$
Parallel RB ($P > v$):	$c_3 + c_4 (u)/P$
Parallel Bresenham:	$c_5 + c_6 (u)/P$

where c_1, c_3, and c_5 represent the preprocessing times. c_2, c_4, and c_6 are constants indicating the amount of time needed to process data per pixel set, disregarding the preprocessing time.

The order of complexity of the parallel RB algorithm is shown to be lower than that of the parallel Bresenham algorithm by examining each algorithm separately. The startup costs will be considered first. When $P \leq v$ in the parallel RB algorithm, the c_1 term is mainly the time needed to find $w_{m, j}, v_{m, j}, S\_ops_{m, j}$, and $D\_ops_{m, j}$. Therefore, c_1 is proportional to two multiplication operations. For the case where $P > v$, c_3 is the time needed to find $m, j, num, pos_i, w_{m, j}, v_{m, j}, S\_ops_{m, j}$, and $D\_ops_{m, j}$. Even though there is slightly more logic than in the case where $P \leq v$, c_3 is once again proportional to two multiplication operations. In contrast, the parallel Bresenham algorithm [6] requires four multiplication/division operations for each processor, i.e., c_5 is about twice the value of c_1 (or c_3). Therefore, the parallel RB algorithm is more efficient during startup. Each of the algorithms sets pixels in time proportional to u/P once the preprocessing work is performed.

Naturally, when implemented, the speed of each algorithm is highly instruction dependent. As a result, optimal time equivalent instructions were substituted for finding the values of m, $b_0^j b_1^j b_2^j ... b_{m-1}^j$, $S\_ops_{m, j}$, and $D\_ops_{m, j}$ in the following implementations in order to show the potential speed of the parallel RB algorithm. With this said, the startup costs for each of the above algorithms were estimated using compiled C on a VAX machine by performing each set of instructions 1,000 times on an arbitrary set of points in the first hexadecimant. The results showed that when $P \leq v$, the startup costs for the parallel RB algorithm were 37.02 ms, significantly lower than the 56.03 ms time of the parallel Bresenham algorithm. When $P > v$, the times were virtually the same. Since the number of instructions performed vary little with respect to the input data for any of the above algorithms, similar times are obtained for almost all sets of (x_s, y_s) and (x_f, y_f). In addition, Table 3, which is reproduced from [7], indicates that the sequential RB algorithm performs more favorably than the sequential Bresenham algorithm for large run lengths of S and D. Therefore, the parallel RB algorithm has an even greater improvement in speed in many instances.

Table 3. Comparison of Bresenham and recursive bisection line algorithm times (measured in milliseconds) for various values of $w - v$ and v using using C without graphics output on a Unix mainframe.

v	$w - v$	Bresenham	RB
30	0	0.83	0.78
30	20	1.16	0.97
30	40	1.16	1.05
30	60	1.4	1.1
30	80	1.7	1.3
30	100	1.9	1.9
30	120	2.4	2.0
30	140	2.8	2.1
30	160	3.5	2.0
30	180	3.5	2.3
30	200	4.9	2.2
100	0	2.4	2.5
100	20	2.7	3.0
100	40	3.5	3.2
100	60	3.1	3.0
100	80	3.3	3.5
100	100	3.5	3.3
100	200	4.6	3.9
100	300	5.5	4.5
500	0	12.3	12.7
500	100	13.3	13.3
500	200	14.3	14.1
500	500	17.8	16.0
500	1000	23.1	25.5

The times above will now be used to make generalizations concerning the speedups and the lengths of lines for which it is beneficial to use the parallel version of each algorithm. First, assume the time complexity of the sequential RB and Bresenham algorithms are as follows:

Sequential RB:	$c_7 + c_8(u)$
Sequential Bresenham:	$c_9 + c_{10}(u)$

where c_7 is a constant related to the amount of work done before the first call to the subroutine for setting pixels is made, and c_8 is a constant representing the amount of work done per pixel set once the subroutine for setting pixels is called. c_9 and c_{10} are constants related to the amount of work done before and during the loop which sets pixels in Bresenham's algorithm. Since $c_8 = c_2$, $c_8 \approx c_4$, and $c_{10} = c_6$, the lengths of the lines for which the times of the parallel algorithms are less than or equal to the times of the serial algorithms are as follows:

$$\text{Parallel RB } (P \leq v): \qquad u \geq \frac{(c_1 - c_7)P}{c_8(P - 1)}$$

$$\text{Parallel RB } (P > v): \qquad u \geq \frac{(c_3 - c_7)P}{c_8(P - 1)}$$

$$\text{Parallel Bresenham:} \qquad u \geq \frac{(c_5 - c_9)P}{c_{10}(P - 1)}$$

Assuming $c_8 \approx c_{10}$ and $c_7 \approx c_9$, the lengths of the lines for which it is beneficial to use the parallel RB algorithm are comparable to those for which it is beneficial to use the parallel Bresenham algorithm. From the previous assumptions, it is also clear that the speedup of the parallel RB algorithm is at least that of the parallel Bresenham when $P \leq v$ and about the same when $P > v$. Again, these comparisons are intended merely as generalizations for a given set of conditions, because the times obtained are dependent upon factors such as the instruction set and the wordsize of the computer.

4 Final Remarks

A new parallel line drawing algorithm is presented and analyzed. Our investigation shows that our parallel RB algorithm is faster than the known Bresenham algorithm. One may suspect that the space required by our algorithm is excessive due to recursive calls. However, it is straightforward to eliminate such calls using Theorems 1 and 2. Perhaps, there will also be advantages if comparable algorithms are developed for drawing other self-similar geometric shapes, such as circles. This is because it is possible that there are fewer variables that must be contended with when a fractal approach is used. It is also possible that any variables present in a fractal algorithm can be calculated more efficiently by only multiplying and dividing by powers of two, as was done in the algorithm presented here. Considering that many powerful parallel computers have been made commercially available and there is an increasing demand in fast graphic algorithms for important applications such as on-line or real-time visualization in large-scale scientific computing, designing efficient parallel algorithms for drawing objects of different shapes is an important and challenging task.

5 References

1. J.E. Bresenham: Algorithm for computer control of a digital plotter. IBM Systems Journal 4(1), 25-30 (1965).
2. J.E. Bresenham: Incremental line compaction. Comp. J. 23(1), 46-52 (1980).
3. J.E. Bresenham: Run length slice algorithm for incremental lines. In: R.A. Earnshaw (ed.): Fundamental Algorithms for Computer Graphics. NATO ASI Series. New York: Springer-Verlag 1985, pp. 59-104.
4. X. Wu and J.G. Rokne: Double-step incremental generation of lines and circles. Comp. Vision Graphics Image Processing 37, 331-334 (1987).
5. G. Casiola: Basic concepts to accelerate line algorithms. Comput. & Graphics 12(3/4), 489-502 (1988).
6. W.E. Wright: Parallelization of Bresenham's line and circle algorithms. IEEE Computer Graphics & Applications 10(5), 60-67 (1990).
7. J.R. Rankin: Recursive bisection line algorithm. Computers & Graphics 15(1), 1-8 (1991).
8. A.V. Aho, J.E. Hopcroft, and J.D. Ullman: The Design and Analysis of Computer Algorithms. Reading, Mass. (USA): Addison-Wesley 1974, pp. 272-273.

6 Appendix: The Detailed Parallel RB Algorithm.

Case Where $P \leq v$ for Subtrees in the First Hexadecimant

P_i:
```
const
        P = number of processors
        i_reverse = not(i[0..log₂P - 1])

var
        x, y:  integer;

procedure bisect(w, v:  integer);
var
        wl, vl, wr, vr, k:  integer;

begin
        if (v > 1) then
                    wl = w div 2
                    vl = v div 2
                    bisect(wl, vl)
                    wr = w - wl
                    vr = v - vl
                    bisect(wr, vr)
        else
                    /* set S operators */
                    for k = 1 to w do
```

```
                        x = x + 1
                        dot_on(x, y)
            end for
            /* set D operator */
            x = x + 1
            y = y + 1
            dot_on(x, y)
      end if
end bisect

procedure recursive_bisect(xs, ys, xf, yf: integer);
var
      u, v, w: integer;

begin
      u = xf - xs
      v = yf - ys
      w = u - v
      /* set x = x_startm, j and y = y_startm, j */
      y = D_ops(v, i, log2P)
      x = S_ops(w, i, log2P) + y + xs
      y = y + ys
      /* calculate wm, j and vm, j */
      if (i_reverse < w[log2P - 1..0]) then
            w = (w div P) + 1
      else
            w = w div P
      end if
      if (i_reverse < v[log2P - 1..0]) then
            v = (v div P) + 1
      else
            v = v div P
      end if
      if (i = 0) then
            dot_on(xs, ys)
      end if
      bisect(w, v)
end recursive_bisect
```

Case Where $P > v$ for Subtrees in the First Hexadecimant

P_i:
const
      P = number of processors

var
      x, y, start_pt: integer;

```
procedure bisect(w, v, num: integer);
var
        wl, vl, wr, vr, k, end_val: integer;

begin
        if (v > 1) then
                wl = w div 2
                vl = v div 2
                bisect(wl, vl)
                wr = w - wl
                vr = v - vl
                bisect(wr, vr)
        else
                end_val = start_pt + w
                /* set S operators */
                while (x < end_val) do
                        dot_on(x, y)
                        x = x + num
                end while
                start_pt = end_val + 1
                y = y + 1
                /* set D operator */
                if (x = end_val) then
                        dot_on(x, y)
                        x = x + num
                end if
        end if
end bisect

procedure recursive_bisect(x_s, y_s, x_f, y_f: integer);
var
        u, v, w, num, pos, m, j: integer;

begin
        u = x_f - x_s
        v = y_f - y_s
        w = u - v
        /* check for when theorems do not hold */
        if (v <> 1) then
                /* calculate m, j, num, and pos_i */
                m = position of most significant "1" bit in v
                j = i mod 2^m
                num = 2^(log_2 P) - m
                pos = i div 2^m
                /* calculate starting x and y coordinate values */
                y = D_ops(v, j, m)
                start_pt = S_ops(v, j, m) + y + x_s + 1
```

```
                    x = start_pt + pos
                    y = y + ys
                    /* calculate wm, j and vm, j */
                    if (not(i[0..m-1]) < w[m-1..0]) then
                              w = (w div 2m) + 1
                    else
                              w = w div 2m
                    end if
                    if (not(i[0..m-1]) < v[m-1..0]) then
                              v = 2
                    else
                              v = 1
                    end if
            else
                    y = ys
                    start_pt = xs + 1
                    x = start_pt + i
                    num = P
            end if
            if (i = 0) then
                    dot_on(xs, ys)
            end if
            bisect(w, v, num)
    end  recursive_bisect
```

A Characterization of Recognizable Picture Languages

Katsushi Inoue
and
Itsuo Takanami

Department of Computer Science and Systems Engineering
Faculty of Engineering
Yamaguchi University
Ube, 755 Japan

Abstract This paper first shows that REC, the family of recognizable picture languages in Giammarresi and Restivo (1991), is equal to the family of picture languages accepted by two-dimensional on-line tessellation acceptors in Inoue and Nakamura (1977). By using this result, we then solve open problems in Giammarresi and Restivo (1991), and show that (i) REC is not closed under complementation, and (ii) REC properly contains the family of picture languages accepted by two-dimensional nondeterministic finite automata even over a one letter alphabet.

1. Introduction

In a pioneering paper [1], Blum and Hewitt introduced the notion of four-way finite automaton moving in a two-dimensional tape, as the natural extention of the one-dimensional two-way finite automaton. Since this work, many investigations about automata on a two-dimensional tape (or picture) have been made [4-8]. See [7] for a recent survey of these investigations.

Recently, Giammarresi and Restivo [3] introduced a new notion of recognizability for two-dimensional (or picture) languages. This notion arose from a characterization of recognizable one-dimensional language in terms of local languages and alphabetic mappings (cf. Theorem 6.1 of [2]).

Giammarresi and Restivo [3] extended these latter concepts in a natural way to two-dimensions and defined recognizable picture languages. In [3], they formulated the following open problems:

(1) Is REC, the family of recognizable picture languages, closed under complementation ?

(2) Does REC properly contain $\mathcal{L}$ [2-NFA], the family of picture languages accepted by two-dimensional nondeterministic finite automata [1,4,6] ?

This paper solves these problems by presenting a characterization of REC. Section 2 contains the formal definitions of REC and two-dimensional on-line tessellation acceptors (2-ota's)[4]. Section 3 shows that REC is equal to the family of picture languages accepted by 2-ota's. By using this result, we show that (i) REC is not closed under complementation, and (ii) REC properly contains $\mathcal{L}$ [2-NFA] even over a one letter alphabet.

2. Preliminaries

2.1. Local Picture Languages

Let Σ be a finite alphabet. A picture (or a two-dimensional tape) over Σ is an $m \times n$ rectangular array of elements of Σ. The pair (m,n) is called the size of the picture [m and n are the numbers of rows and columns of the picture, respectively]. Denote by $\Sigma^{(2)}$ the set of all pictures over Σ and by $\Sigma^{m \, n}$ the set of all pictures of size (m,n). A picture language over Σ is a subset of $\Sigma^{(2)}$. For any picture x of size (m,n), let us denote by $B_{k,r}(x)$, $k \leq m$, $r \leq n$, the set of all subpictures (or blocks) of x of size (k,r). In order to introduce the next definition, we consider pictures over Σ surrounded by a special symbol $* \notin \Sigma$: for any picture $x \in \Sigma^{(2)}$, we denote by b(x) the picture of size (m+2,n+2) obtained by surrounding x with the boundary symbol *.

A picture language $L \subseteq \Sigma^{(2)}$ is local if there exists a subset Q of $(\Sigma \cup \{*\})^{2\times 2}$ such that $L=\{x\in \Sigma^{(2)} \mid B_{2,2}(b(x))\subseteq Q\}$. We write $L=R(Q)$. The following example is given in [3].

Example 1: Let $\Sigma=\{a,b\}$, and let

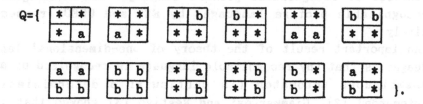

It is easy to see that: $R(Q)=\{$pictures with $n\geq 2$ rows, whose top row is composed by a and other rows are composed by b$\}$.

In the one-dimensional case it is trivially decidable whether a local language is empty. On the other hand, Giammarresi and Restivo [3] showed, by using the undecidability of the "emptiness of the language accepted by a Turing machine" problem, that emptiness problem for local picture languages is undecidable.

2.2. Recognizable Picture Languages

Let Σ and Σ' be two finite alphabets, and $h:\Sigma \to \Sigma'$ a mapping (alphabetic mapping). If x is a picture over Σ, $h(x)$ denotes the picture over Σ' obtained by replacing in x every symbol $\sigma \in \Sigma$ by the corresponding symbol $h(\sigma)$. If L is a picture language over Σ, then $h(L)=\{h(x) \mid x\in L\}$.

In one-dimensional case, it is well known [3] that every recognizable language is the image of a local language by an alphabetic mapping. This naturally suggests us the following definition.

A picture language L over Σ' is <u>recognizable</u> if there exists a finite alphabet Σ, a local language $R(Q)$ over Σ (defined by a finite set $Q\subseteq (\Sigma \cup \{\#\})^{2\times 2}$) and a mapping $h:\Sigma \to \Sigma'$, such that $L=h(R(Q))$. The pair (Q,h) is called a <u>representation</u> of L. A pair (Q,h) is an <u>unambiguous</u> representation of L if, for any picture x∈ L, there exists one and only one picture x' in R(Q) such that h(x')=x. We say that a recognizable picture language L is _unambiguous_ if it admits an unambiguous representation. Let us denote the families of recognizable picture languages and unambiguous recognizable picture languages by REC and UREC, respectively.

An important result of the theory of one-dimensional languages is that any recognizable language is recognized by an unambiguous automaton (in particular a deterministic automaton) [2]. Giammarresi and Restivo [3] showed that in the two-dimensional case it occurs a different situation. That is, they showed that UREC $\subsetneqq$ REC. Other results in [3] about REC are: (i) the emptiness problem for the families REC and UREC is undecidable, (ii) REC is closed under union, intersection, (row and column) concatenation, (row and column) closure operations [7], and (iii) there is a picture language in REC which does not belong to $\mathcal{L}$ [2-NFA], where $\mathcal{L}$ [2-NFA] denotes the family of picture languages accepted by two-dimensional nondeterministic finite automata [1,4,6].

2.3. Two-Dimensional On-Line Tessellation Acceptors

In this section, we recall a two-dimensional on-line tessellation acceptor (2-ota)[4]. Let x be a picture of size (m,n), $m \geq 1, n \geq 1$. If $1 \leq i \leq m$ and $1 \leq j \leq n$, we let $x(i,j)$ denote the symbol in x with coordinates (i,j). Furthermore, if $1 \leq i \leq i' \leq m$ and $1 \leq j \leq j' \leq n$, we let $x[(i,j),(i',j')]$ denote the sub-picture z of x such that (i) the size of z is $(i'-i+1, j'-j+1)$, and (ii) for each k,r $(1 \leq k \leq i'-i+1, 1 \leq r \leq j'-j+1)$, $z(k,r)=x(k+i-1, r+j-1)$. The 2-ota M is an infinite array of identical finite-state machines in two-dimensional space. The input of a picture x to M means that each symbol $x(i,j)$ is placed on the finite-state machine situated at coordinates (i,j) [we call this machine (i,j)-cell] and the boundary symbol "*" is placed on any other cell in M (see Fig.1). When a picture is presented to M, M acts synchronously as follows. First, at

time t=0, each cell in M stays in the quiescent state. At
time t=1, (1,1)-cell enters a stable state corresponding to
the characteristic of x[(1,1),(1,1)]. Generally, at time t=k
(k≥1), each (i,j)-cell such that (i-1)+(j-1)=k-1 enters a
stable state (corresponding to the characteristic of
x[(1,1),(i,j)]) which depends on the symbol x(i,j) and on
the stable states memorized in (i-1,j)-cell and (i,j-1)-
cell. [Until time t=k, each (i,j)-cell such that (i-1)+(j-
1)=k-1 stays in the quiescent state; at time t=k, it enters
a stable state for the first time; and after time t=k, it
stays in the stable state.] M accepts the picture x if and
only if the stable state which (m,n)-cell enters is one of
the accepting states, where (m,n) is the size of x. For example, when a 4×5 picture x as shown in Fig.2 is presented
to a 2-ota M, M acts as follows (see Fig.2): at time t=k
(1≤k≤8), all cells denoted by k enter their stable states
simultaneously. In this case, for example, (3,3)-cell enters
a stable state at time t=5 (which corresponds to the characteristic of x[(1,1),(3,3)]) which depends on the symbol
x(3,3) and on the stable states memorized in (2,3)-cell and
(3,2)-cell. M accepts the picture x if and only if the
stable state which (4,5)-cell enters is one of the accepting
states.

Formally, a 2-ota is defines as $M=(K,\Sigma \cup \{*\},\delta ,q_e,q_0,F)$,
where (1) K is a finite set of states, (2) Σ is a finite
set of input symbols, and "*" is the boundary symbol, not in
Σ , (3) $\delta :K^3\times (\Sigma \cup \{*\})\rightarrow 2^K\cup \{\{q_0\}\}$, where $K'=K-\{q_e,q_0\}$, is
the cell state transition function, (4) $q_e\in K$ is the motive
state, (5) $q_0\in K$ is the quiescent state, (6) $F\subseteq K'$ is a set
of accepting states. K corresponds to the state set of any
one of the finite-state machines in the array. The state of
each (i,j)-cell at time t+1 depends on the states (at time
t) of cells in its neighborhood and on the symbol read by
it, where the neighborhood of (i,j)-cell is the set of
(i,j)-cell, (i-1,j)-cell, and (i,j-1)-cell (see Fig.1). That
is, let $q_{(i,j)}(t)$ be the state of (i,j)-cell at time t. Then
$$q_{(i,j)}(t+1)\in \delta (q_{(i,j)}(t),q_{(i-1,j)}(t),q_{(i,j-1)}(t),a),$$
where a is the symbol in (i,j)-cell.

The motive state q_e plays the role of giving M the motivation to begin to read an input picture. Each element in $K'=K-\{q_e,q_0\}$ is called a stable state. δ must satisfy the following two conditions: For any $a \in \Sigma \cup \{*\}$, and for any $p_i \in K$ ($1 \leq i \leq 3$),

　(i) $\delta(p_1,p_2,p_3,a)=\{q_0\}$ if and only if ($a=*$) or ($p_1=q_0$ and for each i (i=2,3), $p_i \in \{q_e,q_0\}$};

(ii) if $p_1 \in K-\{q_e,q_0\}$ and $a \neq *$, then $\delta(p_1,p_2,p_3,a))=\{p_1\}$.

Condition (ii) means that each cell stays in the stable state regardless of the state of each cell in its neighborhood and of the symbol (except the symbol *) in it, once it enters a stable state.

A state configuration c of M is a mapping from I^2 into K, where I denotes the set of all integers. [As usual, $c(i,j)$ represents the state of (i,j)-cell for each $(i,j) \in I^2$.] The state configuration c of M such that $c(1,1)=q_e$ and $c(v)=q_0$ for each v [$\neq (1,1)$] in I^2 is called the primitive state configuration of M.

We say that M accepts a picture x of size (m,n) if and only if when M begins to read the picture x in its primitive state configuration, (m,n)-cell can enter a stable state in F. [Note that by saying that a picture x of size (m,n) is presented to M, we mean that for each $(i,j) \in A=\{(i,j) \in I^2 \mid 1 \leq i \leq m$ and $1 \leq j \leq n\}$, the symbol $x(i,j)$ is placed on (i,j)-cell and the symbol "*" is placed on each cell not in A.]

It will be easy to see that if a picture x of size (m,n) is presented to M, then (i) at time t ($t \geq 1$), all cells that read the symbols at a distance t-1 from the symbol $x(1,1)$ [that is, all (i,j)-cells such that $(i-1)+(j-1)=t-1$] have attained their ultimate states, and (ii) it is decided just at time m+n-1 whether the picture x is accepted by M.

Let x be a picture of size (m,n) over Σ. A run of M on x is a picture z of size (m,n) over $K-\{q_e,q_0\}$ such that

　(i) $z(1,1) \in \delta(q_e,q_0,q_0,x(1,1))$ and

(ii) for each (i,j) [$\neq (1,1)$] ($1 \leq i \leq m, 1 \leq j \leq n$),

　　　　$z(i,j) \in \delta(q_0,z(i-1,j),z(i,j-1),x(i,j))$,

where $z(0,j)=z(i,0)=q_0$. An accepting run of M on x is a run z of M on x such that $z(m,n) \in F$.

T(M), that is, the picture language accepted by M, is defined as

$T(M)=\{x\in \Sigma^{(2)} \mid$ there is an accepting run of M on x$\}$.

Let us denote by $\mathcal{L}$[2-ota] the family of picture languages accepted by 2-ota's.

3. $\mathcal{L}$[2-ota]=REC

This section investigates a relationship between REC and $\mathcal{L}$ [2-ota], and shows that REC=$\mathcal{L}$[2-ota]. By using this result and our previous results, we solve open problems in [3].

<u>Lemma 3.1.</u> REC$\subseteq \mathcal{L}$[2-ota].

<u>Proof</u>. It is quite easy to see that any local picture language is accepted by some 2-ota. From this and from the fact [4] that $\mathcal{L}$[2-ota] is closed under alphabetic mapping, it follows that REC$\subseteq \mathcal{L}$[2-ota].   Q.E.D.

<u>Lemma 3.2.</u> $\mathcal{L}$[2-ota]$\subseteq$ REC.

<u>Proof</u>. Let $L\in \mathcal{L}$[2-ota] and $M=(K,\Sigma \cup \{*\},\delta,q_\bullet,q_0,F)$ be a 2-ota accepting L. Let $\Gamma =\Sigma \times (K-\{q_\bullet,q_0\})$, and $h_1:\Gamma \rightarrow \Sigma$ and $h_2:\Gamma \rightarrow K-\{q_\bullet,q_0\}$ be alphabetic mappings such that for each $s=(a,q)\in \Gamma$, $h_1(s)=a$ and $h_2(s)=q$. A picture x over Γ is called a <u>computation description picture</u> if $h_2(x)$ is a run of M on $h_1(x)$, and called an <u>accepting computation description tion picture</u> if $h_2(x)$ is an accepting run of M on $h_1(x)$. Let S be the set of all accepting computation description pictures. It will be obvious that $L=T(M)=h_1(S)$. Below, we shall construct from M a subset Q of $(\Gamma \cup \{*\})^{2\;2}$ such that R(Q)=S. Let

$Q_1 =\{$
*	*
*	(a,p)
$\mid (a,p)\in \Gamma$ and $p\in \delta(q_\bullet,q_0,q_0,a)\}$,

$Q_2 = \{$

$\ast$	$\ast$
(a,p)	$\ast$

$\mid (a,p) \in \Gamma \}$,

$Q_3 = \{$

$\ast$	(a,p)
$\ast$	$\ast$

$\mid (a,p) \in \Gamma \}$,

$Q_4 = \{$

(a,p)	$\ast$
$\ast$	$\ast$

$\mid (a,p) \in \Gamma \}$,

$Q_5 = \{$

$\ast$	$\ast$
(b,q)	(a,p)

$\mid (a,p),(b,q) \in \Gamma$ and $p \in \delta (q_0,q_0,q,a)\}$,

$Q_6 = \{$

$\ast$	(b,q)
$\ast$	(a,p)

$\mid (a,p),(b,q) \in \Gamma$ and $p \in \delta (q_0,q,q_0,a)\}$,

$Q_7 = \{$

(b,q)	(a,p)
$\ast$	$\ast$

$\mid (a,p),(b,q) \in \Gamma \}$,

$Q_8 = \{$

(b,q)	$\ast$
(a,p)	$\ast$

$\mid (a,p),(b,q) \in \Gamma \}$,

$Q_9 = \{$

(c,r)	(b,q)
(d,s)	(a,p)

$\mid \begin{array}{l} (a,p),(b,q),(c,r),(d,s) \in \Gamma \\ \text{and } p \in \delta (q_0,q,s,a) \end{array} \}$, and

$Q = \bigcup_{1 \leq i \leq 9} Q_i$.

Let $R' = \{b(x) \mid x$ is a computation description picture$\}$ and $R'' = \{b(x) \mid x \in S\}$. Q_1, Q_2, and Q_3 are used to construct the upper-left hand, the upper-right hand, and the lower-left hand corners of pictures in R', respectively. Q_4 is used to construct the lower-right hand corners of pictures in R". Q_5, Q_6, Q_7, and Q_8 are used to construct the first rows, the first columns, the last rows, and the last columns of pictures in R', respectively. Q_9 is used to construct the inner parts of pictures in R'.

It is easy to see that $R(Q)=S$, and thus $L=T(M)=h_1(R(Q))$. This completes the proof of the lemma.   Q.E.D.

From Lemmas 3.1 and 3.2, we have

Theorem 3.1. REC=$\mathscr{L}$[2-ota].

Giammarresi and Restivo [3] formulated the following two open problems:

(1) Is REC closed under complementation ?
(2) $\mathscr{L}$[2-NFA]$\subsetneqq$REC ?

The following two corollaries solve these problems.

Corollary 3.1. REC is not closed under complementation.

Proof. It is shown in [5] that $\mathscr{L}$[2-ota] is not closed under complementation. From this fact and Theorem 3.1, the corollary follows.   Q.E.D.

Corollary 3.2. $\mathscr{L}$[2-NFA]$\subsetneqq$REC.

Proof. It is shown in [4] that $\mathscr{L}$[2-NFA]$\subsetneqq\mathscr{L}$[2-ota]. From this fact and Theorem 3.1, the corollary follows.   Q.E.D.

Let REC(1) and $\mathscr{L}$[2-NFA](1) be the sub-families of REC and $\mathscr{L}$[2-NFA], respectively, of picture languages over a one letter alphabet. Giammarresi and Restivo [3] conjectured that REC(1)= $\mathscr{L}$[2-NFA](1). The following corollary denies this conjecture, and strengthens Corollary 3.2.

Corollary 3.3. $\mathscr{L}[2\text{-NFA}](1) \subsetneq REC(1)$.

Proof. Let $T=\{x \in \{0\}^{2^n \times n} \mid n \geq 1\}$. It is shown in [6] that $T \notin \mathscr{L}[2\text{-NFA}](1)$. On the other hand, it is shown in Proposition 4.4 of [4] that T is accepted by a 2-ota (in fact, by a deterministic 2-ota). From these facts, Theorem 3.1 and Corollary 3.2, the corollary follows.   Q.E.D.

Acknowledgement We would like to thank the referees of ICPIA'92 for several helpful suggestions concerning this paper.

REFERENCES

[1] M.Blum and C.Hewitt, "Automata on a 2-dimensional tape", IEEE Symp. on Switching and Automata Theory: 155-160 (1967).
[2] S.Eilenberg, "Automata, Languages and Machines", vol.A, Academic Press (1974).
[3] D.Giammarresi and A.Restivo, "Recognizable picture languages", Proc. of the International Colloquium on Parallel Image Processing (edited by M.Nivat,A.Saoudi and P.S.P.Wang), Paris, 3-16 (1991).
[4] K.Inoue and A.Nakamura, "Some properties of two-dimensional on-line tessellation acceptors", Information Sciences 13, 95-121 (1977).
[5] K.Inoue and A.Nakamura, "Nonclosure properties of two-dimensional on-line tessellation acceptors and one-way parallel sequential array acceptors", IEIC Japan Trans.(E), 475-476 (Sept.1977).
[6] K.Inoue and A.Nakamura, "Two-dimensional finite automata and unacceptable functions", Int.J.Comput.Math.Sec.A7, 207-213 (1979).
[7] K.Inoue and I.Takanami, "A survey of two-dimensional automata theory", Information Sciences 55, 99-121 (1991).
[8] A.Rosenfeld, "Picture Languages (Formal Models for Picture Recognition)", Academic Press, New York, 1979.

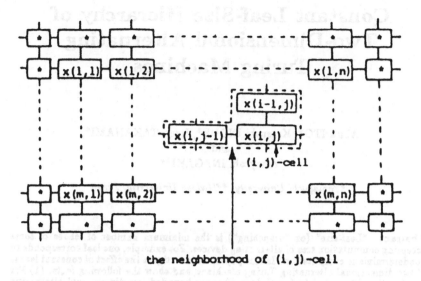

the neighborhood of (i,j)-cell

Fig.1. The input situation of an m×n picture x to a 2-ota
and the neighbourhood of (i,j)-cell.

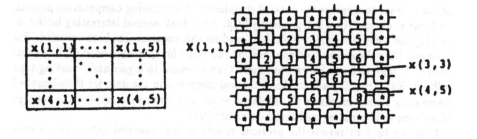

Fig.2. Illustration of the action of a 2-ota.

Constant Leaf-Size Hierarchy of Two-Dimensional Alternating Turing Machines

Akira ITO*, Katsushi INOUE*, Itsuo TAKANAMI*,
and
Yasuyoshi INAGAKI**

*Yamaguchi University, **Nagoya University, Japan

Abstract "Leaf-size" (or "branching") is the minimum number of leaves of some accepting computation tree of alternating devices. For example, one leaf corresponds to nondeterministic computation. In this paper, we investigate the effect of constant leaves of two-dimensional alternating Turing machines, and show the following facts: (1) For any function $L(m, n)$, k leaf- and $L(m, n)$ space-bounded two-dimensional alternating Turing machines which have only universal states are equivalent to the same space bounded deterministic Turing machines for any integer $k \geq 1$, where m (n) is the number of rows (columns) of the rectangular input tapes. (2) For square input tapes, $k+1$ leaf- and $o(\log m)$ space-bounded two-dimensional alternating Turing machines are more powerful than k leaf-bounded ones for each $k \geq 1$. (3) The necessary and sufficient space for three-way deterministic Turing machines to simulate k leaf-bounded two-dimensional alternating finite automata is n^{k+1}, where we restrict the space function of three-way deterministic Turing machines to depend only on the number of columns of the given input tapes.

1. Introduction

As with the other well-known measures for alternation, such as tree-size, tree-width, and so on [1,2], the concept "leaf-size," or sometimes called "branching," was introduced in order to measure the parallel complexity of alternating computation process, by Inoue et al. and King independently [2,3]. After that, several interesting facts concerning the computational complexity based on this measure has been revealed. For example, King shows the "leaf compression theorem" for space-bounded off-line alternating Turing machines which have more than or equal to logarithmic working-tape cells [2]. Similarly, Yamamoto shows the leaf compression theorem for time-bounded alternating Turing machines [4]. Matsuno et al. and Hromkovic applies the concept of leaf-size to alternating multihead automata [5,6].

Now, we turn to review the previous results of leaf-bounded automata on two-dimensional input tapes. For "three-way" alternating Turing machines with $o(\log m)$ space bound, it is known [3] that $(k + 1)$-leaf machines are more powerful than k-leaf ones for each k. Then, a natural question arises: "what about four-way automata?"

If the usage of log space is allowed, a Turing machine can remember the finite coordinates of its finite universal branches. Accordingly, finite leaves never increases the power of four-way nondeterministic (and deterministic) Turing machines which have log or more space [3]. On the other hand, when the machines are disable to use log space, some interesting situations occurs.

Section 3 first show that for four-way alternating Turing machines "with only universal states", the hierarchy collapses to the deterministic class, as with the case of large space bound. (It should be noted that there exists an infinite hierarchy of constant leaves for three-way version of the universal-states machines [7].) By contrast with it, for normal alternating Turing machines using small space bound, a strict hierarchy emerges again. More precisely, it is shown that there exists a set of square tapes accepted by a $k + 1$ leaf-bounded alternating finite automata $(AF(k + 1))$, but not accepted by any k leaf- and $o(\log m)$ space bounded alternating Turing machines. Thus, even the alternating finite automata of two leaves are more powerful than nondeterministic finite automata.

Section 4 investigates the necessary and sufficient space for three-way deterministic Turing machines (3DTs) to simulate $AF(k)$s. It is known [8,9] that n^2 is the necessary and sufficient space for three-way two-dimensional deterministic Turing machines (3DTs) to simulate nondeterministic two-dimensional finite automata, i.e., $AF(1)$s. We will show that n^{k+1} is the necessary and sufficient space for 3DTs to simulate $AF(k)$s, where the space function of 3DTs is restricted to depend on the number of columns but not the number of rows of rectangular input tapes.

2. Definitions

Definition 2.1. Let Σ be a finite set of symbols. A *two-dimensional tape* over Σ is a two-dimensional rectangular array of elements of Σ.

The set of all two-dimensional tapes over Σ is denoted by Σ^{2+}. Given a tape $x \in \Sigma^{2+}$, $l_1(x)$ denotes the number of rows of x and $l_2(x)$ denotes the number of columns of x. If $1 \leq i \leq l_1(x)$ and $1 \leq j \leq l_2(x)$, we let $x(i, j)$ denote the symbol in x with coordinates (i, j). Furthermore, we define

$$x[(i, j), (i', j')],$$

when $1 \leq i \leq i' \leq l_1(x)$ and $1 \leq j \leq j' \leq l_2(x)$, as the two-dimensional tape z satisfying the following:

(i) $l_1(z) = i' - i + 1$ and $l_2(z) = j' - j + 1$,

(ii) for each k, r $[1 \leq k \leq l_1(z)$ and $1 \leq r \leq l_2(z)]$,

$$z(k, r) = x(k + i - 1, r + j - 1).$$

For $x \in \Sigma^{2+}$ with $l_2(x) = n$, the ith row $x[(i, 1), (i, n)]$ of x is simply denoted by $x[i, *]$.

Two-dimensional alternating Turing machines were introduced in [3]. We recall the definition.

Definition 2.2. A *two-dimensional alternating Turing machine* (AT) is a seven-tuple

$$M = (Q, q_0, U, F, \Sigma, \Gamma, \delta),$$

where

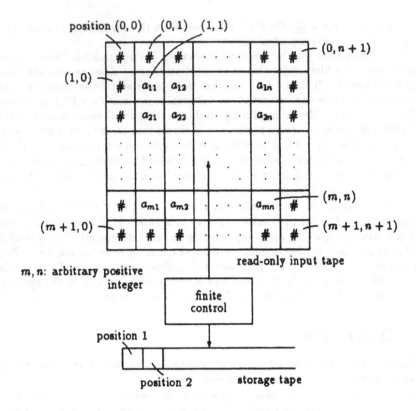

Figure 1. Two-dimentional alternating Turing machine.

(1) Q is a finite set of *states*,

(2) $q_0 \in Q$ is the *initial state*,

(3) $U \subseteq Q$ is the set of *universal states*,

(4) $F \subseteq Q$ is the set of *accepting states*,

(5) Σ is a finite *input alphabet*,

(6) Γ is a finite *storage tape alphabet* ($B \in \Gamma$ is the *blank symbol*), and

(7) $\delta \subseteq (Q \times (\Sigma \cup \{\#\}) \times \Gamma) \times ((Q \times (\Gamma - \{B\}) \times \{$left, right, up, down, no move$\} \times \{$left, right, no move$\})$ is the *next move relation*, where $\# \notin \Sigma$ is the *boundary symbol*.

A state q in $Q - U$ is said to be *existential*. As shown in Fig.1, the machine M has a read-only rectangular input tape with boundary symbols "#"s and one semi-infinite storage tape, initially blank. Of course, M has a finite control, an input head, and a storage tape head. A position is assigned to each cell of the storage tape, as shown in Fig.1. A step of M consists of reading one symbol from each tape, writing a symbol on the storage tape, moving the input and storage heads in specified directions, and entering a new state, in accordance with the next move relation δ. If the input head falls off boundaries of the input tape, then the machine M can make no further move.

Definition 2.3. A *configuration* of an AT $M = (Q, q_0, U, F, \Sigma, \Gamma, \delta)$ is a pair of an element Σ^{2+} and an element of

$$C_M = (\mathbb{N} \cup \{0\})^2 \times S_M,$$

where $S_M = Q \times (\Gamma - \{B\})^* \times \mathbb{N}$ and $\mathbb{N}$ is the set of all positive integers. The first component of a configuration $c = (x, ((i, j), (q, \alpha, k)))$ represents the input to M. The first component (i, j) of the second component of c represents the input head position. The second component (q, α, k) of the second component of c represents the state of the finite control, nonblank contents of the storage tape, and the storage-head position. An element of C_M is called a *semi-configuration* of M and an element of S_M is called a *storage state* of M. If q is the state associated with configuration c, then c is said to be *universal* (*existential, accepting*) configuration if q is a universal (existential, accepting) state. The *initial configuration* of M on input x is

$$I_M(x) = (x, (1, 1), (q_0, \lambda, 1)).$$

Definition 2.4. Given $M = (Q, q_0, U, F, \Sigma, \Gamma, \delta)$, we write

$$c \vdash_M c'$$

and say c' is a *successor* of c if configuration c' follows from configuration c in one step of M, according to the transition rules δ. $\vdash_M^*$ denotes the reflexive transitive closure of $\vdash$. A *computation path* of M on x is a sequence

$$c_0 \vdash_M c_1 \vdash_M \cdots \vdash_M c_n \qquad (n \geq 1).$$

A *computation tree* of M is a nonempty labeled tree with the properties,[†]

(1) each node π of the tree is labeled with a configuration $l(\pi)$,

(2) if π is an internal node (a nonleaf) of the tree, $l(\pi)$ is universal and

$$\{c \mid l(\pi) \vdash_M c\} = \{c_1, \ldots, c_k\},$$

then π has exactly k children $\rho_1, \ldots, \rho_k$ such that $l(\rho_i) = c_i$,

(3) if π is an internal node of the tree and $l(\pi)$ is existential, then π has exactly one child ρ such that

$$l(\pi) \vdash_M l(\rho).$$

An *accepting computation tree* of M on x is a finite computation tree whose root is labeled with $I_M(x)$ and whose leaves are all labeled with accepting configurations. We say that M *accepts* x if there is an accepting computation tree of M on input x. Define

$$T(M) = \{x \in \Sigma^{2+} \mid M \text{ accepts } x\}.$$

[†]In Ref.[3], computation tree itself is defined as a finite (i.e., non-loop) tree.

A deterministic and a nondeterministic two-dimensional Turing machine are special cases of an AT. That is, the former is an AT whose configurations each have at most one successor and the latter is an AT which has no universal states. Similarly, we introduce an alternating two-dimensional Turing machine which has no existential states [7]. By "DT" ("NT","UT") we denote a *deterministic two-dimensional Turing machine* (a *nondeterministic two-dimensional Turing machine*, an *alternating two-dimensional Turing machine with only universal states*).

We next recall the definition of three-way AT.

Definition 2.5. A *three-way two-dimensional alternating Turing machine* (3AT) is an AT $M = (Q, q_0, U, F, \Sigma, \Gamma, \delta)$ such that

$$\delta \subseteq (Q \times (\Sigma \cup \{\#\}) \times \Gamma) \times (Q \times (\Gamma - \{B\}) \times \{\text{left, right, down, no move}\} \times \{\text{left, right, no move}\}).$$

That is, a 3AT is an AT whose input head can move left, right, or down, but not up. "3DT","3NT",and "3UT" are defined similarly.

In this paper, we investigate the properties of ATs whose storage tapes are bounded (in length) to use.

Definition 2.6. Let $L : N \times N \mapsto N$ be a function with two variables m and n. For any AT M, we associate a complexity function SPACE which takes configuration $c = (x, (i,j), (q, \alpha, k))$ to natural numbers. Let $\text{SPACE}(c) = |\alpha|$. We say that M is $L(m,n)$ *space-bounded* if for all m,n and for all x with $l_1(x) = m$ and $l_2(x) = n$, if x is accepted by M, then there is an accepting computation tree of M on input x such that, for each node x of the tree, $\text{SPACE}(l(x)) \leq L(m,n)$. By "AT($L(m,n)$)", we denote an $L(m,n)$ space bounded AT. "DT($L(m,n)$)", "NT($L(m,n)$)", "UT($L(m,n)$)", "3AT($L(m,n)$)", "3DT($L(m,n)$)", and "3UT($L(m,n)$)" are defined similarly.

Especially, AT(0) is denoted by "AF" and called a *two-dimensional alternating finite automaton*. "DF", "NF", and "UF" are defined similarly.

the concept, "leaf-size bounded computation" of alternating Turing machines was introduced in [2,3]. Basically, the leaf-size used by an AT on a given input is the number of leaves of an accepting computation tree with fewest leaves. Leaf-size, in a sense, reflects the minimum number of processes that run in parallel in accepting a given input.

Definition 2.7. Let $Z : N \times N \mapsto N$ be a function. For each tree t, let LEAF(t) denote the number of leaf nodes of t. We say that an AT M is $Z(m,n)$ *leaf-bounded* if for all x with $l_1(x) = m$ & $l_2(x) = n$, if x is accepted by M then there is an accepting tree t of M on x such that $\text{LEAF}(t) \leq Z(m,n)$. By "AT($L(m,n), Z(m,n)$)", we denote a $Z(m,n)$ leaf-size bounded AT($L(m,n)$). "UT($L(m,n), Z(m,n)$)", "AF($Z(m,n)$)", and "UF($Z(m,n)$)" are defined similarly.

In some part of this paper, we concentrate on the properties of ATs whose input tapes are restricted to square ones. In this case, complexity function L or Z has only one variable, conventionally m. By "AT'($L(m)$)" we denote an $L(m)$ space-bounded

AT whose input tapes are restricted to square ones. "DT'$(L(m))$", etc. are defined similarly. The class of sets accepted by AT$(L(m,n))$s is defined as follows.

$$\mathcal{L}[\text{AT}(L(m,n))] = \{T \mid T = T(M) \text{ for some AT}(L(m,n)) \ M\}.$$

$\mathcal{L}[\text{AT}'(L(m))]$, etc. are defined similarly.

Definition 2.8. Let $g : \mathbb{N} \mapsto \mathbb{N}$ be a function and x be a two-dimensional tape with $l_2(x) = n$. For each j $(1 \leq j \leq l_1(x)/g(n))$, we call

$$x[((j-1)g(n)+1, 1), (jg(n), n)]$$

the jth $g(n)$-*block of* x, when $l_1(x)$ is divided by $g(n)$. We simply denote it by $x[\text{block}_{g(n)}(j)]$.

Here, We give some mathematical notations. For any set A, "$\mathcal{P}(A)$" denotes the power set of A and "$m\text{-}\mathcal{P}_i(A)$" denotes the set of multisets consisting of i elements from A. We assume that any function is a mapping from $\mathbb{N}$ to $\mathbb{N}$. As usual, "$\log n$" is the abbreviated form of the function $\max\{1, \lceil \log_2 n \rceil\}$.

3. Hierarchy based on Constant Leaf-Size

In this section, we investigates a fundamental question concerning leaf-size: Is $k+1$ leaves better than k? We first show that in case of alternating Turing machine with only universal states (UT), no hierarchy exists for any space bound.

Theorem 3.1. *For any* $k \in \mathbb{N}$ *and any function* $L(m,n)$,

$$\mathcal{L}[\text{UT}(L(m,n), k)] = \mathcal{L}[\text{DT}(L(m,n))].$$

Proof. Given a k leaf-size bounded UT M and an input tape x, a DT M' performs a depth-first-search on the computation tree of M on x without any extra cells of the working tape: Normal tree-search method needs one stack for backtracking. Instead, M' adopts only the forward tracking from the root to each leaf and uses finite internal memories in the finite control. Note that since M has constant leaves, the branching structure of universal configurations of M on x is also constant. After each traversal of a path and finding out its leaf is labeled with an accepting configuration, M' adds the newly obtained information about the tree structure into a memory cell of the finite control. Then, M begins to walk from the root to the next leaf, whose route can be specified by referring to the memories of the finite control. When the whole travel have been done and if M is surely k leaf-bounded, M' enters an accepting state. Note that M' accepts exactly $T(M)$ and that M' is L space bounded iff M is L space bounded. □

Note that there exists a set of square tapes accepted by "three-way" k leaf-bounded UF, but not accepted by any "three-way" k leaf- and $o(\log m)$ space-bounded AT [7].

Corollary 3.1. *For any* $k \in \mathbb{N}$, $\mathcal{L}[\text{UF}(k)] = \mathcal{L}[\text{DF}]$.

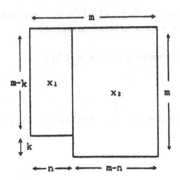
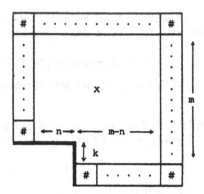

Figure 2. (m,n,k)-chunk   Figure 3. $x(\#)$

With contrast to four-way universal machines, we can show that there exists an infinite hierarchy of $o(\log m)$ space-bounded two-dimensional alternating Turing machines based on leaf-size. To this end, we have to give several preliminaries at first.

Let Σ be a finite alphabet. For each k,m,n ($k \geq 1, m \geq k+1, 1 \leq n \leq m-1$), an (m,n,k)-*chunk* over Σ is a pattern x over Σ as shown in Fig.2, where $x_1, x_2 \in \Sigma^{2+}$, $l_1(x_1) = m - k$, $l_2(x_1) = n$, $l_1(x_2) = m$ and $l_2(x_2) = m - n$. For an (m,n,k)-chunk x, we denote by $x(\#)$ the pattern (obtained from x by surrounding x by the boundary symbols $\#$'s) as shown in Fig.3. Suppose that a two-dimensional automaton enters or exits the pattern $x(\#)$ only at the face designated by the bold line in Fig.3. Then, for any (m,n,k)-chunk x, the number of entrance points to $x(\#)$ is $n+k+3$ and the number of exit points from $x(\#)$ is $n+k+1$. Let $Ent(m,n,k)$ $(Ext(m,n,k))$ denote the set of those entrance (exit) points. Further, the set of all non-boundary points of $x(\#)$ is denoted by $Int(m,n,k)$.

Let M be an $AT^s(L(m),k)$ whose input alphabet is Σ. Let s,r be the number of states of finite control and the number of working tape alphabet of M, respectively. We denote by $St(M,m)$ the set of available storage states of M on input tape x with $l_1(x) = l_2(x) = m$. Then, it is clear that $|St(M,m)| = sL(m)t^{L(m)}$. For any $(m,n,k+2)$-chunk x and any $AT^s(L(m),k)$ M, define the mapping

$$M_x : Ent(m,n,k+2) \times St(M,m)$$
$$\mapsto \quad \mathcal{P}\left(\bigcup_{i=1}^{k} m\text{-}\mathcal{P}_i\left(Ext(m,n,k+2) \times St(M,m)\right)\right)$$

as follows (without no confusion, we regard semi-configurations of M as configurations of M on x):

$M_x(P,q) = \{V_1, V_2, \ldots, V_J\}$

$\iff$ There exist J finite computation trees $t_1, t_2, \ldots, t_J$ of M on $x[v]$ for some $v \in \{0,1\}^{2+}$ such that for each $1 \leq i \leq J$, (1) the root of t_i is labeled with the

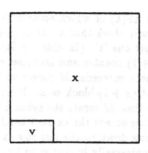

Figure 4. $x[v]$

semi-configuration (P, q), (2) all internal nodes of t_i are labeled with semi-configuration (P', q') for some $P' \in Int(m, n, k + 2)$ and $q' \in St(M, m)$, and (3) the leaves of t_i are labeled with (possibly overlapped) semi-configurations $c_1, c_2, \ldots, c_d$, where $\{c_1, c_2, \ldots, c_d\} = V_i$ $(d \le k)$.

For any $(m, n, k + 2)$-chunks x and y, we say that x and y are M-*equivalent* if for each $P \in Ent(m, n, k + 2)$ and each $q \in St(M, m)$, $M_x(P, q) = M_y(P, q)$. For any $(m, n, k + 2)$-chunk x and any tape $v \in \Sigma^{2+}$ with $l_1(v) = 1$ & $l_2(v) = n$, let $x[v]$ be the two-dimensional tape over Σ as shown in Fig.4. Then, the following fact will be easily verified.

Proposition 3.1. *Let M be an $AT(L(m), k)$ whose input alphabet is Σ. Let x, y be two M-equivalent $(m, n, k + 2)$-chunk over Σ, then for any $v \in \Sigma^{2+}$ with $l_1(v) = k + 2$ & $l_2(v) = n$, $x[v] \in T(M)$ if and only if $y[v] \in T(M)$.*

In other words, M cannot distinguish two M-equivalent chunks. We are now ready to prove the key lemma.

Lemma 3.1. *For each $k \in N$, define*

$$T(k) = \{x \in \{0, 1\}^{2+} \mid \exists m \ge 2[l_1(x) = k \cdot m \ \&$$
$$\exists i(1 \le i \le m - 1)[$$
$$x[block_k(i)] = x[block_k(m)] \ ] \ \&$$
$$\forall i(1 \le i \le l_1(x))[$$
$$(each \ row \ x[i, *] \ of \ x \ has \ exactly \ one \ '1' \ ) \ ]]\}$$

and

$$T^s(k) = \{x \in T(k) \mid l_1(x) = l_2(x)\}.$$

Then, (1) $T(k + 1) \in \mathcal{L}[AF(k)]$,
(2) *if* $L(m) = o(logm)$, $T^s(k + 2) \notin \mathcal{L}[AT^s(L(m), k)]$.

Proof. (1): We construct an AF(k) M which accepts $T(k+1)$ as follows. Given an input tape $x \in \{0,1\}^{2+}$, M firstly check that x has m $(k+1)$-blocks for some $m \geq 2$ and that each row of x has just one '1'. (In order to locate itself within a $(k+1)$-block of x, M uses a mod($k+1$) counter and increases or decreases the counter at each vertical step.) If this check succeeds, M moves to the position of the symbol '1' on the first row of the last $(k+1)$-block of x. From this position, M begin to move up looking for '1'. Each time M meets the symbol '1' on the first row of some $(k+1)$-block, it guesses whether or not the current block is equal to the last block. If so, it moves from the first row down to the last row of this block. On the lth row in the block ($2 \leq l \leq k$), it universally branches to two machines, one to continue descending and the other to turn its motion to the horizontal direction looking for '1' on the lth row. On the $k+1$st row in the block, M only turn its motion to the horizontal direction looking for '1'. Each machine, say M_l ($2 \leq l \leq k+1$), which have reached the symbol '1' on the lth row begins to move down for row-by-row check of two $(k+1)$-blocks equality. In the last $(k+1)$-block of x, machine M_l ($2 \leq l \leq k+1$) enters an accepting state if and only if the symbol of the lth row in the last block is '1'. It is clear that $T(M) = T(k+1)$ and M is k leaf-bounded.

(2): Suppose to the contrary that there exits an AT$^s(L(m), k)$ M accepting $T^s(k+2)$, where $L(m) = o(\log m)$. Without loss of generality, we assume that when M accepts a given input tape x, it enters an accepting state at the leftmost position of the last row of x. For each $n \geq 1$, let

$$V(n) = \{x \in T^s(k+2) \mid l_1(x) = l_2(x) = (k+2)\cdot(n^{k+2}+1) \ \&$$
$$x[(1, n+1), (l_1(x), l_2(x))] \in \{0\}^{2+}\},$$
$$Y(n) = \{x \in \{0,1\}^{2+} \mid l_1(x) = 1 \ \& \ l_2(x) = n\}, and$$
$$Row(x) = \{y \in Y(n) \mid \exists i (1 \leq i \leq m-1)[\ x[i, *] = y\ ]\}$$

Let $m = (k+2)\cdot(n^{k+2}+1)$ for readability and let $eqv_M(x)$ denote the M-equivalent class of an $(m, n, k+2)$-chunk x over $\{0,1\}$ (See Section 2 for their definitions). Then, the following proposition holds.

Proposition 3.2. *For any $(m, n, k+2)$-chunks x and y over $\{0,1\}$,*
if $Row(x) \neq Row(y)$ then $eqv_M(x) \neq eqv_M(y)$.

Proof. Suppose to the contrary that $Row(x) \neq Row(y)$ but $eqv_M(x) = eqv_M(y)$. Without loss of generality, we assume $\rho \in Row(x)$, $\rho \notin Row(y)$ for some $\rho \in Y(n)$. Consider M on two tapes $x[\rho]$ and $y[\rho]$. Since $x[\rho] \in V(n) \subseteq T^s(k+2)$, M accepts $x[\rho]$. Then, from Proposition 3.1, it follows that M also accepts $y[\rho]$, which is a contradiction. (Note that $y[\rho] \notin T^s(k+2)$.) $\qquad \Box$

Proof of Lemma 3.1 (continued). From

$$|m\text{-}\mathcal{P}_i(C)| = \binom{|C|+i-1}{i} \leq |C|^i,$$

we have

$$\left| \mathcal{P}(\bigcup_{i=1}^{t} m\text{-}\mathcal{P}_i(C)) \right| = 2^{O(|C|^k)}.$$

Therefore, letting $C = Ext(m, n, k + 2) \times St(M, m)$ and $C' = Ent(m, n, k + 2) \times St(M, m)$, the number of M-equivalent classes $|\{equ_M(x) \mid x \in V(n)\}|$ is at most

$$e(n) = (2^{O(|C|^k)})^{|C'|} = 2^{O(|C|^k \cdot |C'|)}.$$

By taking log, we have

$$\log e(n) \leq d \cdot |C|^k \cdot |C'|$$
$$= d \cdot (s(n + k + 3)L(m)t^{L(m)})^k \cdot s(n + k + 5)L(m)t^{L(m)}$$

for some constant d. Taking logarithm again, we have

$$\log \log e(n) \leq k \log(n + k + 3) + \log(n + k + 5)$$
$$+ (k + 1)\{L(m)\log t + \log L(m) + \log s\} + \log d$$
$$\leq (k + 1)\log(nk5)$$
$$+ (k + 1)\{L(m)\log t + \log L(m) + \log s\} + \log d$$
$$\leq (k + 1)\log n + d'L(m) + d''$$

for some constant d' and d''. On the other hand, letting $R(n) = \{row(x) \mid x \in V(n)\}$, we have

$$|R(n)| = \binom{n^{k+2}}{1} + \binom{n^{k+2}}{2} + \cdots + \binom{n^{k+2}}{n^{k+2}}$$
$$= 2^{n^{k+2}} - 1$$

Taking logarithm twice, we have

$$\log \log(|R(n)| + 1) = (k + 2)\log n.$$

Now, it should be noted that

$$L(m) = o(\log m) = o(\log n),$$

since $\log m = \log(n^{k+2} + 1) + \log(k + 2) \leq c \cdot \log n$ for some constant c, which is the logarithm of both sides of $m = (k + 2) \cdot (n^{k+2} + 1)$. From these facts, the inequality

$$\log \log(|R(n)| + 1) < \log \log e(n)$$

holds for large n. Hence, $\log(|R(n)| + 1) < \log e(n)$ for large n, and finally we get $|R(n)| < |R(n)| + 1 < e(n)$ for a large n. For such an n, it follows that there exist two M-equivalent $(m, n, k + 2)$-chucks x and y such that $Row(x) \neq Row(y)$, which contradicts Proposition 3.2. We have finished the proof of Lemma 3.1. $\square$

From Lemma 3.1, we get the desired result.

Theorem 3.2. *For each $k \in \mathbb{N}$, if $L(m) = o(\log m)$, then*

$$\mathcal{L}[AT'(L(m), k)] \subsetneq \mathcal{L}[AT'(L(m), k + 1)].$$

This answers the open question posed in [3].

Corollary 3.2. *For each $k \in \mathbb{N}$,*

$$\mathcal{L}[AF'(k)] \subsetneq \mathcal{L}[AF'(k + 1)].$$

4. Simulation of Constant Leaf-Size Alternating Finite automata by Three-Way Deterministic Turing Machines

In this section, we investigate the necessary and sufficient space for three-way deterministic Turing machines (3DTs) to simulate k leaf-bounded alternating finite automata (AF(k)s). This quantity does not give the definite characterization of AF(k)s, but in some extent it captures their internal capability for recognizing two-dimensional tapes.

Lemma 4.1. $\mathcal{L}[\mathrm{AF}(k)] \subseteq \mathcal{L}[3\mathrm{DT}(n^{k+1})]$.

Proof. The simulation method of AF(k) by 3DT is an extension of that of NF by 3DT in [8]. Suppose an AF(k) M and input tape x is given. Without loss of generality, we assume that it begins to move at the leftmost boundary symbol of the bottom boundary row of x, i.e., position $(l_1(x) + 1, 0)$, and that when M accepts an input x, it enters an accepting state at position $(l_1(x) + 1, 0)$. We also assume that when M enters a universal state, it always branches into at least two machines (i.e., deterministic transitions should appear in existential states).

Let $C = Q \times \{0, 1, \ldots, n+1\}$, where Q is the set of states of the finite control of M and $n = l_2(x)$. For each i $(0 \le i \le l_1(x) + 1)$, we define the mapping

$$g_i^{\uparrow} : C \mapsto \mathcal{P}\left(\bigcup_{l=1}^{k} m\text{-}\mathcal{P}_l(C)\right)$$

as follows.

$g_i^{\uparrow}(q, j) = \{V_1, V_2, \ldots, V_J\}$

⟺ There exist J finite computation trees $t_1, t_2, \ldots, t_J$ of M on x such that for each i $(1 \le i \le J)$, (1) the root of t_i is labeled with the configuration $(x, ((i-1, j), q))$, (2) all internal nodes of t_i are labeled with configurations $(x, ((i', j'), q'))$ for some i' $(0 \le i' \le i - 1)$, $(q', j') \in C$, and (3) the leaves of t_i are labeled with (possibly overlapped) configurations $(x, ((i, j_1), q_1))$, $(x, ((i, j_2), q_2)), \ldots, (x, ((i, j_d), q_d))$, where $\{(q_1, j_1), (q_2, j_2), \ldots, (q_d, j_d)\} = V_i$ $(d \le k)$.

From $|m\text{-}\mathcal{P}_i(C)| \le |C|^i = O(n^i)$, we have

$$\left| \mathcal{P}\left(\bigcup_{i=1}^{k} m\text{-}\mathcal{P}_i(C)\right) \right| = 2^{O(n^k)}.$$

Therefore, $O(n) \times O(n^k) = O(n^{k+1})$ space suffices in order to record one table $g_i^{\uparrow}$. Roughly speaking, a 3DT M' constructs $g_{i+1}^{\uparrow}$ by using $g_i^{\uparrow}$ on the ith row of x and decides whether or not M accepts x by using $g_{m+1}^{\uparrow}$ on the bottom boundary row of x.

set $g_0^{\uparrow}(c) = \emptyset$ for all $c \in C$;
from $i = 0$ to $l_1(x) + 1$ do the following:
    move to the ith row (If $i = 0$ then, assume #s as the input symbols on the first row);

/* construction of $g_{i+1}^{\dagger}$ from $g_i^{\dagger}$ */
for each $(q, j) \in C$ do the following:
    initialise a working list $H = \{\{(q, j)\}\}$;
    if q is an existential state then go to ① else go to ②;
    repeat ① and ② below k times:
    ① /* processing of existential parts of the computation trees */
      for each multi-set v in H do the following:

        Let u be the subset of v each element of which has an existential state as
its first component and does not have a check-mark (check-mark indicates the
marked element being a configuration which M will enter after going down
to the $i + 1$st row). For each $c_l = (q', j') \in u$ $(1 \leq l \leq |u|)$, initialise a
working list $H_l = \{\}$ and begin to find computation paths of M on x each
of whose start nodes is labeled with $(x, ((i, j'), q'))$, or $(x, ((i - 1, j'), q'))$ if
it is a temporary E-element $temp_E(q', j')$ (its meaning should be explained
later), and whose goal nodes are labeled with some universal configurations
or configurations which M will enter after going down to the $i + 1$st row. As
described in the proof of Theorem 3.3 in [10], the path search is performed
by all the possible one step simulations for unexamined configurations of M.
During the search, if M will enter some universal configuration $(x, ((i, j''), q''))$,
add the pair (q'', j'') to the list H_l, unless any overlap occurs. If M would
go down to the $i + 1$st row at the j''th column and would be the state p''
after that, then put a check-mark on (q'', j'') and add the pair to the list H_l.
When M will go up to the $i - 1$st row at the j''th column and will enter state
q'', then refer to the table $g_i^{\dagger}(q'', j'')$ and continue to research only for one-
tuple elements in $g_i^{\dagger}(q'', j'')$; for two- or more tuple element $\{d_1, d_2, \ldots, d_K\}$ in
$g_i^{\dagger}(q'', j'')$, add a new type element $temp_U[d_1, d_2, \ldots, d_K]$, called a temporary
U-element, to the corresponding goal list.

        At the end of all path-search, if some goal node list H_l $(1 \leq l \leq |u|)$
is empty, remove v from H. Finally, insert in H all multisets of the form
$v' = \{c_1, c_2, \ldots, c_J\} \cup (v - u)$, where $J = |u|$ and $c_l \in H_l$ for each l $(1 \leq l \leq J)$.
Remove the old element v from H. Proceed to the step ②.

    ② /* processing of universal parts of the computation trees of M */
      for each multi-set v in H do the following:

        Let u be the subset of v each element of which has a universal state as
its first component and does not have a check-mark or which is a temporary
U-element. For each $c_l = (q', j') \in u$ $(1 \leq l \leq |u|)$, initialise a working list
$G_l = \{\}$ and begin to perform depth-first search of the computation tree of
M whose root is labeled with the configuration $(x, ((i, j'), q'))$, or if c_l is a U-
element $temp_U[(q_1, j_1), (q_2, j_2), \ldots, (q_K, j_K)]$, assume one pseudo-root labeled
with a universal configuration whose sons are labeled with $(x, ((i, j_1), q_1))$,
$(x, ((i, j_2), q_2)), \ldots$, and $(x, ((i, j_K), q_K))$. During the search, if it is found that
M enters a loop or the tree has more than k leaves, then stop the search and
remove v from H. Otherwise, if M will enter some universal configuration
$(x, ((i, j''), q''))$, add the pair (q'', j'') to the leaf-node list G_l regardless of any
overlap, and backtrack to its predecessor. If M would go down to the $i + 1$st
row at the j''th column and would be the state p'' after that, then put a check-
mark on (q'', j'') and add the pair to the list G_l and backtrack. When M will go
up to the $i - 1$st row at the j''th column and will enter state q'', then refer to the
table $g_{i-1}^{\dagger}(q'', j'')$: if $g_{i-1}^{\dagger}(q'', j'')$ has only one element, which means that no
existential branch goes out from the configuration $(x, ((i - 1, j''), q''))$, continue

to research from this element; if $g_{i-1}^{\downarrow}(q'',j'')$ has more than one element, which means that some existential branch goes out, add the element $temp_E(q'',j'')$, called temporary E-element, to the corresponding list G_l.

At the end of all tree-search for u, if it is found out that the number of the elements of multi-set $v' = \{G_1 \cup G_2 \cdots G_J \cup (v-u)\}$, where $J = |u|$, exceeds k, then remove v from H. Otherwise, insert v' in H and remove the old element v from H. Proceed to the step ①.

/* final step for construction of $g_{i+1}^{\downarrow}(q,j)$ */

copy to $g_{i+1}^{\downarrow}(q,j)$ the multi-sets in H which consist of the elements all having check-marks;

/* decision about the acceptance of M at the bottom boundary row */
initialize a working list $H = \{\{(q_0,0)\}\}$, where q_0 is the initial state of M;
repeat ① and ② above k times;
if there exist an element v in H such that each pair of v is the form $(q_f,0)$, where q_f is an accepting state of M, then accepts x. Otherwise, rejects x.

Note that by one execution of Step ②, the number of elements of each multi-set $v \in H$ increases at least one if v has at least one pair whose first component is a universal state. It is obvious that $T(M') = T(M)$. Now, we consider the amount of space used by the procedure. Concerning the description length of working lists, $\|H_l\| \le O(n^k) \cdot \log|C| = O(n^k \log n)$, $\|G_l\|$, $\|u\|$, $\|v\| \le k \cdot \log|C| = O(\log n)$, and $\|H\| \le O(n^k) \cdot \|v\| = O(n^k \log n)$. The stack height in step ② is at most $k \cdot \log|C| = O(\log n)$. Thus, the total space used by M' is never beyond $O(n^{k+1})$. □

Lemma 4.2. *If* $L(n) = o(n^k)$, *then* $T\langle k \rangle \notin \mathcal{L}[3DT(L(n))]$, *where* $T\langle k \rangle$ *is the set defined in Lemma 4.1.*

Proof. Easy modification of the proof of Lemma 4.1 (2) in [10]. □

From Lemma 4.1 and Lemma ??, we get the desired result.

Theorem 4.1. *For each* $k \in N$, *the necessary and sufficient space for 3DTs to simulate AF(k)s is* n^{k+1}.

Corollary 4.1 ([8],[9]). *The necessary and sufficient space for 3DTs to simulate NFs is* n^2.

Note that the 3DT space corresponding to two-dimensional deterministic finite automata (DF) is $n \log n$ [8,9].

5. Discussion

In the first half of this paper, we have investigated the constant leaf-size hierarchy of four-way Turing machines using small space. Table.1 summarizes the results concerning constant leaf-size hierarchy of two-dimensional Turing machines.

In the latter half, we have investigated the necessary and sufficient space for three-way deterministic Turing machines (3DTs) to simulate constantly leaf-bounded alternating finite automata (AF(k)s). The polynomial space of simulating 3DTs illustrates

space bound	AT	UT	3AT	3UT
less than log	Y	N	Y	Y
more than or equal to log	N	N	N	N

Table 1. Existence of constant leaf-size hierarchy

the moderate power level of AF(k)s for picture recognition device. For example, it is known [3] that an mn leaf-bounded alternating finite automaton can recognize the connected pictures, but unknown [11] whether or not a deterministic or nondeterministic finite automata can do. Thus, it is an interesting question to ask whether or not an AF(k) can recognize the connected pictures for some k.

References

[1] W.L.Ruzzo, Tree-size bounded alternation, *J. Comp. Sys. Sci.* **21**, pp.218–235 (1980).

[2] K.N.King, Measures of parallelism in alternating computation trees, *Proc. 13th ACM Symp. on Theory of Comp.*, pp.189–201 (1981).

[3] K.Inoue,I.Takanami,and H.Taniguchi, Two-dimen-sional alternating Turing machines, *Theoret. Comput. Sci.* **27**, pp.61–83 (1983).

[4] H.Yamamoto, Leaf reduction theorem on time- and leaf-bounded alternating Turing machines, *IEICE Trans. Inf. & Syst.* **E75-D**, No.1, pp.133–140 (Jan. 1992).

[5] H.Matsuno, K.Inoue, I.Takanami, and H.Taniguchi, Alternating simple multi-head automata, *Theoret. Comput. Sci.* **36**, No.2–3, pp.291–308 (Mar.1985).

[6] J.Hromkovic,K.Inoue,and I.Takanami, Lower bounds for language recognition on two-dimen-sional alternating multihead machines, to appear in *J. Comp. Sys. Sci.*

[7] A.Ito, K.Inoue, I.Takanami, and H.Taniguchi, Two-dimensional alternating Turing machines with only universal states, *Inform. and Control* **55**, pp.193–221 (1983).

[8] K.Morita, H.Umeo, and K.Sugata, Accepting Abilities of Offside-free Two-dimensional Marker Automata — The simulation of Four-way Automata by Three-way Tape-bounded Turing Machines, *The Technical Reports of the Institute of Electronics and Communication Engineers of Japan* Vol. **AL79**, No.19, pp.1–10(1979).

[9] K.Inoue and I.Takanami, A Note on Deterministic Three-Way Tape-Bounded Two-Dimen-sional Turing Machines, *Information Sciences* **20**, pp.41–55 (1980).

[10] A.Ito,K.Inoue,and I.Takanami, The simulation of two-dimensional one-marker automata by three-way Turing machines, *Inter. J. of Patt. Recog. & Art. Intel.* 3, No.3&4, pp.393–404 (1989).

[11] A.Rosenfeld, *Picture Languages — Formal models for Picture Recognition,* Academic Press (1979).

Shape Recovery and Error Correction Based on Hypothetical Constraints by Parallel Network for Energy Minimization

Koh Kakusho*, Seiichiro Dan*, Norihiro Abe** and Tadahiro Kitahashi*

* The Institute of Scientific and Industrial Research, Osaka University
Ibaraki, Osaka 567, JAPAN
** Faculty of Computer Science and Systems Engineering, Kyushu Institute of Technology
Iizuka, Fukuoka 820, JAPAN

Abstract. Shape recovery from a monocular image with errors is addressed. Shape recovery necessitates the use of additional plausible constraints on typical structures and features of the objects in an ordinary scene. We propose an hypothesization and verification method for 3D shape recovery based on geometrical constraints peculiar to man-made objects. One difficulty with this method lies in the mutual dependency between proper assignment of constraints to the regions in a given image and recovery of a consistent 3D shape. Another lies in dealing with the error by preliminary processes to extract 2D geometrical features from a real image. A concurrent mechanism has been implemented which is based on energy minimization using a parallel network for relaxation, and is capable of maintaining consistency between constraint assignment and shape recovery. The error in an input image is also corrected through the process of shape recovery.

1 Introduction

Recovery of the three dimensional (3D) scene from a monocular image has been studied since the early stages of computer vision research. The information conveyed by an image is insufficient to reconstruct the 3D shapes of objects in an image. This implies that shape recovery from an image necessitates the use of additional constraints on typical structures and features of the objects in an ordinary scene.

Many methods for shape recovery have been successfully developed by employing specific constraints as well as general ones. But even if a method is effective for one class of scenes, it might be useless for another class. A robust vision system should be able to adaptively choose a set of optional constraints which are effective for solving each of the partial problems involved in shape recovery from a given image. We call these optional constraints "hypotheses", because they are not universal and are not always useful either. Hypothetical use of constraints instead of conventional assertive use of them could make a vision system more flexible.

This hypothesis-based paradigm provides an alternative to the domain dependency of the conventional approach. However, it essentially includes the so-called *chicken and egg problem*. That is, shape recovery requires proper constraints, but the constraints to be chosen depend on the 3D shape to be recovered. This scheme is realized by the interaction between the following two processes; 1) assigning hypotheses to the regions of the image through evaluation of global consistency of the resultant 3D shape and 2) recovering the 3D shape based on the assigned hypotheses. Parallel and interactive mechanisms, such as neural networks, will be more effective for implementing this strategy than sequential mechanisms, because the information is spatially distributed in an image and we can not find any significant ordering of procedures in general.

Another problem is concerned with the error by preliminary processes to extract 2D geometrical features from a real image. It is hardly possible to extract the data free from input noise and processing errors from the real image. For the robustness of the system, we must provide our system with the ability to correct such noise and errors.

In this paper, we will discuss a solution of these problems in terms of an energy minimization process. The process is carried out by a parallel network for relaxation. The network is obtained by extending a Hopfield-type neural network [1]. The implementation solves the two main problems; 1) a mechanism for self-assignment of hypotheses, and 2) correction of the error in the image through shape recovery.

2 A Hypothesis-Based Method

2.1 Hypotheses

The feasibility of shape recovery methods depends upon the constraints placed on a scene and the input conditions of the image. When a scene consists of polyhedral objects and the images of their edges are clear, geometrical regularities such as rectangularity and symmetry of surfaces can be employed as powerful constraints to recover the 3D shape [2, 3]. Furthermore, certain heuristics which properly deal with *horizontal*, *vertical* and other gravity-based environmental labels can make shape recovery more efficient.

Among these geometrical and environmental constraints, we focus on the following three constraints.

1) *Rectangularity Hypothesis*
"The surface corresponding to a region is a rectangle."

2) *Horizontality Hypothesis*
"The surface corresponding to a region is environmentally horizontal (perpendicular to the gravity vector)."

3) *Verticality Hypothesis*
"The surface corresponding to a region is environmentally vertical (parallel to the gravity vector)."

These constraints are named *hypotheses* because they are not absolute constraints. These hypotheses are optionally used for recovering the 3D surface corresponding to each region of an image.

2.2 Domain Space

One of our problems is to find a suitable mechanism for shape recovery on the basis of these hypotheses. For the sake of simplicity, we will restrict objects to polyhedra consisting only of quadrangular planar surfaces, so that each surface can satisfy at least one of the proposed hypotheses. More hypotheses could be used without losing generality in order to extend the domain space.

The object images are assumed to be free from occlusions. We assume that both the 2D locations of all the vertices in an input image and all the pairs of labels of bordering regions in the image are given.

2.3 Hypothesization and Verification

The unique 3D shape of a line drawing image can not be recovered from the image unless some hypotheses and their proper assignments to the regions are taken into consideration. One problem here is that the proper assignments are unknown before recovery. The hypothesization and verification scheme can be used for solving this problem. In this scheme, if a consistent 3D shape can be recovered based on the hypotheses tentatively assigned to the regions, this assignment is regarded as adequate, and if not, it is revised (see Fig.1).

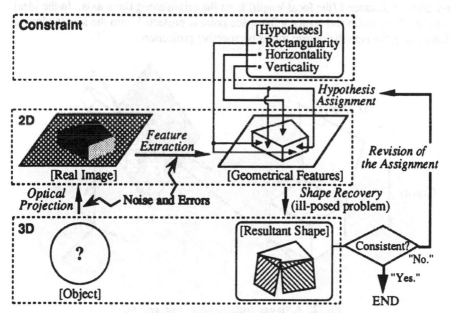

Fig.1. Hypothesization and Verification Strategy

A measure of *appropriateness* should be defined in order to achieve the appropriate hypothesis assignment. The following five conditions are used. The more conditions the assignment satisfies, the more appropriate it is assumed to be.

1) The recovered 3D shape of a region must meet all the hypotheses assigned to the region.
2) The assignment of the hypotheses must be unambiguous, which is to say, the assignment of each hypothesis should be binary.
3) The horizontality hypothesis and the verticality hypothesis must not be assigned to the same region.
4) At least one hypothesis should be assigned to each region.
5) The horizontality hypothesis should not be assigned to adjoining regions simultaneously (we assume that each edge in the image corresponds to a *roof-edge* of a 3D object).

The condition 1) describes the interactive relationship between hypothesis assignment and shape recovery.

3 The Formulation

Prior to discussing the implementation of the hypothesization and verification strategy, we will begin by formalizing the shape recovery problem.

We will use a viewer-centered coordinate system in a gravity-dependent environment (see Fig.2). The origin, O, is at the center of the lens. The screen is parallel to the x-y plane at distance f (the focal length) from the origin along the z-axis. In the ideal case free from noise and errors, a 3D space point is projected onto the screen along a line passing through the origin; this is perspective projection.

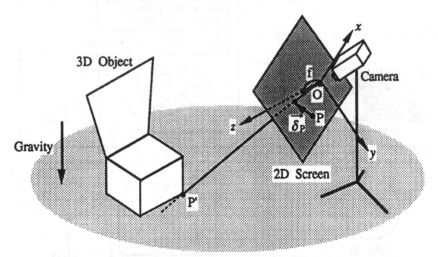

Fig. 2. A viewer-centered coordinate system.

A 2D image is obtained by projecting all the points on a 3D object to the screen. What *shape recovery* means in this paper is determining the 3D point corresponding to a given point in a 2D image using environmental and/or geometrical hypotheses. Throughout this paper, P stands for a point in an input image and P' stands for its hypothesized 3D point.

Considering the error included in the image, the location of P' are calculated by the following equation;

$$\overrightarrow{OP'} = k_P\left(\overrightarrow{OP} + \overrightarrow{\delta_P}\right) \qquad (1)$$

where $\overrightarrow{\delta_P}$ denotes the vector correcting the location of the point P, which we call the *correction vector* of P, and k_p is the ratio of $\overrightarrow{OP'}$ to $\left|\overrightarrow{OP} + \overrightarrow{\delta_P}\right|$, which we call the *depth parameter* of P.

As we assume that the focal length f is given, the 3D location of the point P' is uniquely determined by the depth parameter and the correction vector of P from equation (1). The set of possible depth parameters and correction vectors of all the observable points in an input image serves as the problem space (or state space), and corresponds to the set of all possible 3D shapes associated with the image.

Objects are assumed to be polyhedral. It follows from this that the 3D shapes of objects can be determined by their vertices only. Consequently, our *shape recovery* problem is reduced to determining the depth parameters and the correction vectors of all the vertices of regions in the image.

The horizontality and verticality hypotheses come from the assumption that we deal with man-made objects in the real world with gravity. The direction of the gravity vector in this viewer-centered coordinate system is vital to the application of these hypotheses. In this paper, the 3D direction of the gravity vector is given a priori.

4 Energy Functions

4.1 Energy Functions Associated with the Constraints

First we define energy functions which correspond to the constraints used for shape recovery. Each energy function defined below will be employed as a term of a global energy function for the hypothesization and verification in the following section.

As we assume that objects are polyhedra, we need a constraint to ensure that surfaces are flat. This constraint must always be satisfied by the resultant shape. The energy function for the flatness constraint is defined below.

We denote dot product and cross product of vectors $\vec{v_1}$ and $\vec{v_2}$ by $\vec{v_1} \cdot \vec{v_2}$ and $\vec{v_1} \times \vec{v_2}$, respectively. Note that P' represents an hypothesized 3D point which corresponds to the 2D point P (see section 3).

Definition 1 (*Rectangularity and Parallelism of Two Vectors*)
Let the angle between the vectors $\vec{v_1}$ and $\vec{v_2}$ be $\theta(\vec{v_1}, \vec{v_2})$. Then $\text{rect}(\vec{v_1}, \vec{v_2})$ and $\text{para}(\vec{v_1}, \vec{v_2})$ are defined as follows:

$$\text{rect}(\vec{v_1}, \vec{v_2}) = \cos^2\left\{\theta(\vec{v_1}, \vec{v_2})\right\}$$

$$= \frac{\left(\vec{v_1} \cdot \vec{v_2}\right)^2}{\left|\vec{v_1}\right|^2 \left|\vec{v_2}\right|^2}$$

$$(2)$$

$$\mathrm{para}(\vec{v_1}, \vec{v_2}) = \sin^2\left\{\theta(\vec{v_1}, \vec{v_2})\right\}$$
$$= 1 - \cos^2\left\{\theta(\vec{v_1}, \vec{v_2})\right\}$$

$$(3)$$

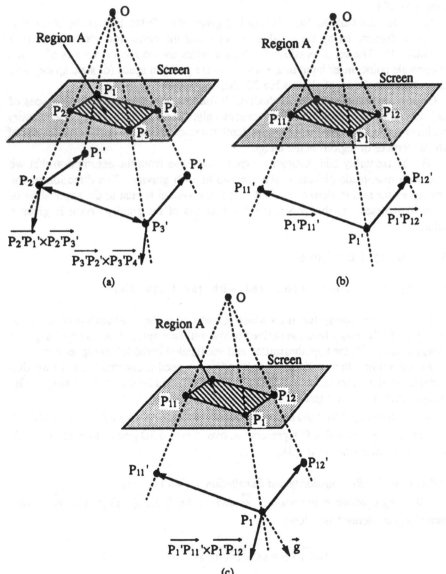

Fig. 3. Meaning of Energy Functions.

Definition 2 (*Flatness of Surface*)

Let P_2 and P_3 be given vertices that are adjacent to each other on a region A and let their adjacent vertices other than P_2 and P_3 be P_1 and P_4. The energy of the flatness constraint E_F is defined as follows:

$$E_{FA} = \sum para(\overrightarrow{P_2'P_1'} \times \overrightarrow{P_2'P_3'}, \overrightarrow{P_3'P_2'} \times \overrightarrow{P_3'P_4'}) \qquad (4)$$

$$E_F = \sum_A E_{FA} \qquad (5)$$

where $\sum$ in the first formula denotes the sum over all vertex pairs adjacent to each other on A and $\sum$ in the second formula denotes the sum over all regions in an image.

At each vertex of region A, construct a vector defined as the normal vector of the plane determined by the 3D locations of the vertex and its two adjacent vertices of region A. The minimum of the function E_{FA} corresponds to the situation in which the two normal vectors at two adjacent vertices of the region A are parallel to each other. In Fig.3(a), $\overrightarrow{P_2'P_1'} \times \overrightarrow{P_2'P_3'}$ at P_2 and $\overrightarrow{P_3'P_2'} \times \overrightarrow{P_3'P_4'}$ at P_3 are one pair of 3D normal vectors of region A.

The three hypotheses that we proposed earlier (i.e. rectangularity, horizontality and verticality) are optional while the flatness constraint is required. The energy functions representing the three hypotheses are defined below.

Definition 3 (*Rectangularity, Horizontality and Verticality of a Plane*)

Let a given vertex of the region A and its two adjacent vertices on A be P_i, P_{i1} and P_{i2}, respectively. Then the energy of the rectangularity hypothesis E_{RA}, of the horizontality hypothesis E_{HA}, and of the verticality hypothesis E_{VA} with respect to A are defined as follows:

$$E_{RA} = \sum_{i=1}^{4} rect(\overrightarrow{P_i'P_{i1}'}, \overrightarrow{P_i'P_{i2}'}) \qquad (6)$$

$$E_{HA} = \sum_{i=1}^{4} para(\overrightarrow{P_i'P_{i1}'} \times \overrightarrow{P_i'P_{i2}'}, \vec{g}) \qquad (7)$$

$$E_{VA} = \sum_{i=1}^{4} rect(\overrightarrow{P_i'P_{i1}'} \times \overrightarrow{P_i'P_{i2}'}, \vec{g}) \qquad (8)$$

where $\vec{g}$ is the gravity vector which is known a priori.

A minimum of E_{RA} corresponds to the situation in which all the 3D angles of every region in the image are right angles. In Fig.3(b), $\overrightarrow{P_1'P_{11}'}$ and $\overrightarrow{P_1'P_{12}'}$ sharing P_1' are one of the pairs of 3D vectors which are required to be perpendicular to each other. A minimum of E_{HA} corresponds to the situation in which the normal vector at each vertex of the region A is parallel to the gravity vector $\vec{g}$. In Fig.3(c), $\overrightarrow{P_1'P_{11}'} \times \overrightarrow{P_1'P_{12}'}$

is the normal vector at vertex P_1. Similarly a minimum of E_{VA} corresponds to the situation in which the normal vector at each vertex of A is perpendicular to $\vec{g}$.

4.2 Global Energy Function for Hypothesis-Based Shape Recovery

We define the energy function E which stipulates the interaction between hypothesis assignment and shape recovery as follows:

$$
\begin{aligned}
E = \quad & c_1 E_F \\
+ \sum \{ \quad & c_2 (E_{RA} r_A + c_R(1 - r_A)) + c_2 (E_{HA} h_A + c_H(1 - h_A)) + c_2 (E_{VA} v_A + c_V(1 - v_A)) \\
& + c_3 r_A (1 - r_A) + c_3 h_A (1 - h_A) + c_3 v_A (1 - v_A) \\
& + c_4 h_A v_A \\
& + c_5 (1 - r_A)(1 - h_A)(1 - v_A) \qquad \qquad \qquad \qquad \qquad \} \\
+ \sum & c_6 h_A h_B + c_7 \sum_i \left(\delta_P^{x\,2} + \delta_P^{y\,2} \right)
\end{aligned}
$$

$$(9)$$

where r_A, h_A, v_A denote the assignment of the rectangularity hypothesis, the horizontality hypothesis and the verticality hypothesis to the region A, respectively. That is, $r_A = 1$ means that the rectangularity hypothesis is assigned to region A, and $r_A = 0$ means that the hypothesis is not assigned to the region. The variables δ_P^x and δ_P^y denote the x-component and the y-component of the correction vector $\vec{\delta_P}$, respectively.

The first term E_F in E corresponds to the flatness constraint defined by equation (5). The first $\sum$ denotes the sum over all regions in the image. The first term in this summation implies that it selectively takes the lower value of $E_{RA} r_A$ or $c_R (1 - r_A)$ and corresponds to the condition 1) given in the section 2.3. The second and third terms mean the similar operation. The fourth term expresses the logical mutual inhibition of the selection of both r_A and its inverse $1 - r_A$ by means of setting its minimum points at $r_A = 1$ or 0. This term corresponds to the condition 2). The fifth and sixth terms do the similar operation concerning h_A and v_A, respectively. The seventh term and the eighth term correspond to the condition 3) and 4), respectively. The former represents the mutual inhibition of both cases $h_A = 1$ and $v_A = 1$ occurring at the same time by making its value maximum in that case. The eighth term does the opposite; it inhibits the simultaneous occurrence of $r_A = 0$, $h_A = 0$ and $v_A = 0$. The second $\sum$ denotes the sum over all pairs of the two adjoining regions A and B, and this corresponds to the condition 5). The last term of E represents that orthogonally projected point of each vertex to the screen must coincide with the corresponding point in the image. The constant parameter c_1 is fidelity of the recovered 3D shape to the flatness constraint. The values of c_R, c_H and c_V control the strictness of the representation of the condition 1), and c_2, ..., c_6 are constants which determine fidelity of the process to each condition. The parameter c_7 represents certainty of an input image.

5 A Parallel Network for Relaxational Energy Minimization

A 3D shape based on the three hypotheses can be recovered by minimizing the energy defined in section 4.2. Here, it should be emphasized that the energy function is completely defined by local relationships among the variables. This locality enables the minimization process to be carried out in a massively parallel way by a relaxation network. In other words, massive parallelism of computation becomes possible thanks to this definition of the energy function. In the massively parallel computation, the amount of computation weakly depends on the complexity in an input image.

Consistent shape recovery by energy minimization can be performed by an iterative search which starts off with a poor interpretation and progressively improves it by reducing the energy function that measures the extent to which the current interpretation violates the constraints.

In this section, we discuss a parallel network to minimize the energy function.

5.1 Representation of the Problem Space in the Network

First we map the problem space of shape recovery into a parallel network. The network consists of three kinds of units: *interpretation units, correction units* and *hypothesis units.* The correction units are classified as *x-correction units* and *y-correction units*. A pair of them and an interpretation unit is dedicated to each vertex of regions in an image. The output value of an interpretation unit denotes the depth parameter of the vertex corresponding to the unit. The output values of a x-correction unit and a y-correction unit denote the x-component and the y-component of the correction vector of the vertex corresponding to those units. The internal activation levels and output values of an interpretation unit and a correction unit are continuous. They range within the interval $[0, \infty)$. The output values of the units are exactly equal to the internal level of the units.

The interpretation units and the correction units represent the 3D shape. Different 3D shapes of the same input image correspond to different patterns of activity over the group of the units. As a special case, the input image itself is represented when the output values of all the interpretation units are 1.0 and those of all the correction units are 0.0.

The hypothesis units represent the assignment of the hypotheses. They are classified as *rectangularity units, horizontality units* or *verticality units*. A triple of these hypothesis units is associated with each region of the image. The output value of a hypothesis unit has the range $[0, 1]$ and follows a sigmoid function of the internal activation level of the unit. The output values of the rectangularity unit, horizontality unit and verticality unit, associated with region A represent r_A, h_A and v_A, respectively.

The structure of the network depends on the input image. A different network must be built for a different image. However, this does not mean that the network is to be made by hand. The topology and the dynamics of the network are automatically formed from input data.

5.2 The Update Rules

The energy function (9) is a function of the outputs of all the units. Each possible state of activity of the network is associated with an energy level. The rule used for updating the state must be chosen so that this energy keeps falling.

Hopfield et al. have defined the energy of a network, which is relaxationally minimized by means of the *steepest descent method* through the update of the network [1]. Although Hopfield's networks may not be complete in the sense that there is no guarantee that they will reach global minima, their parallelism and interactiveness will be quite helpful for construction of vision systems. Here, we would like to find the optimum or at least an optimal shape for a given image. For this reason, we are going to make use of a parallel network for relaxation, similar to a Hopfield network for solving our problem. As our energy function is more complex, we can not naively adopt a Hopfield network. We will focus on the computational basis of a Hopfield network, i.e. the *steepest descent method*. Let $\Delta k_P(t)$ designate the change in k_P at time t, then the update rule for the interpretation units is defined as follows:

$$\Delta k_P(t) = -\alpha \frac{\partial E}{\partial k_P(t)} \qquad (10)$$

The update rule for the x-correction units is defined as follows:

$$\Delta \delta_P^x(t) = -\beta \frac{\partial E}{\partial \delta_P^x(t)} \qquad (11)$$

The update rules for the y-correction units is the same as those of the x-correction units. The update rule for the rectangularity units is defined as follows:

$$\Delta u_{RA}(t) = -\gamma \frac{\partial E}{\partial r_A(t)} \qquad (12)$$

$$r_A(t) = \frac{1}{1 + \exp(-u_{RA}(t))} \qquad (13)$$

where $u_{RA}(t)$ and $r_A(t)$ denote the internal level and the output value of the rectangularity unit related to the region A at time t. The update rules for the horizontality units and the verticality units are the same as those of the rectangularity units.

5.3 Structure of the Network

The structure of this network is illustrated in Fig.4. A *ball* represents an interpretation unit or a correction unit, and a *cylinder* represents an hypothesis unit. A *branch* settled between two units represents interaction between them, which is to say, they send their output values to each other without weighting. All the units send their outputs to their neighboring units only, and this locality of the connections of the structure is essential to realize massively parallel processing on this network. Automatic performance of the hypothesization and verification strategy results from the interactions among the hypothesis units and the triples of an interpretation unit and two correction units. On the one hand, each hypothesis unit updates its output value so that the hypotheses are appropriately assigned to the regions in the image according to a tentative 3D shape represented by the output values of the interpretation units and the correction units. On the other hand, each interpretation unit or correction unit updates its output value so that a globally consistent 3D shape is recovered based on the assigned hypotheses represented by the output values of the hypothesis units.

A unit's job is, following equation (10) - (13), to compute the change of its output as a function of the separate inputs which it receives from each of the individual units, and to update its output which is sent to its neighbors. We need more intricate processing units than those of a Hopfield network in order to realize the update. It is not significant that the units in our network are not neuron-like because we can construct a neural network which is functionally equivalent to our network and entirely consist of neuron-like units. The equivalent network is obtained by replacing each unit in the original network with a feedforward layered neural network which consists of neuron-like units and performs the same mapping from input to output as equation (10) - (13).

It has been proved that a three-layered neural network can represent arbitrary continuous mappings [4]. A procedure for designing a three-layered neural network which approximates an arbitrary continuous mapping has also been presented [5]. These results might be applied to construction of the layered neural networks in our example. In this paper, however, we do not go into the details of the equivalent network, because our focuses are on the issue of processes for the hypothesization and verification strategy and error correction through the processes in the framework of relaxational energy minimization, and on massively parallel computation by networks.

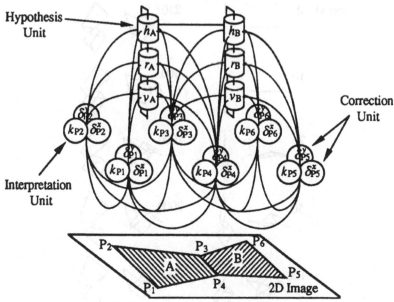

Fig.4. The Structure of the Network.

6 Experimental Results

6.1 Shape Recovery from a computed image free from error

The parallel network for a computed image free from error was simulated on a digital computer. In this case we did not use the correction units. It means that the correction vectors of all the vertices were *zero vectors* throughout the simulation. Fig.5(a) is a computed perspective projection of a chair-like object, which includes one vertical trapezoid face corresponding to region A, two rectangular faces corresponding to regions B and C, and one vertical face corresponding to region D. The images in Fig.5(b) are parallel projections of the recovered shape to $x+y-z=0$, x-z plane, y-z plane and x-y plane in the coordinate system showed in Fig.2.

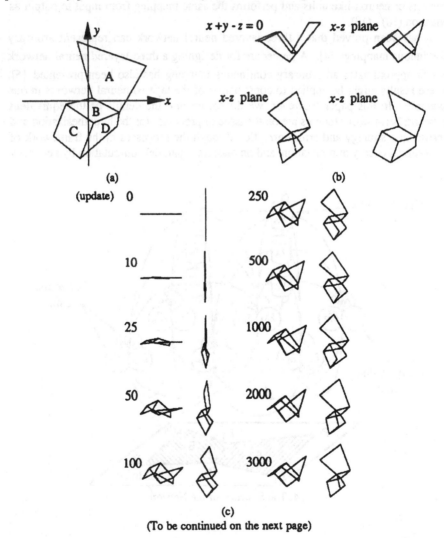

(a)

(b)

(c)

(To be continued on the next page)

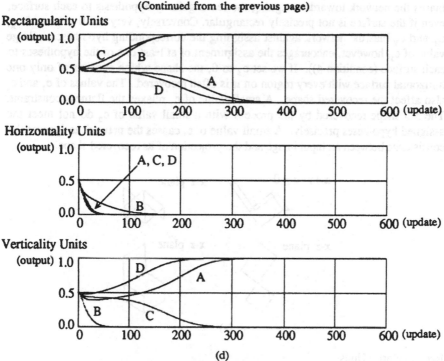

(Continued from the previous page)

(d)

Fig.5. Shape recovery from an ideal image.

As initial conditions, we set the output value of all the interpretation units to 1.0 and all the hypothesis units to 0.5. The interpretation units were updated five times for every update of the hypothesis units to avoid hasty assignment of hypotheses. The 3D shape is gradually built up from the input image through the updates of the interpretation unit (see Fig.5(c)). At the same time, the assignment of hypotheses is determined through the updates of the hypothesis units. Fig.5(d) shows the output values of the rectangularity units, the horizontality units and the verticality units related to regions A, B, C and D. After 3000 updates of the interpretation units and 600 updates of the hypothesis units, the rectangularity hypothesis and the verticality hypothesis were assigned to regions B, C and regions A, D, respectively, and the 3D shape was recovered based on the assigned hypotheses.

This result was obtained with $c_R = 0.005$, $c_H = c_V = 0.01$, $c_1 = 1.0$, $c_2 = 2.0$, $c_3 = 2.5$, $c_4 = 3.0$, $c_5 = 1.5$, $c_6 = 0.01$.

6.2 Recovery of a Coarse 3D Shape by Top-down Approach

The values of the parameters in the energy function E strongly affect the recovered shape and the assignment of the hypotheses. The constant parameters c_R, c_H and c_V represent penalties for failing to assign the rectangularity hypothesis, the horizontality hypothesis, and the verticality hypothesis, respectively. Heavy penalties force the network to assign the hypotheses to each region. For example, a large value of c_R

biases the network towards assigning the rectangularity hypothesis to each surface, even if the surface is not precisely rectangular. Conversely, very small values of c_R, c_H, and c_V bias the network against assigning the corresponding hypotheses; a large value of c_S, however, encourages the assignment of at least one of the hypotheses to each surface (condition 4)). If we set c_6 to 0, the shape which consists of only one horizontal surface with every region on it is often recovered. The values of c_1 and c_2 also affect the recovered shape. A small value of c_1 relaxes the flatness constraint. The 3D shape recovered by the process with a small value of c_2 do not meet the assigned hypotheses precisely. A small value of c_7 causes the network to ignores the consistence between an input image and the projection of its recovered shape.

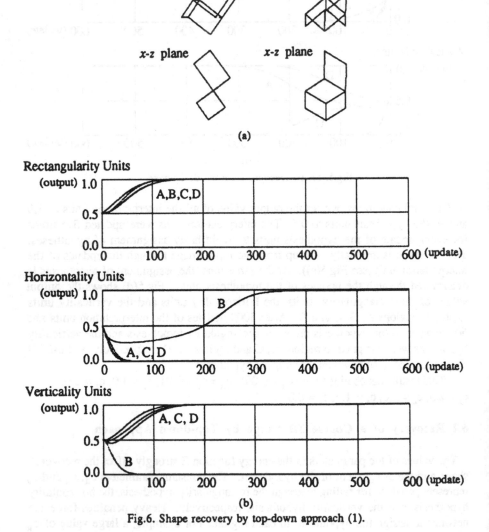

(a)

(b)

Fig.6. Shape recovery by top-down approach (1).

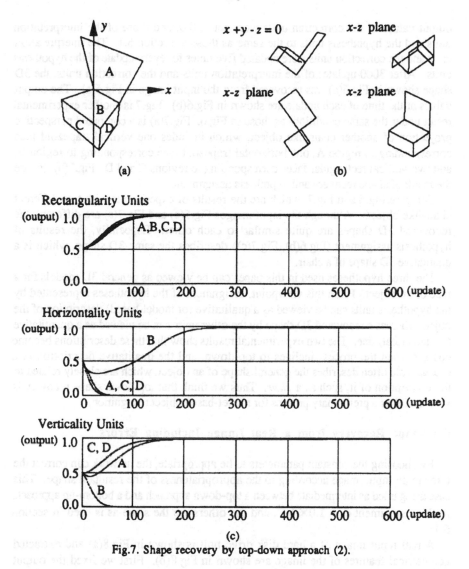

Fig.7. Shape recovery by top-down approach (2).

It follows that we can modulate the inclination of the process to *top-down* or *bottom-up* by changing the values of constant parameters. For example, large values of c_R, c_H, c_V and a small one of c_7 will get the process to incline toward top-down, and the inverse case will lead it toward bottom-up. The experimental result described in section 6.1 shows the extreme of bottom-up shape recovery. In the experiment, the network can not represent the recovered 3D shape whose back-projection to the screen deviate from the original input image. It corresponds to the case with $c_7 = \infty$.

In this section, we will investigate the other case: top-down shape recovery. In the following experiments, we used $c_R = c_H = c_V = 0.2$, $c_7 = 1.0 \times 10^{-16}$, and set the other constant parameters to the same values as those used in section 6.1. The initial output values of the correction units were set to 0.0, and those of the interpretation units and the hypothesis units to the same as those in section 6.1. The interpretation units and the correction units were updated five times for every update of the hypothesis units. After 3000 updates of the interpretation units and the correction units, the 3D shape shown in Fig.6(a) was recovered from the input image in Fig.5(a). The output values at the time of each update are shown in Fig.6(b). Fig.7 is another experimental result under the same conditions as those in Fig.6. Fig.7(a) is a computed perspective projection of another chair-like object, which includes one vertical trapezoid face corresponding to region A, one horizontal trapezoid face corresponding to region B, and two vertical rectangular faces corresponding to regions C and D. Fig.7(b), (c) are the result of shape recovery and hypothesis assignment.

Compare Fig.6 and Fig.7, which are the results of experiments with two different chair-like objects. Although the input images Fig.5(a) and Fig.7(a) are different, the recovered 3D shapes are quite similar to each other. Especially, the results of hypothesis assignment (Fig.6(b), Fig.7(c)) describes the same 3D shape, which is a qualitative 3D shape of a chair.

The three hypotheses used in this paper can be viewed as general 3D models for a face of an object. From this viewpoint, assignment of the hypotheses represented by the hypothesis units can be viewed as a qualitative (or model-based) description of the object whereas a recovered 3D shape by the other units can be viewed as a quantitative (or numerical) one. The two experimental results show that these descriptions become coarser when the process inclines to top-down, and the qualitative description in a coarse scale often describes the general shape of an object, which are closely related to the conception of it, such as a *chair*. Thus we think that top-down shape recovery is very useful as a preliminary process for model-based object recognition.

6.3 Shape Recovery from a Real Image Including Errors

By choosing the constant parameters to be appropriate, the network can correct the error in an input image according to the appropriateness of the resultant shape. This case is regarded as intermediate between a top-down approach and a bottom-up approach. In this experiment $c_7 = 1.0 \times 10^{-8}$, and the others are the same as is used in section 6.1.

A real input image of a hard disk drive unit is shown in Fig.8(a) and extracted geometrical features of the image are shown in Fig.8(b). First we fixed the output values of the correction units, and allowed the interpretation units and the hypothesis units to run freely (*bottom-up phase*). All the hypothesis units made one update, each time all the interpretation units made ten updates. After 5000 updates of the interpretation units and 500 updates of the hypothesis units, the rectangularity hypothesis was assigned to all the regions and the horizontality hypothesis to the region A, and the verticality hypothesis to the region B, C. The tentative 3D shape recovered as the result of this phase is shown in Fig.8(c). Then we fixed the output of all the hypothesis units, and let the interpretation units and correction units run freely (*top-down phase*). In this phase, correction units update their outputs so that the errors in the image are corrected. After 5000 updates of the interpretation units and the correction units, the revised 3D shape was obtained as the result of this phase (see

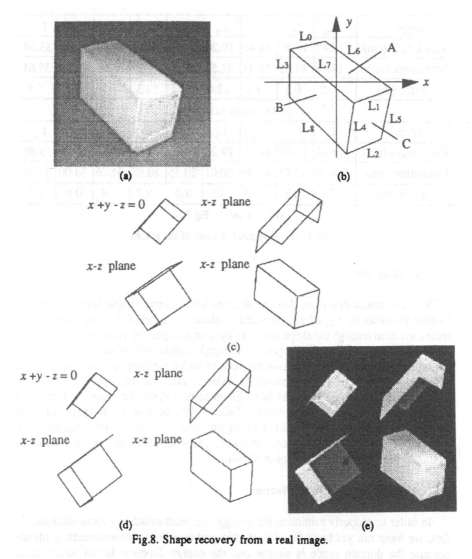

(a)

(b)

$x + y - z = 0$ x-z plane

x-z plane x-z plane

(c)

$x + y - z = 0$ x-z plane

x-z plane x-z plane

(d)

(e)

Fig.8. Shape recovery from a real image.

Fig.8(d)). Fig.8(e) is generated by mapping the gray level of each pixel of the input image in Fig.8(a) to the corresponding point in the recovered shape in Fig.8(d). Table.1 numerically describes the result. The estimation values are calculated from the results shown in Fig.8(c), (d). As we cannot get absolute values with this method, the figures are normalized by the length of edge L_0. So far as this instance is concerned, the error ratio of estimation after error correction is no more than 10% for every edge.

In all the experiments showed in 6.1 - 6.3, different values of α, β and γ mainly affect convergence speed. Larger values of them enable rapid convergence but often give harmful oscillation. Experimentally, we have found favorable convergence for $\alpha = 2.0 \times 10^{-4}$, $\beta = 0.5$, $\gamma = 0.01$.

Edge	L_0	L_1	L_2	L_3	L_4	L_5	L_6	L_7	L_8
Real length (cm)	14.40	14.40	14.40	19.20	19.20	19.20	33.80	33.80	33.80
Estimation (cm)	14.40	14.35	14.14	21.58	21.52	21.56	36.06	35.99	35.61
Error (%)	0.0	0.3	1.8	12.4	12.1	12.3	6.7	6.5	5.4

(a) The result in Fig.7(a)

Edge	L_0	L_1	L_2	L_3	L_4	L_5	L_6	L_7	L_8
Real length (cm)	14.40	14.40	14.40	19.20	19.20	19.20	33.80	33.80	33.80
Estimation (cm)	14.40	13.85	13.25	20.01	20.55	20.97	34.26	34.00	34.04
Error (%)	0.0	3.8	8.0	4.2	7.0	9.2	1.4	0.6	0.7

(b) The result in Fig.7(b)

Table 1. The numerical figures of the results.

7 Discussion

We have concluded from these experiments with some simple images that it is feasible to realize the hypothesization and verification method for shape recovery and error correction through the shape recovery by means of parallel networks.

It should be noted that our proposal in this paper can be viewed as hypothesis-based reasoning or constraint-based reasoning, both of which are now receiving the interest of many AI researchers. In most cases, these techniques are applied to symbolic information. Little attention has been given to non-symbolic, iconic information including error when considering these techniques. We believe that this paper suggests an approach to processing this kind of information, although it only scratches the surface of the true visual perception problem. Many problems still remain unsolved. In this section, we will discuss some of them.

7.1 Avoidance of Local Minima

In order to properly minimize the energy, we must avoid any local minima. In fact, we have not yet been faced with this problem in our experiments, probably because the domain space is simple and the energy function is not so intricate. However, the problem of avoiding local minima becomes serious when the number of hypotheses to be employed and the complexity of the input images increase. Our energy function guarantees that good interpretations are minima, but does not guarantee that bad interpretations are not minima. Bad interpretations will have higher energy than good ones, so a global optimization technique, if one exists, will be less likely to settle into them. Although this is an general problem with the energy minimization methods, there are few solutions to this problem which are both practical and satisfactory. The most popular one is the statistical approach called *simulated annealing* [6]. We are now trying to incorporate this technique.

7.2 Extension of the Domain Space

Shape recovery on the hypotheses of rectangularity, horizontality and verticality is possible as long as the proper assignment of the hypotheses supplies sufficient constraints to recover the 3D shape from an input image. In other words, if shape recovery with the proper assignment of the hypotheses is still an underconstrained problem, it is not well-posed and the 3D shape can not be recovered. For example, bottom-up approach is impossible to recover the 3D shape based on the three hypotheses from a 2D image of an object which does not have rectangular, horizontal or vertical surfaces. Although top-down approach can recover a coase 3D shape of the object, additional hypotheses are necessary if the fine 3D shape is required.

We employed only three hypotheses: the rectangularity hypothesis, the horizontality hypothesis, and the verticality hypothesis. This does not mean that our approach is bound to these hypotheses. Any number of constraints of any kind are available if they can be represented in the form of energy functions. For example, symmetry in the shape of a surface might be used in human visual perception. In fact, a shape recovery method based on the symmetry constraint has been proposed [7]. We are currently trying to make use of it.

If we employ many hypotheses, their plausible assignment may not be unique and various 3D shapes may be admissible. In order to avoid this ambiguity, we must use some plausible relationships among the hypotheses based on the kinds of structures that typically appear, which involves higher knowledge about the scene than the hypotheses. For example, the relationships of parallelism, perpendicularity and symmetry among component surfaces of an object might be useful to deal with man-made objects. The hypothesis assignment and the restriction based on it are interactive and this interaction can also be realized by our approach. Hence, when we try to employ more hypotheses, what we must consider is how to represent the constraints as energy functions.

Acknowledgment

The authors wishes to thank Mr. Yukihiko Shimizu, who was one of their colleagues at the Institute of Scientific and Industrial Research, Osaka University and now is in Information Systems Research Center, Canon Inc., for help with programming and the experiments.

References

1. J.J.Hopfield and D.W.Tank: "Neural" Computation of Decisions in Optimization Problems. Biological Cybernetics 52, 141-152 (1985)
2. R.Horaud: New Methods for Matching 3-D Objects with Single Perspective Views. IEEE Transactions on Pattern Analysis and Machine Intelligence PAMI-9, 401-412 (1987)
3. Y.Shirai: Three-Dimensional Computer Vision. Berlin: Springer 1987

4. B.Irie, S.Miyake: Capabilities of Three-layered Perceptrons. Proceedings of IEEE Annual International Conference on Neural Networks 1, 641-648 (1988)
5. H.Kawahara, T.Irino: A Procedure for Designing 3-Layer Neural Networks Which Approximate Arbitrary Continuous Mapping: Applications to Pattern Processing. IEICE Techical Report MBE88-54, 47-54 (1989) (in Japanese)
6. S.Kirkpatrik, C.D.Gelatt and Jr.M.P.Vecchi: Optimization by Simulated Annealing. Science 220, 671-680 (1983)
7. F.Ulupinar and R.Nevatia: Constraints for Interpretation of Line Drawings under Perspective Projection. CVGIP Image Understanding 53, 88-96 (1991)

Use of Gradated Patterns in Associative Neural Memory for Invariant Pattern Recognition

Kazukuni Kobara Taiho Kanaoka Koukichi Munechika [*]
Yoshihiko Hamamoto and Shingo Tomita

Department of Computer Science and Systems Engineering,
Faculty of Engineering, Yamaguchi University
Ube, 755 Japan

[*]Information Processing Center, Yamaguchi University
Ube, 755 Japan

Abstract

Distortion invariant pattern recognition is interesting problem
from the biological and technological point of view, however, it has
not yet been solved by neural networks in satisfactory way. This
paper investigates a associative neural network system to improve
the recalling accuracy for distortion patterns. On a neural network
of perceptron type with feedback, error back-propagation algorithm
and energy function are used for learning process and recalling
process respectively. By using gradated patterns as learning and
unknown patterns, it is shown that recalling accuracy become higher
than using original pattern themselves.

1. Introduction

Invariant pattern recognition is one of the hardest problems in the
theory of perception. Several approaches by using the neural networks for the problem are reported, however, it has not yet been

solved satisfactory way. Though multilayered networks using the back-ward error propagation model is well known as a method for the invariant pattern recognition, it requires a large number of training passes and an exhaustive training set for achieving high recognition accuracy[1]. Graph matching method using a energy function in neural network needs exhaustive calculations[2]. And higher-order neural network[3,4], by which topological information in patterns can be extracted, is not so good for distortion unknown patterns. On the other hand, a model of Hopfield network[5] is attractive as a model of human associative memory, however, it can't recall correctly for distortion patterns such as slightly shifted or slightly scaled ones.

This paper examines a associative neural network system by using gradated patterns to improve a recalling accuracy for distortion patterns. Two layered neural network with feedback(2LNF) is used as associative neural memory. In learning process, we use error backpropagation algorithm and use energy of the network in recalling process. It is shown from fundamental experimental results that, by using the gradated patterns, the recalling accuracy in the 2LNF become higher than usual associative memory.

2. Layered Neural Network with Feedback

LNF(Layered Neural Network with Feedback) can be considered as a model that output of last layer in perceptron model is feedbacked to first layer, or Hopfield model with hidden layer. In the subsequent discussion, for simplicity, we examine the 2LNF(two Layered Neural Network with Feedback), however, it is easy to extend the discussion to multi LNF.

2.1 2LNF

In this section, we describe 2LNF whose illsutrative example is shown in Figure 1. At first each neuron of first layer receives a input from outside and its output is input to each neuron of second layer. Further output of second layer is sent to first layer. And the behavior like this manner is continued. Output pattern can be seen at first layer everytime. We use the following notations.

m : the number of neurons of first layer

n : the number of neurons of second layer

I_{ri} : sum of inputs to ith neuron of rth layer

u_{ri} : output of ith neuron of rth layer

θ_{ri} : threshold of ith neuron of rth layer

w_{rij}: connectivity strength from ith neuron of rth layer
　　　　to jth neuron of other layer

Input-output relationships for ith neuron of rth layer are given as follows.

$$I_{ri} = \Sigma_j \ w_{\bar{r}ji} u_{\bar{r}j} - \theta_{\bar{r}i} \qquad (1)$$

$$u_{ri} = f(I_{ri}) \qquad (2)$$

where $\bar{r}$ denotes other layer for rth layer, and f denotes input-output function for which we use sigmoid function $f(x) = 1/(1+\exp(-x))$.

2.2 Learning Process

In learning process, we regard the 2LNF as 3 layers perceptron network that is, first, second and third layers correspond to first, second and first layer respectively. Then, the error back propagation is used so that output patterns same as input patterns are appear at the first layer.

2.3 Energy in the 2LNF

In recalling process, to evaluate the state of overall 2LNF we introduce the following energy E.

$$E = E_{12} + E_{21} \qquad (3)$$

,where

$$E_{pr} = -\sum_i \sum_j u_{pi} w_{pij} u_{rj} + \sum \theta_{rj} u_{rj} \qquad (4)$$

In the equation (3), Epr denotes the energy of the network whose information flow is from pth layer to rth layer. Then, as well known, we can derive the following.

$$\frac{dE_{12}}{dt} \leq 0 \ , \qquad \frac{dE_{21}}{dt} \leq 0 \qquad (5)$$

Thus $E_{12}(E_{21})$ never increase as long as a $u_{21}(u_{12})$ changes and $u_{12}(u_{21})$ doesn't change. Since both E_{12} and E_{21} work increasingly each other, the state transition of the system halts when the change of energies E_{12} and E_{21} can not be seen.

3. A System of 2LNF

In the system of 2LNF, gradated patterns are used as input and output patterns. The gradated patterns are made by gradating transformation of binary patterns and gradated patterns can be transformed to binary patterns by inverse gradating transformation. In this section, gradating transformation, inverse gradating transformation and how to input and output of pattern is described.

3.1 Gradating Transformation

Let $P = [p_{i,j}]_{1 \leq i,j \leq q}$ be a binary image pattern. And let

$X = (x_1, x_2, \ldots, x_m)$ $(m = q^2)$ be a binary pattern obtained from P, where $p_{i,j} = x_{(i-1)q+j}$ and $p_{i,j}$ denotes the grey level of position (i,j) of P.

At first we get a gradatation pattern $Y = (y_1, y_2, \ldots, y_n)$ from X by the following equation.

$$Y = NX \qquad (6)$$

where

$$N = \begin{bmatrix} (1,1) & (2,1) & \cdots & (m,1) \\ (1,2) & (2,2) & \cdots & (m,2) \\ & & & \\ & & & \\ (1,m) & (2,m) & \cdots & (m,m) \end{bmatrix} \qquad (7)$$

where $(i,j) = \exp(-d(i,j)^2 / \rho)$ and d(i,j) denotes the Euclid distance between ith and jth pixel in the pattern X, and ρ denotes the parameter of gradation rate. Figure 2 shows a pixel with grey level of 1 and its gradated pattern. Further, y_i is normalized to some value between zero and one as follows.

$$z_k = 1 - \exp(-y_k) \qquad (0 \leqq z_k \leqq 1) \qquad (8)$$

Inverse gradating transformation from gradated pattern Y to binary pattern X is obtained easily by the following equation.

$$y_k = \log(1 - z_k) \qquad (9)$$

$$X = N^{-1}Y \qquad (10)$$

where N^{-1} denotes the inverse matrix of N. A example of binary pattern and its gradated pattern is shown in Figure 3.

4. Simulation and Discussion

In order to examine the effect of utilizing the gradated patterns to recalling accuracy, we tried to simulate for the following four types of input and output procedure.

(T-1) GG processing

Gradated patterns are used in learning and recalling (Fig.ure 4.1).

(T-2) BG processing

Gradated patterns are used in learning. And in recalling process binary unknown patterns are recalled(Figure 4.2).

(T-3) GB processing

Binary patterns are used in learning. After unknown gradated patterns are inputted, binary patterns are recalled(Figure 4.3).

(T-4) BB processing

Binary patterns are used for all proceses(Figure 4.4).

For the several types of input output procedure stated above, simulation is carried out for the following items.

(I-1) Recalling accuracy for gradating parameter ρ s.

(I-2) Recalling accuracy for the number of learning categories.

We adopt the number of neurons of first and second layer to 256. And the following some types of learning patterns are used.

(P-1) Ten numerical patterns from 0 to 10(Figure 5).

(P-2) Ten gradated patterns of (P-1).

(P-3) Five numerical patterns from 0 to 4(Figure 5).

(P-4) Five gradated patterns of (P-3).

In recalling process the following types of unknown patterns are used.

(R-1) Twenty patterns of 8 bits difference with binary learning
patterns(Figure 6).

(R-2) Twenty patterns with one bits shifted from the position
of original patterns in (P-3) to left, right, upper or lower
direction.

(R-3) Twenty gradated patterns of (R-1).

(R-4) Twenty gradated patterns of (R-2).

At learning process, the initial values of connectivity strength w_{1ij}
and w_{2ij} are set to random values between -0.3 and 0.3. Learning is
stopped when, for each neuron in last layer, the difference between
the expected training output and actual output become under a
value.

The results of recalling accuracy are shown in Table 1 and Figure
7. Recalling accuracy decreased when the number of learning
categories increase. It can be seen that there may be optimum value
of ρ. However recalling accuracy is depend on the type of input-
output procedure and the type of recalling pattern, it showed the
highest value between ρ =2.5 and 3.5.

In the case that pattern is apart from the position of learning
pattern, overlapping between input pattern and learning pattern becomes small if ρ is too small, and overlapping between input pattern
and other different learning patterns becomes large if ρ is too
large. Thus it is seen there exists optimum ρ . When ρ is set optimum value, it is seen that GG, BG and GB processings using gradated
patterns showed higher recalling accuracy than BB processing.

In a recalling simulation for one pixel shifted unknown patterns,
recalling accuracy is 50% in BB processing but 95% in GG processing for five learning categories, and for ten learning categories 45
% accuracy is obtained in BB processing but 90% in GG processing.
However, for two pixel shifted unknown patterns, remarkable effect
of gradating can not be seen. One of reasons of this result is that
input unknown pattterns are too far from learning patterns.

Simulation results for rotated, scaled up and scaled down patterns are shown in Figure 8 and Figure 9. Comparatively small different patterns from learnig pattern shown in Figure 8.1 and Figure 8.2 are recalled correctly by both GG and BB processing. However much more different patterns from learning patterns shown in Figure 8.3 to Figure 10.2 are recalled correctly by GG processing.

5. Conclusions

A associative neural network system using gradated patterns is proposed to improve the recalling accuracy for distortion patterns. From the fundamental simulation results, it was shown that the recalling accuracy for using the gradated patterns as learning and unknown patterns was much higher than using the binary patterns themselves.

However, in the neural associative memory, the problem that decrease of recalling accuracy become serious for distortion patterns is not yet solved sufficiently.

Detail consideration about the effect of the using gradated patterns in neural associative memory is a problem in the near future.

References

[1] K.Yamada, H.Kami, J.Tukumo and T.Tenma, "Handwritten numerical recognition by multi-layered neural network with improved learning algorithm", Proc. Joint Int.Conf.on Neural Networks, Washington.D.C., June 18-22, pp.259-266, 1989.

[2] E.Bienenstock and C.Von der Malsburg, "A neural network for invariant pattern recognition", Europhys.Lett.,4(1),pp.121-126, 1987.

[3] G.L.Giles and T.Maxwell, "Learning, invariance, and generalization in higher-order neural networks", Applied Optics, 26,pp.4972-4978, 1987.

[4] M.B.Reid, L.Spirkovska and E.Ochoa, "Simultaneous position, scale and rotation invariance pattern classification using third oder neural networks", Int.J.of.Neural Networks, 1,pp.154-159, 1989.

[5] J.J.Hopfield, "Neural networks and physocal system with emergent collective computational abilities", Proceeding of the National Academy of Sciences USA 79, 2254-2258, 1982.

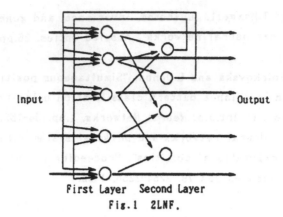

First Layer   Second Layer
Fig.1   2LNF.

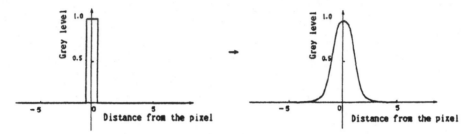

Fig.2.1   Before gradating.   Fig.2.2   After gradating.

Fig.2   A concept of gradating transformation for a pixel.

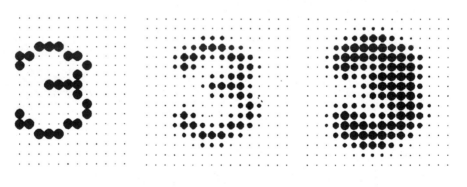

ρ =3 ρ =7.5

binary pattern   gradated pattern

Fig.3   A example of binary pattern and its gradated pattern.

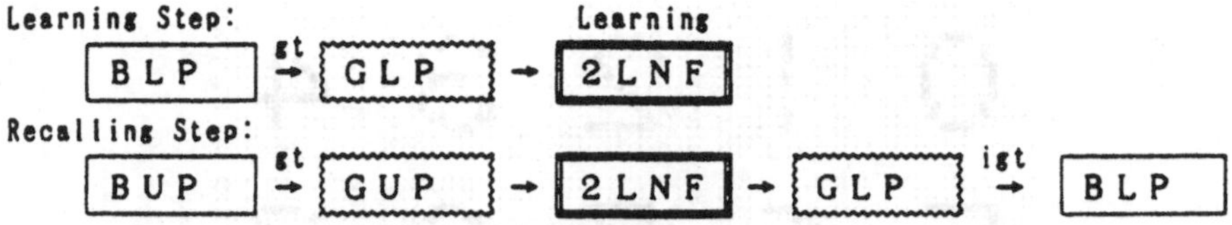

Learning Step:

BLP $\overset{gt}{\to}$ GLP → 2LNF

Learning

Recalling Step:

BUP $\overset{gt}{\to}$ GUP → 2LNF → GLP $\overset{igt}{\to}$ BLP

Fig.4.1 GG processing.

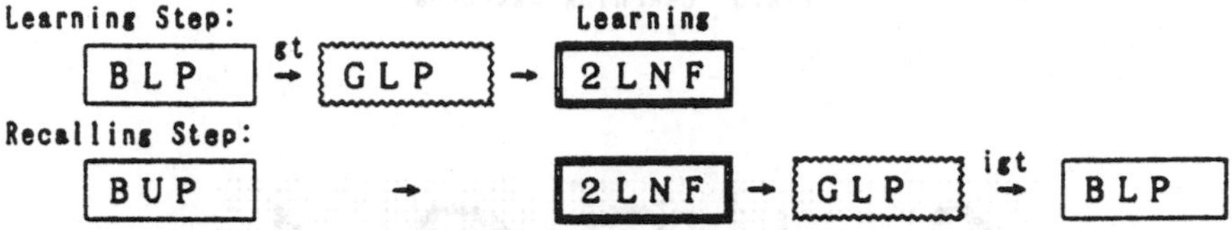

Learning Step:

BLP $\overset{gt}{\to}$ GLP → 2LNF

Learning

Recalling Step:

BUP → 2LNF → GLP $\overset{igt}{\to}$ BLP

Fig.4.2 BG processing.

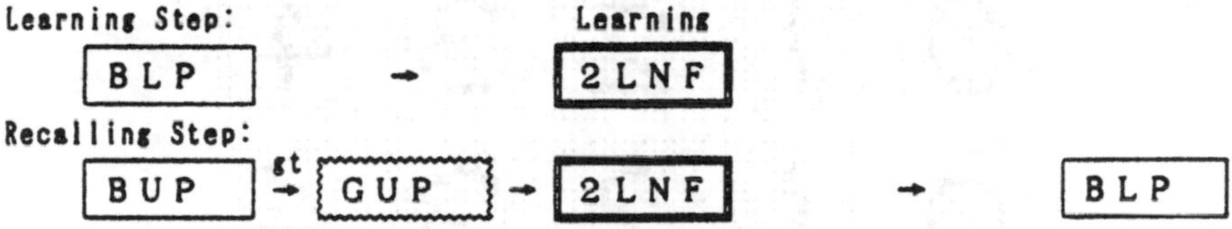

Learning Step:

BLP → 2LNF

Learning

Recalling Step:

BUP $\overset{gt}{\to}$ GUP → 2LNF → BLP

Fig.4.3 GB processing.

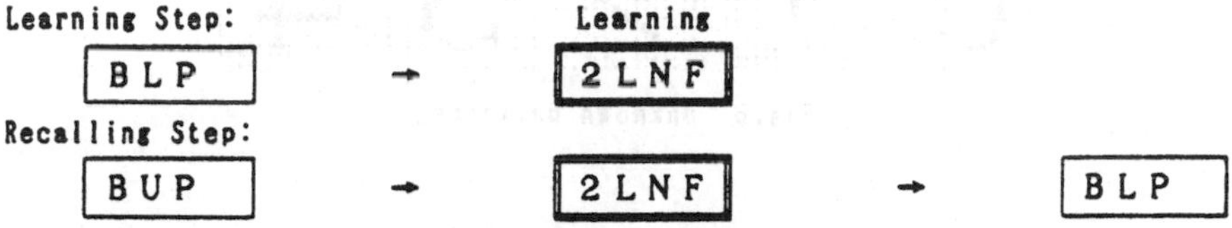

Learning Step:

BLP → 2LNF

Learning

Recalling Step:

BUP → 2LNF → BLP

Fig.4.4 BB processing.

Fig.4   Input - Output Procedures.
(Notations; gt : gradating transformation
              igt: inverse gradating transformation
              BLP: Binary Learning Pattern
              BUP: Binary Unknown Pattern
              GLP: Gradated Learning Pattern
              GUP: Gradated Unknown Pattern   )

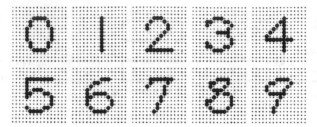

Fig.5 Learning patterns.

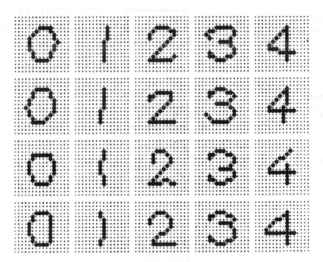

Fig.6 Unknown patterns.

Table.1 Results of GG processing and BB processing (ρ =3).

unknown patterns	Recalling accuracy					
	8 bit difference		1 pixel shifted		2 pixel shifted	
number of learning patterns	5	1 0	5	1 0	5	1 0
BB processing	1 0 0 %	1 0 0 %	5 0 %	4 5 %	4 5 %	1 5 %
SS processing	1 0 0 %	1 0 0 %	9 0 %	9 0 %	4 5 %	3 0 %

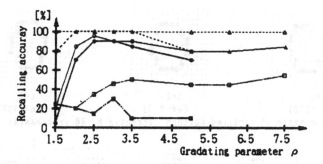

Fig.7.1 GG processing.

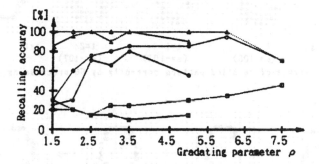

Fig.7.2 BG processing.

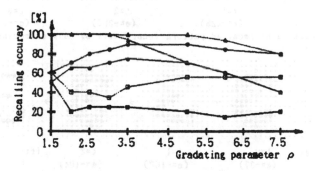

Fig.7.3 GB processing.

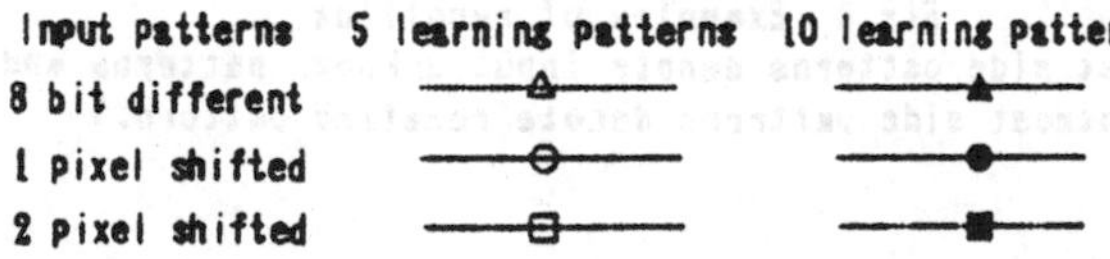

Input patterns	5 learning patterns	10 learning patterns
8 bit different	—△—	—▲—
1 pixel shifted	—⊖—	—●—
2 pixel shifted	—□—	—■—

Fig.7 Relation between gradating parameter ρ and recalling accuracy.

192

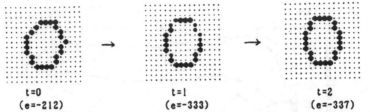

t=0　　　　　　　　　t=1　　　　　　　　　t=2
(e=-212)　　　　　　(e=-333)　　　　　　(e=-337)

Fig.8.1　A example of recalled pattern correctly by BB proccessing for(R-1).

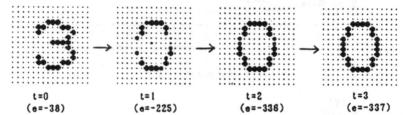

gt　　　t=0　　　　t=1　　　　t=2　　　igt
(e=-105)　　(e=-106)　　(e=-107)　　　　　(ρ =5).

Fig.8.2　A example of recalled pattern correctly by GG proccessing for(r-3)

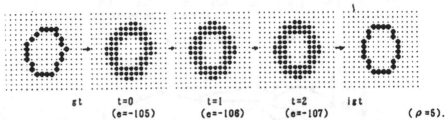

t=0　　　　　　t=1　　　　　　t=2　　　　　　t=3
(e=-38)　　　　(e=-225)　　　　(e=-336)　　　　(e=-337)

Fig.8.3　A example of recalled pattern incorrectly by BB proccessing for(R-2).

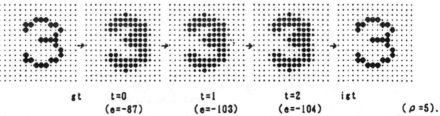

gt　　　t=0　　　　t=1　　　　t=2　　　igt
(e=-87)　　(e=-103)　　(e=-104)　　　　　(ρ =5).

Fig.8.4　A example of recalled pattern correctly by GG proccessing for(R-4)

Fig.8　Examples of recalling
(Leftmost side patterns denote input unknown patterns and
rightmost side patterns denote recalled pattern.)

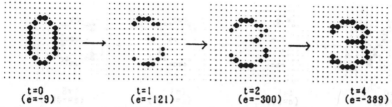

t=0
(e=-9)

t=1
(e=-121)

t=2
(e=-300)

t=4
(e=-389)

Fig.9.1 A example of recalled pattern incorrectly by BB proccessing for scaled down pattern.

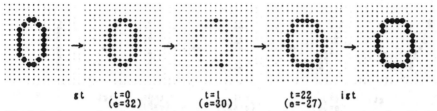

gt

t=0
(e=32)

t=1
(e=30)

t=22
(e=-27)

igt

Fig.9.2 A example of recalled pattern correctly by GG proccessing (ρ=2) for scaled down pattern.

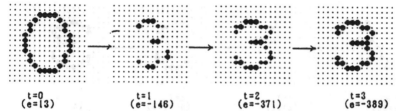

t=0
(e=13)

t=1
(e=-146)

t=2
(e=-371)

t=3
(e=-389)

Fig.9.3 A example of recalled pattern incorrectly by BB proccessing for scaled up pattern.

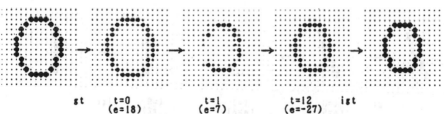

gt

t=0
(e=18)

t=1
(e=7)

t=12
(e=-27)

igt

Fig.9.4 A example of recalled pattern correctly by GG proccessing (ρ=2) for scaled up pattern.

Fig.9 Examples of recalling.
(Leftmost side patterns denote input unknown patterns and rightmost side patterns denote recalled pattern.)

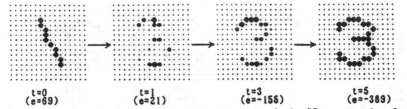

Fig.10.1 A example of recalled pattern incorrectly by BB proccessing for rotated pattern.

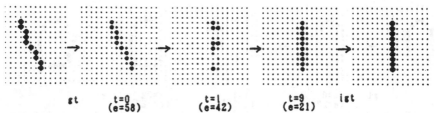

Fig.10.2 A example of recalled pattern correctly by GG proccessing (ρ =2) for rotated pattern.

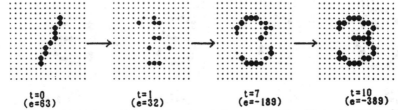

Fig.10.3 A example of recalled pattern incorrectly by BB proccessing for rotated pattern.

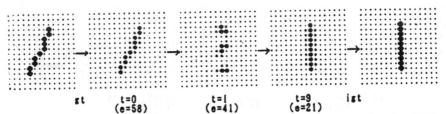

Fig.10.2 A example of recalled pattern correctly by GG proccessing (ρ =2) for rotated pattern.

Fig.10 Examples of recalling.
(Leftmost side patterns denote input unknown patterns and rightmost side patterns denote recalled pattern.)

Context-sensitivity of Puzzle Grammars

P. Laroche (*), M. Nivat (*) and A. Saoudi (**)

(*) Université Paris VII
L.I.T.P.
2, place Jussieu
75 251 Paris Cedex 05, France
e-mail : laroche@litp.ibp.fr

(**) Université Paris XIII L.I.P.N.
Institut Galilée
Av. J.B. Clément
93 430 Villetaneuse, France
e-mail : as@lipn.univ-paris13.fr

Juin 1992

Abstract

We study some properties of array grammars, called puzzle grammars, introduced in [NS89].We give a new method, using puzzle grammar, for generating the set of rectangles. We prove the the emptiness problem for puzzle grammar is undecidable. We show that the non-overlapping problem for puzzle grammar is decidable.

1 Introduction

The classical syntax theory, introduced by Chomsky and Schutzenberger has many applications such as in compiling, computationnal linguistics and pattern recognition. This theory has been extended to two-dimensionnal case by Rosenfeld ([Ro 71], [Ro 73] and [Ro 79]), Siromoney et al. ([SKK 92]), Nivat et al.([NS 89], and [NSSSD 92]) Wang ([wa 89]) and Yamamoto et al.([YMS 89]).

Two-dimensionnal array grammars have been used for generating picture languages. These grammars are defined by a set of two-dimensionnal rules together with a rewriting strategy such as parallel rewriting or sequential rewriting. For more details about other classes of array grammars and their corresponding array automata, see [RO 79] and [Wa 89].

A new model of array grammars, called puzzle grammars, has been introduced by Nivat et al.([NS 89] and [NSSSD 92]) as a basic model for generating arrays. In this paper, we examine some properties of puzzle grammars such as their power for generating the set of all rectangles, the undecidability

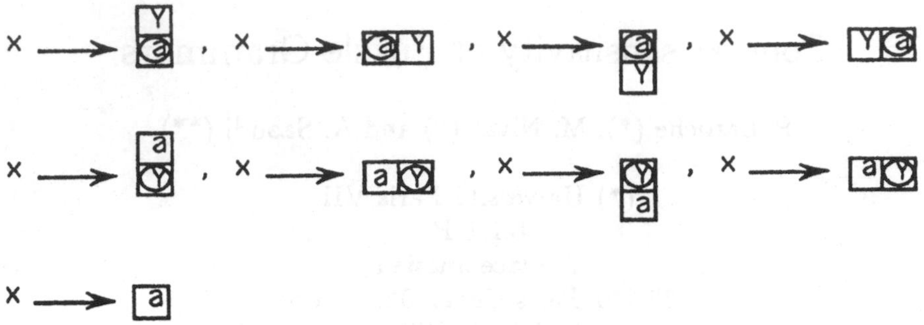

Figure 1: Forms of the rules of a basic puzzle grammar

of the emptiness problem, and the decidability of overlapping problem. We give a new method, inspired by Yamamoto et al.[YMS 89], for generating the set of rectangles. We present a new proof of the undecidability of the emptiness problem byreduction to the well known halting problem for Turing machines. We examine the power of restricting the class of puzzle grammars to those with non-overlapping and then consider the non-overlapping problem for puzzle grammars. We show that the non-overlapping problem for puzzle grammar is decidable, where the non-overlapping problem is the problem to check whether or not a puzzle grammar contains an overlapping.

We prove that puzzle grammars with non-overlapping cannot generate the set of rectangles.

2 Definitions and notations

In this section we introduce the basic notions of puzzle grammars. For a more detailed presentation, and a comparison with array grammars, see [NS 89] and [NSSSD 92].

Definition 2.1 *A basic puzzle grammar is a structure $G=< N, T, R, S >$ where N and T are finite sets of symbols; $N \cap T = \emptyset$. Elements of N are called nonterminals and of T, terminals. $S \in N$ is the start symbol or the axiom. R consists of rules of the forms shown in figure 1, for sets $N = \{X, Y\}$ and $T = \{a\}$.*

Derivations begin with S written in a unit cell, in the two-dimensional plane, with all other cells containing the blank symbol #, not in $N \cup T$. In a derivation step, denoted by $\Rightarrow$, a nonterminal X is replaced by the right hand member of a rule, whose left hand side is X. In this replacement, the circled symbol of the right side of the rule used occupies the cell of the replaced symbol and the non-circled symbol of the right side occupies the cell to right or to the left, or above, or bottom the cell of the replaced symbol depending on the type of the rule used. The replacement is possible and defined only if the cell to be filled in by the non-circled symbol contains a blank symbol.

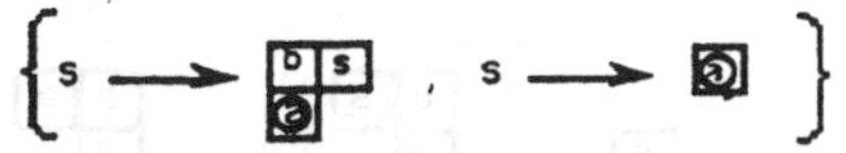

$$\left\{ \text{s} \longrightarrow \boxed{\begin{smallmatrix} \text{b} & \text{s} \\ \text{\textcircled{a}} & \end{smallmatrix}} \quad , \quad \text{s} \longrightarrow \boxed{\text{\textcircled{a}}} \right\}$$

Figure 2: The rewriting rules of Gstairs.

Definition 2.2 *A picture over an alphabet A is a connected, digitized finite array of elements of A.*

Definition 2.3 *A Context-free puzzle grammar (CFPG) is a structure $G = < N, T, R, S >$ where N, T and S are defined as before, and R is the set of rewriting rules of the form $X \to \alpha$, where α is a finite connected array of one or more cells, each cell containing either a non-terminal or a terminal symbol, with the symbol in one of the cells of α being circled.*

As before, a derivations begin with S written in a unit cell in the two-dimensional plane with all other cells containing the blank symbol #. In a direct derivation step, a non-terminal X in a cell is replaced by the right side α of the rule $X \to \alpha$. In this replacement, the circled symbol occupies the cell with symbol X and the remaining symbols of α occupying their respective relative positions with respect to the circled symbol of α. Again, the rewriting by $X \to \alpha$ is possible and defined only when the cells to be filled in by the non-circled symbols of α contain the blank symbol #.

We define precisely the notion of derivation by a sequence of pictures over $N \cup T$, each picture depending on its predecessor. In this formalism, a cell is described by a triple, giving its value in $N \cup T$ and its position in the two-dimensional plane. This triple is an element of $U = \{N \cup T\} \times Z^2$. We define the mapping π, from 2^U onto the set of pictures, $\pi(K)$, with $K \subset U$. $\pi(K)$ equals the two-dimensional plane with each cell containing a #, except the cells in K, with the origin of the plane arbitrary chosen, for example, the lowest, and the leftmost of the cells. If a coordinate is given twice in K, that is, two triples of K have the same second and third values, $\pi(K)$ is not defined.

Now, for a CF PG $G = < N, T, R, S >$, we define a mapping δ, from U to 2^{2^U} as follow. Let $X \to \alpha$ be a rule of R, where α is formed of n cells of symbols $s_1, ..., s_n$. We associate a distance (d_{x_i}, d_{y_i}) with each s_i, relative to the circled symbol. $d_{x_i}, d_{y_i} \in Z$ where $|d_{x_i}|$ is the number of cells in the horizontal direction and $|d_{y_i}|$ in the vertical direction between s_i and the circled symbol. Let $m = \max_{i=1 \text{ to } n} \{|d_{x_i}|, |d_{y_i}|\}$. $\forall d_x, d_y \in [-m, m], \{(s_1, d_{x_1}, d_{y_1}), ..., (s_n, d_{x_n}, d_{y_n})\} \in \delta((X, d_x, d_y))$ and for all other $u \in U, \delta(u) = \emptyset$.

We give an example in order to bring out the notions defined.

Example : Let $G_{stairs} = < S, a, b, R, s >$ be the puzzle grammar described in figure 2.

For G_{stairs}, m has the value 1. δ is defined as follows. From the first rule, $\forall d_x, d_y \in [-1, 1], \{(a, 0, 0), (b, 0, 1), (s, 1, 1)\} \in \delta((s, d_x, d_y))$, and from the second rule, $\forall d_x, d_y \in [-1, 1], \{(s, 0, 0)\} \in \delta((s, d_x, d_y))$.

Figure 3: Two derivations in G_{stairs} : $P_0 \Rightarrow P_1 \Rightarrow P_2$ and $P_0 \Rightarrow P_1 \Rightarrow P_3$

Then, a derivation begins with the picture $\pi(\{(S,0,0)\})$, and a derivation step is possible from a picture P $=\pi(Q)$ if there is an element $c = (a_c, x_c, y_c)$ of Q, and a set $C' \in \delta((a_c, 0, 0))$ (undefined if a_c is not a nonterminal) such that $Q' = Q\backslash\{c\} \cup \{(a, x_c + d_x, y_c + d_y)|(a, d_x, d_y) \in C'\}$, and P' $= \pi(Q')$ is the picture derived. We denote this transformation by $P \Rightarrow P'$, and the equivalent in 2^U, $Q \to Q'$. We say then that P' directly derive from P.

For example, $\{(s,0,0)\} \to \{(a,0,0),(b,0,1),(s,1,1,)\} = Q$ (we call P_1 the image of Q by π) . With c $= (s,1,1)$, we get $C' = \{(a,0,0),(b,0,1),(s,1,1,)\}$, $Q' = \{(a,0,0),(b,0,1)\}\cup\{(a,1+d_x,1+d_y)|(a,d_x,d_y) \in C'\} = \{(a,0,0),(b,0,1), (a,1,1),(b,1,2),(s,2,2)\}$ $(Q \to Q'$ and $P_2 = \pi(Q'))$ or $C'' = \{(a,0,0)\}$, and $Q'' = \{(a,0,0),(b,0,1),(a,1,1,)\}$ $(Q \to Q''$ and $P_3 = \pi(Q''$)). This derivations are shown in figure 3

If all the elements of a subset Q of U are in T, then no derivation is possible, and $\pi(Q)$ is a terminal picture (in the example P_3 is a terminal picture).

Definition 2.4 *A derivation in a puzzle grammar is finite sequence of pictures $P_0, P_1, \ldots, P_n$, P_0 being the image by π of (S,0,0), such that $\forall i < n, P_i \Rightarrow P_{i+1}$. So, $P_0 \overset{*}{\Rightarrow} P_n$, and P_n is called the derived picture.*

The language of a puzzle grammar G, denoted L(G), is the set of pictures over T (i.e., not containing nonterminals) that are obtained from the start symbol by a finite derivation.

3 Puzzle grammars generating rectangles

[NS 89] and [NSSSD 92] do not explicitly examine overlapping in the rewriting process of puzzle grammars. Here we bring out the context-sensitivity of puzzle grammars when the overlapping feature is present. We give a method, inspired from the one given by Yamamoto et al. [YMS 89] for generating the set of rectangles. We associate to any Turing machine a puzzle grammar which generates a rectangle iff the Turing Machine halts and thereby prove that the emptiness problem for puzzle grammars is undecidable.

There are many forms in which the rules of a basic puzzle grammar can be written, which do not modify their generating power. If, from a nonterminal, there is only one sequence of derivation steps possible, we can give the result of this sequence as the product of the initial symbol (See Figure 4 for an example).

Conversely, it is possible to give a presentation like the one in Definition 2.1, if the rules allowed are $X \to \alpha$, where α is a finite connected array of one or more cells, which satisfies the following property, (See Figuree 5).

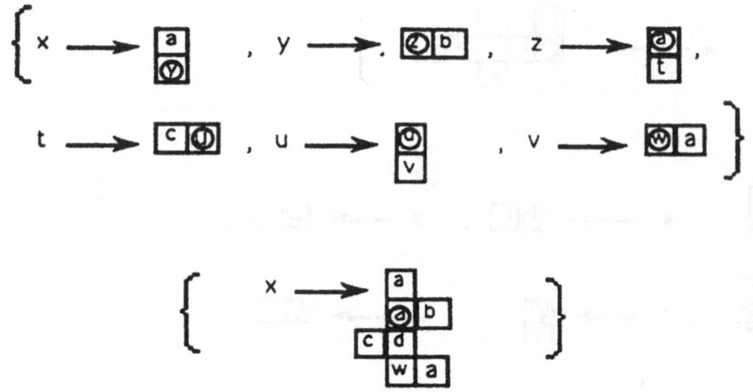

Figure 4: The two possible presentations for a basic puzzle grammar. The two subsets of rules are equivalent is there is no other rule in R using y, z, t, u or v in its left side.

We consider paths on cells of α, defined as sequences of neighbouring cells. Then it is possible to modify the form of the rule if either i) each cell has a terminal symbol, except for one which has a nonterminal symbol N, and there is a path of cells from the circled symbol to N, and all other cells with terminal symbol, not on this path, are just adjacent to cells in this path, or ii) each cell has a terminal symbol, and there is a path from the circled symbol to some terminal symbol having a similar property.

Its easy to see how we can switch from one presentation to the other. The transformation in Figure 5 constructs an equivalent set of basic rules from one rule.

3.1 A basic puzzle grammar generating the set of rectangles

Definition 3.5 *Let RECT be the set of all the solid upright rectangles over one letter alphabet* $\{a\}$.

We denote the height and the width of a rectangle $x \in RECT$ as h(x) and w(x) respectively.

Let $RECT^-$ be the subset of RECT containing only the rectangles of odd height and even width, with both greater than two :

$$RECT^- = \{x \mid x \in RECT \text{ and } h(x) = 2i + 1, w(x) = 2j + 2, i, j \in \{1, 2, \ldots\}\}$$

Lemma 1 *The grammar* G_{RECT-} *given below generates* $RECT^-$.

$G_{RECT-} = \langle N = \{s, i, j, r, l, e, f, x, y\}, T = \{a\}, Q, s \rangle$. Figure 6 shows Q. In all the following figures we draw cells containing the letter "a" in gray, because our main concern is in overlapping, and these properties are not affected by terminal alphabet.

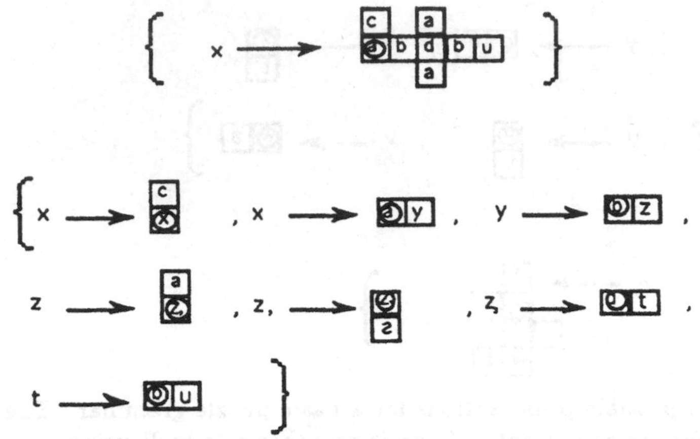

$$\left\{ \quad x \longrightarrow \begin{array}{c} \boxed{c}\ \boxed{a} \\ \boxed{\circlearrowleft}\boxed{b}\boxed{d}\boxed{b}\boxed{u} \\ \boxed{a} \end{array} \right\}$$

$$\left\{ x \longrightarrow \begin{array}{c}\boxed{c}\\\boxed{\otimes}\end{array} \quad,\quad x \longrightarrow \boxed{\circlearrowleft}\boxed{y} \quad,\quad y \longrightarrow \boxed{\circlearrowright}\boxed{z} \quad, \right.$$

$$z \longrightarrow \begin{array}{c}\boxed{a}\\\boxed{\circlearrowleft}\end{array} \quad,\quad z, \longrightarrow \begin{array}{c}\boxed{\circlearrowright}\\\boxed{z}\end{array} \quad,\quad z \longrightarrow \boxed{\circlearrowright}\boxed{t} \quad,$$

$$\left. t \longrightarrow \boxed{\circlearrowright}\boxed{u} \right\}$$

Figure 5: If x_1, y, z_1, z_2, z_3, t are not in $N \cup T$, replacing the first subset of R by the second does not modify the language generated by the grammar.

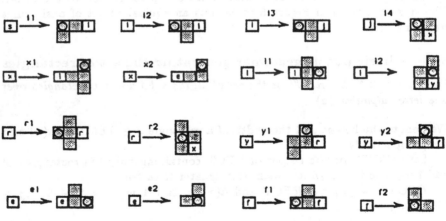

Figure 6: The set of rules Q of Grect-

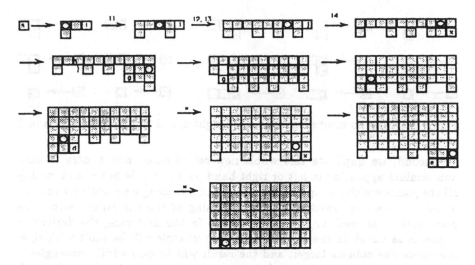

Figure 7: A derivation of a rectangle in G_{RECT-}

The process of derivation lay down successive layer of terminal symbols, from right to left and then from left to right, until a final layer completes a rectangle. Figure 7 shows different steps of the derivation of a rectangle.

Each right part of a rule is a piece which can fit only in a specific place in a rectangle. The pieces of the production parts of rules l_1, l_2, r_1 and r_2 are horizontal dominos, with a square upside, which "tests" pieces already laid, and another downside, which will guide the following layer.

After using the rule i_1 to rewrite the axiom, i can be rewritten any number of time with i_2, defining any even number of columns for the future rectangle. Then the rules i_3 and i_4 are used, the growth to the right stops, and the main part of the process begins.

The pieces at the right and left extremities of rows bear a square downside on the left, but the pieces inside bear it on the right. So, at each step, only one rule on two is selected between l_1 and l_2, in the other sense r_1 or r_2, to rewrite l (respectively r).

Thus, $L(G_{RECT-}) = RECT^-$. □

Proposition 1 *The set set of rectangles (i.e., RECT) can be generated by a basic puzzle grammar.*

Proofs sketch : We construct the grammar G_{RECT}, which generate RECT, by modifying G_{RECT-}.

First, we need to construct rectangles of any height : we duplicate rules $x_2, y_2, e_1, e_2, f_1, f_2$, and modify them by adding a row of cells downside each piece, and transform each symbol e and f of these rules to e' and f'. So, when the derivation comes to its end, the last layer uses e, f or e', f' rules, giving a rectangle of height respectively odd or even, still greater than 2.

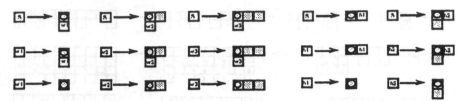

Figure 8: Rules to create rectangles of height 1 and 2 and width 1, 2 and 3

Second, we duplicate this whole new set of rules, add quotes to each nonterminal appearing in left or right hand part of a rule but s, and modify all the pieces which can appear to the right of a rectangle by adding a column of cells: when s is rewritten at the beginning of the derivation, either the piece with i_1 is used, or the one with i_1'. In the first case, the derivation proceeds as usual, in the second case, the rectangle will be built with right side piece one column larger, and the result will be odd width rectangles.

Third, we add the rules needed to create rectangles of width 1, 2 and 3, and height 1 or 2. They are shown in figure 8. □

3.2 The emptiness problem

In this section, we give a new proof that the emptiness problem is undecidable, for CF-puzzle grammars and basic puzzle grammars. This is already proved by Nivat et al.[NSSSD 92] using general properties of array grammar. This new method is direct, and can be seen as a good example of proof technique on puzzle grammars.

Problem 1 *(emptiness problem) : Given a basic puzzle grammar G, is L(G) nonempty ?*

To prove the undecidablity of the emptiness problem for puzzle grammars, we need to recall some basic definitions.

Let us first give a definition of a Turing Machine:

Definition 3.6 *: A Turing Machine $\mathcal{M}$ is a 5-uple (Q, Σ, F, q_0, q_f), where :*

1. *Q is a set of state,*

2. *Σ is the tape alphabet,*

3. *$\mathcal{F}$ is the set of transition, $\mathcal{F} \in Q \times \Sigma \times Q \times \Sigma \times \{G, D\}$,*

4. *q_0 is the initial state,*

5. *q_f isthe final state.*

A configuration of $\mathcal{M}$ is a word $\alpha q \beta$ where $\alpha, \beta \in \Sigma^*$ and $q \in Q$. The reading head of $\mathcal{M}$ points to the first letter of β.

A configuration $d = \alpha_d q_d \beta_d$ is directly obtained from another one $c = \alpha_c a q_c b \beta_c$, denoted by $c \to d$, if $\exists t = (q_c, b, q_d, e, G)$ such that $\alpha_d = \alpha_c$ and $\beta_d = ae\beta_c$, or $\exists t = (q_c, b, q_d, e, D)$ such that $\alpha_d = \alpha_c ae$ and $\beta_d = \beta_c$.

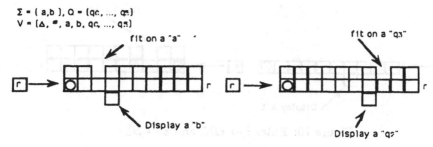

$\Sigma = \{ a,b \}, Q = \{q_0, ..., q_3\}$
$V = \{\Delta, \#, a, b, q_0, ..., q_3\}$

fit on a "a"

fit on a "q3"

$r \rightarrow$

r

$r \rightarrow$

r

Display a "b"

Display a "q2"

Figure 9: Description of rules $r \rightarrow B_b^a r$ and $r \rightarrow B_{q_2}^{q_3} r$

A successful computation of $\mathcal{M}$ is a sequence of words $(c_1, \ldots, c_k)$, where c_i is the ith configuration of the machine, such that $q_0 \in c_1, \forall i < k, c_i \rightarrow c_{i+1}$, and $q_f \in c_k$. We can suppose each c_i has the same length n (by adding # symbols on the right if necessary).

Theorem 1 *The emptiness problem for puzzle grammar is undecidable.*

Proof. We reduce the halting problem for Turing Machine to the emptiness problem for puzzle grammar.

Let $\mathcal{M}$ be a Turing machine. We define a puzzle grammar $G_{\mathcal{M}}$ such that $\mathcal{M}$ halts if and only if $L(G_{\mathcal{M}})$ contains a picture.

A / Definition of the puzzle grammar associated to $\mathcal{M}$

B / Core of the proof.

A/Definition of the puzzle grammar associated to $\mathcal{M}$

Input : $\mathcal{M} = (Q, \Sigma, F, q_0, q_f)$.

Output : a basic puzzle grammar $G_{\mathcal{M}} = < N, T, s, R >$, where

N = {s, i, j, x, y, r, r', l, l', e}, T = {a}, and R is computed from $\mathcal{F}$, as explained below.

The right hand sides of rules are bricks, which code information on their sides, like the position of a brick was indicated by a cell at right or left in G_{RECT} : upside, it is the place it fits in, downside what it describes. Here, we code upside and downside a value in $V = \Sigma \cup Q \cup \{\Delta\} \cup \{\#\}$, Δ is a new symbol appended to left and right end to show the border. Thus, we can describe with a row of bricks a sequence of values in V which is an instantaneous description of $\mathcal{M}$. If x and y are elements of V, we denote by B_y^x a brick which fits under x and shows y downside. We use the simplest of the possible codes : the brick has always a length equal to $|V|$, and a cell is present at the ith place to code the ith value of the set V. To test the place where it fits, there a row of cells upside, but for the ith place, if the brick fits in the ith symbol.

To illustrate the use of this notation of bricks, we show four rules with a right hand part which is a brick, in figure 9 and 10.

Proof. Now, we describe how we create the set R of rules of $G_{\mathcal{M}}$ from $\mathcal{M}$. These rules are grouped in 7 parts :

1/ The initialization rules :

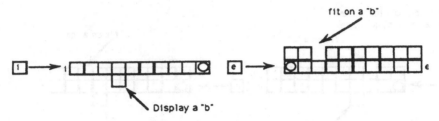

Figure 10: Rules $i \to iB_b^-$ and $e \to B_-^b e$

(i1) $S \to iB_\Delta^-$

(i2) $\forall s \in \Sigma, i \to iB_s^-$

(i3) $i \to jB_{q0}^-$

(i4) $\forall s \in \Sigma, j \to jB_s^-$

(i5) $j \to B_\Delta^-$
$\quad\quad B_\Delta^\Delta l$

2/ The copy rules :

(c1) $\forall s \in \Sigma, r \to B_s^s r$ and $r' \to B_s^s r'$

(c2) $\forall s \in \Sigma, l \to lB_s^s$ and $l' \to l'B_s^s$

3/ the computation rules :

(p1) $\forall(q_a, s_a, q_b, s_b, D) \in \mathcal{F}, r \to B_{s_b}^{q_a} B_{q_b}^{s_a}\, r$

(p2) $\forall(q_a, s_a, q_b, s_b, G) \in \mathcal{F}, r \to B_{q_b}^{q_a} B_{s_a}^{s_b} r'$

4/ the right to left turning rule :

(t1) $r \to B_\Delta^\Delta$
$\quad\quad\quad lB_\Delta^\Delta$

(t2) $r' \to B_\Delta^\Delta$
$\quad\quad\quad l'B_\Delta^\Delta$

5/ the left to right rules that copy states :

(c3) $\forall q \in Q, l \to lB_q^q$

(c4) $\forall q \in Q, \forall s \in \Sigma, l' \to lB_q^s B_s^q$

6 / the left to right turning rule :

(t3) $l \to B_\Delta^\Delta$
$\quad\quad\quad B_\Delta^\Delta\, r$

(t4)$l \to B_\Delta^\Delta$
$\quad\quad\quad B_\Delta^\Delta e$

7 / the final rules :

(f1) $\forall s \in \Sigma, e \to B_s^s e$

(f2) $e \to B^{q_f} e.$

(f3) $\forall s \in \Sigma, e \to B_-^\Delta$

B/ Core of the proof

For any Turing Machine T, $G_\mathcal{M}$ generates a picture iff $\mathcal{M}$ halts.

Lemma 2 *If a rectangle code on its bottom a configuration of a machine $\mathcal{M}$, and the nonterminal r is at the extreme left, $G_\mathcal{M}$ can generate two layers of bricks, the last coding a new configuration iff the new configuration derives directly of the former one.*

The rules can be used only in a precise order : when the nonterminal r is at the extreme left, only rules of type (c1) can be used : they create the beginning of a new line identical to the one upside, they just copy the letter of Σ, from left to right ; when a rule (c1) can no longer apply, this means we reached the head : one of the rules (p1) of (p2) is chosen, and a new nonterminal is created : if it is r, then the head had moved to right, and the grammar just finish the line with (c1) rules, and comes back to left with (c2) and (c3) rules ; but if the head moved to left, then r' is created, which will later caused the copy process to move the state brick one columns to left on the second row : the line is completed in the same manner, but when the derivation proceeds to left, the nonterminal l' is created, so that when reaching a state brick, a rule (c4) is used, and afterwards, as l is generated, the come-back procedure is the same as for the right move case.

If the two layers can be successfully created, then the new configuration derives directly from the former one by the application of the rule of $\mathcal{M}$ used in the (p1) or (p2) rewriting. If no configuration can be derived from the one described on the bottom of a rectangle, it means no rule (p1) or (p2) can be chosen, i.e., no rule of $\mathcal{M}$ can be applied when the machine is in the state coded with the letter after it under the head.

If a rectangle codes c on its bottom, and $c \to c'$, then it is possible to derive a rectangle coding c' on its bottom. $\square$

First Claim : If $G_{\mathcal{M}}$ generates a picture, T halts.

Let W be a picture created by $G_{\mathcal{M}}$. It is clear that it is a rectangle. To generate the first line of the rectangle, the derivation has created bricks coding letter of Σ, then a brick coding q_0, then other letters. So the first line of $G_{\mathcal{M}}$ gives an initial configuration, that we name c_1.

Using the lemma, we know that after 2k layers, the derived picture is still a rectangle with a bottom coding a configuration c_k, and c_k is derivable from c_1.

A last layer can be successfully created only if the state q_f is in the last configuration c_k because the rule (f2) has to be used in this last part. So q_f is the current state in the final configuration, this means that $(c_1, \ldots, c_k)$ is a successful computation of $\mathcal{M}$.

Second Claim : If $\mathcal{M}$ halts, then $G_{\mathcal{M}}$ generates a picture.

As $\mathcal{M}$ halts, there is a sequence $(c_1, \ldots, c_k)$ of instantaneous descriptions, from which we can extract a sequence of derivation steps for $G_{\mathcal{M}}$, by reading it from left to right on a row, and right to left on the next one, and so on until the last layer. First, the rules (i1), (i2) (i3) (i4) and then (i5) can be used without restriction to make grow a first line, coding exactly c1 on its bottom. Then, the lemma says that we are sure that we can derive a new rectangle from one with c_i on the bottom, until we reach c_k, and c_k contains q_f, so a picture can be generated.

4 Non overlapping grammar

If a grammar never induces overlapping, and so is such that all rules can apply without testing the neighborhood, it has no context-sensitivity, and can not generate RECT

4.1 Derivation trees

For context-free string grammars, it is usual to represent the derivation process by a tree. A derivation tree has an interior node labeled by some non-terminal X, and the children of the node are labeled form left to right by the elements of the right hand side a of a production of the form $X \to \alpha$.

In the case of two dimensional patterns, one need to have a structure for representing a derivation. For this, we construct a tree for representing a derivation by associating to each symbol the relative displacement with its direct father. As for the string case, the tree gives no information on the order of the rewriting.

We assume in the following that the grammars are under Chomsky Normal Form, that is, each right hand side of the rewiting rules contains two nonterminal symbols or a terminal symbol. Any context-free or basic puzzle grammar can be put under this form [NS 90] ; the main idea of this transformation is shown in the beginning of the third section. We denote by $\mathcal{D}$ the free monoid generated by 1, 2.

$\mathcal{D} = \{1, 2\}^* = \{f = f_1 \ldots f_n \mid n \geq 0; f_1, \ldots f_n \in \{1, 2\}\}$.

A subset D of $\mathcal{D}$ is said to be left factorial (cf [BN 85]) if for every element $fg \in D$ we have also $f \in D$. We call A the alphabet $\{ h, \bar{h}, v, \bar{v} \}$, where we use h to denote a movement to the right, $\bar{h}$ to the left, v to the top, $\bar{v}$ to the bottom, and $U = \{N \cup T\} \times \{A \cup e\}$. A derivation tree T relative to G is a partial function from D to U, defined on a left-factorial set. We define $D_G^{x,l}$, the set of derivation trees relative to G with root labeled (x,l), by the following properties :

- $\forall l \in A \cup \epsilon, \forall a \in T, (a, l) \in D_G^{a,l}$

- $\forall l \in A, \forall r \in R, \forall a \in T,$ if $r = x \to a$, then
$$\begin{array}{c} (x, l) \\ | \\ (a, \epsilon) \end{array} \in D_G^{x,l}$$

- $\forall l \in A \cup \epsilon, \forall r \in R,$ if $r = x \to \alpha$, with y the circled symbol of a, and z the symbol placed in a direction l_z from the circled symbol,
$$t_1 \in D_G^{y,\epsilon}, t_2 \in D_G^{z,l_z} \text{ then } \begin{array}{c} (x, l) \\ \bigwedge \\ t_1 \ t_2 \end{array} \in D_G^{x,l}$$

$D_G^{s,\epsilon}$ is the set of the derivation trees relative to G (s is the axiom) .

To each node w of T labeled $(v, l) \in U$, we associate another value from $\{N \cup T\} \times A^*$, denoted $global_T(w)$, giving its relative displacement with the initial axiom.

$$global_T(\dot{w}) = (v, \sum_{f \in FG(w)} \delta(T(f))$$

FG(w) denote the set of the left-factors of w, and δ is mapping from A to Z^2, which associates to each letter a value expressing the displacement in a two-dimensionnal plane : $\delta(h)=(1,0)$, $\delta(\bar{h})=(-1,0)$, $\delta(v)=(0,1)$, $\delta(\bar{v})=(0,-1)$ and $\delta(\epsilon)=(0,0)$.

For a given T, $global_T$ is also a mapping from D to $\{N \cup T\} \times A^*$, that is, a tree defined on the same set that T, but with nodes labeled with absolute coordinate. Fr(T) denotes the set of leaves of a tree T.

Definition 4.7 *The image P by π of $global_T(Fr(T))$ is called the derived picture of the trees T, and T is a derivation tree of P. By extension, we denote $\pi(global_T(Fr(T))$ by $\pi(T)$.*

Definition 4.8 *A grammar G is said non overlapping (NO) if for all derivation tree T relative to G, $\pi(T)$ is defined (i.e., no cell on the frontier had been defined twice).*

Proposition 2 *A non overlapping basic puzzle grammar cannot generate RECT.*

Proof. Assume that there is a puzzle grammar G without overlapping, G=< N, T, R, S >, under Chomsky Normal Form, generating RECT.

We denote by M the number of possible labels in a derivation tree relative to G, $|N \cup T| \times |A \cup e|$. If a given symbol is used twice in the same branch of the derivation tree of a pattern, then the part of the tree between the positions of that symbol can be repeated any number of times and generate another picture, as G is non-overlapping and so no coordinate conflict can appear.

We denote by $Rect_{x,y}$ a picture which is a rectangle of width x and height y.

First, we prove that there is at least one symbol used in the derivation of a rectangle big enough that can not be used for other sizes.

Lemma 3 *If a label (X,l), with $X \in N$ is used in a node and in the subtree originating from this node, in a derivation tree Ti of $Rect_{2^i,m_i}$, $i > M$, $m_i > 0$ it cannot be used in any label of a derivation tree of $Rect_{2^j,m_j}$, with $j > i$ and $m_j > m_i$.*

There is always a label (X,l) that is present two times in a branch of a derivation tree of $Rect_{2^i,m_i}$, $i > n$, $m_i > 0$, because there is at least one branch of length greater than M. Let T_i be a derivation tree of $Rect_{2^i,m_i}$.

T_i is divided into three sub-trees : T_{i1}, Ti truncated at the level of the first use of (X,l), T_{i2} the subtree originating from this point, but without the part extending after the second use, which is called T_{i3}. T_{i2} can be repeated any

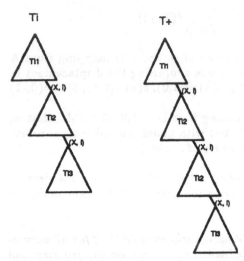

Figure 11: T_i and $T+$

number of times, that is inserted at the place of the node labeled (X,l). By iterating the second tree component, and then adding T_{i3}, we get derivation trees of pictures (because the grammar is non-overlapping). In figure 11, $T+$ is a derivation tree relative to G, created by iterating once T_{i2}. T_{i2} has a fixed number of leaves, and thus the picture representing this part of a tree has a vertical section, smaller or equal to 2^i.

Now we consider another rectangle, $Rect_{2^j,m_j}$, with $j > i$ and $m_j > m_i$. Let T_j be a derivation tree of this rectangle.If (X,l) appears in the derivation tree T_j , we divide T_j in two parts, the tree truncated a (X,l), T_{j1}, and the subtree originating from (X,l), T_{j2}. (X,l) can then be replaced any number of time t with the tree T_{i2} before using T_{j2} derivation tree, and for t big enough, it is clear the resulting picture is not a rectangle. So X can not be in the derivation tree of another rectangle $Rect_{2^j,m_j}$ if j is superior to i and m_j greater than m_i.

Precisely, if the relative displacement between two occurences of (X,l) in a branch of T_j is only horizontal (vertical), say d square to the left (bottom) for example, after t replacement of (X,l) with T_{i2}, such that $t \times d$ is larger than the diameter of $\pi(T_{j1})$ and $\pi(T_{j2})$, a vertical section of the picture at a distance $t \times d + 1$ from a cell of $\pi(Tj1)$ will give only cells of T_{i2}, with height limited to 2^i cells. An exemple of this process is shown in figure 12.

If the relative displacement between two occurences of (X,l) in the resulting rectangle is along the two axes, the repeated derivation of T_{i2} would give a linear growing of the area of the picture with t, but the growing of the area of a rectangle would have to be quadratic with the increasing size of a diagonal, so the resulting pictures can not be rectangles for each t.

In both cases, the result of the appearance of (X,l) in a derivation tree of a rectangle of other width and height is the possibility for the grammar G to generate picture other than rectangle. This is impossible as G generates

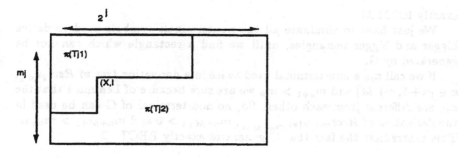

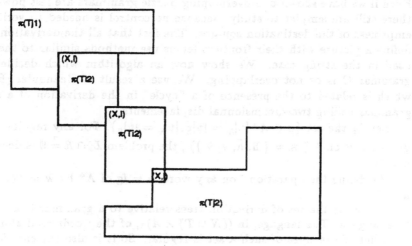

Figure 12: T_{i2} used to replace (X,l) in T_j

exactly RECT.□

We just have to eliminate all the nonterminals symbols used to derive bigger and bigger rectangles, until we find a rectangle which can not be generated by G.

If we call n_{tk} a non-terminal used twice in a derivation tree of $Rect_{2^k,m_k}$, $k \in [n+1, n+M]$ and $m_{k+1} > m_k$ we are sure because of Lemma 1 that the n_{tk} are different from each other. So, no non-terminal of G can be used in the derivation of $Rect_{2^{n+|M|+1},m_{n+M+1}}$, $m_{n+M+1} > 0$ and $m_{n+M+1} > m_{n+M}$. This contradicts the fact that G generates exactly RECT. □

4.2 Deciding if a grammar is non overlapping

Even if we have shown non-overlapping puzzle grammars are less powerfull, there still are simpler to study, because no control is needed to verify the emptiness of the destination squares. The fact that all the derivation trees define a picture with their frontiers let us use methods similar to the ones used in the string case. We show now an algorithm which decides if a grammar G is or not overlapping. We use a result of Beauquier [Be 91] which is related to the presence of a "cycle" in the derivation of a string grammar coding two-dimensionnal displacements.

Let L_b the set $\{u \in A \mid |u|_h = |u|_{\bar{h}}, |u|_v = |u|_{\bar{v}}\}$. For any regular string grammar R on A ($A = \{h, \bar{h}, v, \bar{v}\}$), the problem $L_b \cap R \stackrel{?}{=} \emptyset$ is decidable [Be 91] .

We define the operation ‾ on any word $w = fg$ of A^* by $\bar{w} = \bar{f}\bar{g}$, $\bar{\bar{h}} = h$ and $\bar{v} = v$.

It is clear the set of derivation trees relative to a grammar is a regular tree langage. The langage, in $((N \cup T) \times A)*$, of the words read along the branches of the trees of such a set is regular. So it is also the case for the projection of these words on A^*. Then the words created by reading the letter from $A \cup \epsilon$ on the labels of the nodes from the root along any branch of the tree define a regular string language. We call $L_l(r)$ the language of the words read from under a node X of the tree, for the X derived with the rule r, and beginning with the left child, $L_r(r)$ the language generated beginning with the right child, and $L(r)$ the union of $L_l(r)$ and $L_r(r)$.

Problem 2 : *Given a puzzle grammar G, is G non overlapping ?*

Theorem 2 *The problem II is decidable.*

Proof

First, we show the equivalence : A puzzle grammar G is non-overlapping iff $\forall r \in R, L_l(r)\overline{L_r(r)} \cap L_b = \emptyset$

First Claim : If a puzzle grammar G is not non-overlapping, then $\exists r \in R, L_l(r)\overline{L_r(r)} \cap L_b \neq \emptyset$.

G being not non-overlapping, there is one derivation tree T relative to G with two leaves n_1 and n_2 at the same global coordinates. We call w_{n_1} and w_{n_2} the words read on the branches leading from the root to these leaves.

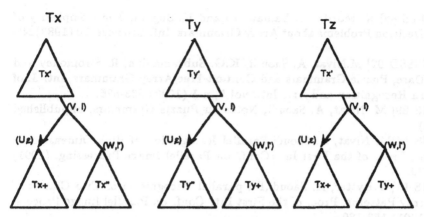

Figure 13: The buiding of the tree T_z

They have the property $|w_{n_1}|_h - |w_{n_1}|_{\bar{h}} = |w_{n_2}|_h - |w_{n_2}|_{\bar{h}}$, $|w_{n_1}|_v - |w_{n_1}|_{\bar{v}} = |w_{n_2}|_v - |w_{n_2}|_{\bar{v}}$. Thus, it is clear that $w_{n_1}\overline{w_{n_2}} \in L_b$. So if we call r_S the rule used to rewrite S in this tree, $L_l(r_S)\overline{L_r(r_S)} \cap L_b \neq \emptyset$.

Second Claim : If $\exists r \in R, L_l(r)\overline{L_r(r)} \cap L_b \neq \emptyset$ then G is not non-overlapping. We call $l_x \in L_l(r)$ and $l_y \in L_r(r)$ two word satisfiying $l_x\overline{l_y} \in L_b$, and V the left side nonterminal of r. l_x appears in a derivation tree relative to G, T_x, and l_y in a tree T_y.

Now we build a new tree T_z from T_x and T_y, like on the Figure 13. As this new tree respect the constraints defined in section 4.1, it is a derivation tree relative to G. As $l_x\overline{l_y} \in L_b$, there is somewhere in T_x+ and T_y+ two nodes at the same coordonate. So G is not non-overlapping, because $\pi(T_z)$ is not defined.

To decide problem II is equivalent to decide if $\forall r \in R, L_l(r)\overline{L_r(r)} \cap L_b = \emptyset$. As $L_l(r)$ and $\overline{L_r(r)}$ are regular language, so is $L_l(r)\overline{L_r(r)}$, and the number of rules being finite, the Problem II is decidable. $\square$

5 Bibliography

[Be 91] Danielle Beauquier, *An undecidable problem about rationnal sets and contour words of polyominoes*, Inf. Process. Letters. Vol 37 n 5 (March 1991), 257-263.

[BN 85] D. Beauquier, M. Nivat, About rationnal subset of algebras of infinite words, in Automata, Languages and programming (W. Brauer ed.), LNCS, n 192, Springer-Verlag, Berlin (1985) 33-42.

[CW 78] C.R. Cook and P.S.P. Wang, Chomsky hierarchy of isotonic array grammars and languages, Computer Graphics and Image Proccesing 8(1978) 144-152.

[IS 91] K. Inoue, I. Sakuramoto, M. Sakamoto and I. Takanami, Two Topics concerning Two-dimensionnal Automata Operating in Parallel, Proc. of the First Int. Conf. on Parallel Image Processing, (1991) 231-262.

[MYS 83] K. Morita, Y. Yamamoto and K. Sugata, The Complexity of some Decision Problems about Array Grammars, Inf. Sciences, 30 (1983)241-262.

[NSSSD 92] M.Nivat, A. Saoudi, K.G. Subramanian, R. Siromoney and V.R. Dare, Puzzle Grammars and Context-Free Array Grammars, Int. J. of Pattern Recognition and Art. Int., vol 5, n 5 (1992) 663-675.

[NS 89] M. Nivat, A. Saoudi, Notes on Puzzle Grammars, unpublished (1989).

[NS 91]M. Nivat, A. Saoudi,Parallel Recognition of High Dimensionnal Images, Proc. of the First Int. Conf. on Parallel Image Processing, (1991) 123-130.

[RS 91]W.Rytter, A. Saoudi,Pn parallel Recognition of Two Classes of 2D-Array Patterns, Proc. of the First Int. Conf. on Parallel Image Processing, (1991) 153-160.

[Ro 71] A. Rosenfeld, Isotonic Grammars, Parallel Grammars and Picture Languages, in Machine Intelligence (B. Meltzer and D. Michie, Eds), Univ. of Edinburgh Press (1971)

[Ro 73] A. Rosenfeld, Array Grammars normal forms, Informations and Control 23(1973)173-182

[Ro 79] A. Rosenfeld, Picture languages : formal models for picture recognition, Academic Press 1979.

[YMS 89] Y. Yamamoto, K.Morita and K. Sutaga, Context-sensitivity of two-dimentionnal regular array grammars, Int. J. of Pattern Recognition and Art. Int. vol. 3&4 (1989)295-320.

Parallel Generation and Parsing of Array Languages Using Reversible Cellular Automata

Kenichi Morita and Satoshi Ueno

Yamagata University, Yonezawa-shi 992, Japan

Abstract. We propose a new system of generating array languages in parallel, based on a partitioned cellular automaton (PCA), a kind of cellular automaton. This system is called a PCA array generator (PCAAG). The characteristic of PCAAG is that a "reversible" version is easily defined. A reversible PCA (RPCA) is a backward deterministic PCA, and we can construct a deterministic "inverse" PCA that undoes the operations of the RPCA. Thus if an array language is generated by an RPCA, it can be parsed in parallel by a deterministic inverse PCA without backtracking. We also define two subclasses of PCAAG, and give examples of them that generate geometrical figures.

1 Introduction

To treat two-dimensional patterns in the framework of formal language theory, several types of array generating systems have been proposed until now [1, 2, 7, 8, 9, 10, 11, 12, 13, 15, 16]. An isometric array grammar (IAG) [2, 8] is one of such systems, and many subclasses have been introduced and investigated. For example, a monotonic array grammar (MAG) [2, 8], a context-free array grammar (CFAG) [1, 13], and a regular array grammar (RAG) [1, 13] form a Chomsky-like hierarchy in IAG. It has been shown that, RAG, the lowest subclass in the hierarchy, has relatively high generating ability [14]. However, membership (or parsing) problem becomes NP-complete even for RAGs because of their "isometric" nature [3]. As another subclass of IAG, a uniquely parsable array grammar (UPAG) has been proposed [15, 16]. UPAG has a favorable property that parsing can be done without backtracking. Especially, array languages generated by monotone terminating UPAGs (MTUPAG), a subclass of UPAG, can be parsed in linear time [15].

Here, we propose a new system of generating array languages based on a non-deterministic partitioned cellular automaton (PCA). It is called a "PCA array generator" (PCAAG). This system is somewhat different from the framework of IAG, but has some similarity to the grammar formalism. Further, by imposing the constraint of "reversibility" to it, we can get a system having the property of unique parsability like UPAG.

PCA was introduced to investigate computing ability of reversible cellular automata [4, 5, 6]. It is a kind of cellular automaton where each cell is partitioned into the equal number of parts to the neighborhood size, and the information stored in each part is sent to only one neighboring cell. PCAAG generates an array language by operating the underlying PCA nondeterministically from a specified initial configuration.

One advantage of using PCA, besides its parallelism, is that reversible version of PCA (RPCA) can be easily defined. An RPCA is a "backward deterministic" PCA. Thus, if an array is generated by an RPCA, it can be parsed deterministically by retracing its movement backward. We show that, in PCA, reversibility of a local transition function is equivalent to that of a global transition function. Therefore, it is easier to design an RPCA than to design a reversible one in the framework of usual CA.

In this paper, we first give basic definitions of PCA, PCAAG, and three subclasses of PCAAG, i.e., a reversible PCAAG (RPCAAG), a monotonic RPCAAG (MRPCAAG), and a bounded transmission signal RPCAAG (BRPCAAG). We show some examples of these systems that generate geometrical figures. We then define an inverse PCA (IPCA) that undoes the operations of a PCA, and show that an array language generated by an RPCAAG can be parsed by a corresponding deterministic IPCA in parallel.

2 Definitions and Preliminaries

Definition 2.1 *A nondeterministic two-dimensional 5-neighbor partitioned cellular automaton* (PCA) *is a system defined by*

$$P = (\mathbf{Z}^2, C, U, R, D, L, f_P)$$

where $\mathbf{Z}$ is the set of all integers, C, U, R, D, and L are non-empty finite sets of center, up, right, down, and left internal states, respectively, and $f_P : C \times D \times L \times U \times R \rightarrow 2^{C \times U \times R \times D \times L}$ is a mapping called a *local function.* □

In the following,

$$f_P(c, d, l, u, r) = \{(c_1, u_1, r_1, d_1, l_1), \cdots, (c_n, u_n, r_n, d_n, l_n)\},$$

is sometimes represented by

$$(c, d, l, u, r) \rightarrow (c_1, u_1, r_1, d_1, l_1), \cdots, (c_n, u_n, r_n, d_n, l_n)$$

or by the following figure, and called a *rule* of P.

We regard f_P to denote the set of all rules of P whose righthand sides are nonempty, for convenience.

A *configuration* over $C \times U \times R \times D \times L$ (or of P) is a mapping $\alpha : \mathbf{Z}^2 \to C \times U \times R \times D \times L$. The set of all configurations over $C \times U \times R \times D \times L$ is denoted by $\text{Conf}(C \times U \times R \times D \times L)$, i.e.,

$$\text{Conf}(C \times U \times R \times D \times L) = \{\alpha \mid \alpha : \mathbf{Z}^2 \to C \times U \times R \times D \times L\}.$$

Let CENTER: $C \times U \times R \times D \times L \to C$ be the projection function that picks out the element of C from the quintuple in $C \times U \times R \times D \times L$ (the projections UP, RIGHT, DOWN, and LEFT are also defined similarly). The *global function* F_P : $\text{Conf}(C \times U \times R \times D \times L) \to 2^{\text{Conf}(C \times U \times R \times D \times L)}$ of P is defined as follows.

$$
\begin{aligned}
F_P(\alpha) = \{\alpha' \mid \forall i, j \in \mathbf{Z} \\
[\, \alpha'(i,j) \in f_P(\text{CENTER}(\alpha(i,j)), \text{DOWN}(\alpha(i,j+1)), \\
\text{LEFT}(\alpha(i+1,j)), \text{UP}(\alpha(i,j-1)), \text{RIGHT}(\alpha(i-1,j)))\,]\,\}
\end{aligned}
$$

Let α and α' be configurations of P. If $\alpha' \in F_P(\alpha)$, we say α' *is derived from* α in P, and denote it by $\alpha \underset{P}{\Rightarrow} \alpha'$. Reflexive and transitive closure of the relation $\underset{P}{\Rightarrow}$ is denoted by $\underset{P}{\overset{*}{\Rightarrow}}$. Further, we use the notation $\alpha \underset{P}{\overset{n}{\Rightarrow}} \alpha'$ to denote that there exist $\alpha_1, \alpha_2, \cdots, \alpha_{n-1}$ such that $\alpha \underset{P}{\Rightarrow} \alpha_1 \underset{P}{\Rightarrow} \alpha_2 \underset{P}{\Rightarrow} \cdots \underset{P}{\Rightarrow} \alpha_{n-1} \underset{P}{\Rightarrow} \alpha'$.

Definition 2.2 A PCA $P = (\mathbf{Z}^2, C, U, R, D, L, f_P)$ is called *locally reversible* if

$$\forall s_1, s_2 \in C \times D \times L \times U \times R \ [\, s_1 \neq s_2 \to f_P(s_1) \cap f_P(s_2) = \emptyset\,],$$

and is called *globally reversible* if

$$\forall \alpha_1, \alpha_2 \in \text{Conf}(C \times U \times R \times D \times L) \ [\, \alpha_1 \neq \alpha_2 \to F_P(\alpha_1) \cap F_P(\alpha_2) = \emptyset\,].$$

$\square$

In [4] it is shown that the notions of local and global reversibility are equivalent for deterministic one-dimensional case. It is also easy to prove the lemma for nondeterministic two-dimensional case.

Lemma 2.1 A PCA P is globally reversible iff it is locally reversible.

Proof. We first show the "if" part. Suppose P is locally reversible but not globally reversible. Then there are configurations α, α_1 and α_2 such that $[\alpha \in F_P(\alpha_1) \cap F_P(\alpha_2)] \wedge [\alpha_1 \neq \alpha_2]$. By the definition of F_P,

$$
\begin{aligned}
\forall i, j \in \mathbf{Z} \, [\, \alpha(i,j) \in \\
f_P(\text{CENTER}(\alpha_1(i,j)), \text{DOWN}(\alpha_1(i,j+1)), \\
\text{LEFT}(\alpha_1(i+1,j)), \text{UP}(\alpha_1(i,j-1)), \text{RIGHT}(\alpha_1(i-1,j)))\,]\,\} \\
\forall i, j \in \mathbf{Z} \, [\, \alpha(i,j) \in \\
f_P(\text{CENTER}(\alpha_2(i,j)), \text{DOWN}(\alpha_2(i,j+1)), \\
\text{LEFT}(\alpha_2(i+1,j)), \text{UP}(\alpha_2(i,j-1)), \text{RIGHT}(\alpha_2(i-1,j)))\,]\,\}
\end{aligned}
$$

holds. However, since $\alpha_1 \neq \alpha_2$, there must be $i^*, j^* \in \mathbb{Z}$ such that

$$\text{CENTER}(\alpha_1(i^*, j^*)) \neq \text{CENTER}(\alpha_2(i^*, j^*)) \lor \text{UP}(\alpha_1(i^*, j^*)) \neq \text{UP}(\alpha_2(i^*, j^*)) \lor$$
$$\text{RIGHT}(\alpha_1(i^*, j^*)) \neq \text{RIGHT}(\alpha_2(i^*, j^*)) \lor \text{DOWN}(\alpha_1(i^*, j^*)) \neq \text{DOWN}(\alpha_2(i^*, j^*)) \lor$$
$$\text{LEFT}(\alpha_1(i^*, j^*)) \neq \text{LEFT}(\alpha_2(i^*, j^*)).$$

This contradicts the assumption that P is locally reversible.

Next we show the "only if" part. Suppose P is globally reversible but not locally reversible. Then there are $u_1, u_2 \in C$, $v_1, v_2 \in D$, $w_1, w_2 \in L$, $x_1, x_2 \in U$, $y_1, y_2 \in R$ that satisfy

$$[u_1 \neq u_2 \lor v_1 \neq v_2 \lor w_1 \neq w_2 \lor x_1 \neq x_2 \lor y_1 \neq y_2]$$
$$\land [f_P(u_1, v_1, w_1, x_1, y_1) \cap f_P(u_2, v_2, w_2, x_2, y_2) \neq \emptyset].$$

Let α_1 and α_2 be distinct configurations such that

$$\begin{aligned}
\text{CENTER}(\alpha_1(0, 0)) &= u_1, & \text{CENTER}(\alpha_2(0, 0)) &= u_2, \\
\text{DOWN}(\alpha_1(0, 1)) &= v_1, & \text{DOWN}(\alpha_2(0, 1)) &= v_2, \\
\text{LEFT}(\alpha_1(1, 0)) &= w_1, & \text{LEFT}(\alpha_2(1, 0)) &= w_2, \\
\text{UP}(\alpha_1(0, -1)) &= x_1, & \text{UP}(\alpha_2(0, -1)) &= x_2, \\
\text{RIGHT}(\alpha_1(-1, 0)) &= y_1, & \text{RIGHT}(\alpha_2(-1, 0)) &= y_2,
\end{aligned}$$

and all the other parts of α_1 and α_2 are the same.

$F_P(\alpha_1) \cap F_P(\alpha_2) \neq \emptyset$ holds for such α_1 and α_2, and this contradicts the assumption of global reversibility of P. □

In the following, a globally or locally reversible PCA is called simply "reversible" and denoted by RPCA.

We now extend the projection CENTER to $\text{CENTER}^* : \text{Conf}(C \times U \times R \times D \times L) \to \text{Conf}(C)$ as follows.

$$\forall (i, j) \in \mathbb{Z}^2 \; [\, \text{CENTER}^*(\alpha)(i, j) = \text{CENTER}(\alpha(i, j)) \,]$$

From now on, we use CENTER instead of CENTER^* for convenience.

Let Σ be a nonempty finite set of symbols. A two-dimensional *word* over Σ is a two-dimensional finite connected array of symbols in Σ. The set of all words over Σ is denoted by Σ^{2+} (the empty word is not contained in Σ^{2+}). The #-*embedded* array of a word $w \in \Sigma^{2+}$ is an infinite array over $\Sigma \cup \{\#\}$ obtained by embedding w in two-dimensional infinite array of #s, and is denoted by $w_\#$.

Definition 2.3 *PCA array generator* (PCAAG) is a system defined by

$$G = (A, T, P, S, \$, \#)$$

where A and T are nonempty finite sets of array symbols and transmission symbols, respectively, S ($\in T$) is a start symbol, $\$$ ($\in T$) is an end symbol,

$\#$ ($\notin A \cup T$) is a blank symbol, and $P = (\mathbb{Z}^2, C, U, R, D, L, f_P)$ is a PCA that satisfies the following conditions.

(a) $C = A \cup \{\#\}$, $U = R = D = L = T \cup \{\#\}$
(b) $(\#, \#, \#, \#, \#) \rightarrow (\#, \#, \#, \#, \#)$
   (State $(\#, \#, \#, \#, \#) \in C \times U \times R \times D \times L$ is called a *quiescent state*.)
(c) $\forall (c, d, l, u, r) \in C \times D \times L \times U \times R$
   $[\,(d = \$) \vee (l = \$) \vee (u = \$) \vee (r = \$) \; \rightarrow \; f_P(c, d, l, u, r) = \emptyset \,]$

$\square$

Array symbols and transmission symbols in PCAAG roughly correspond to terminal and nonterminal symbols in the usual string grammar. In the following, we omit to write the rule $(\#, \#, \#, \#, \#) \rightarrow (\#, \#, \#, \#, \#)$, and indicate the symbol $\#$ by a blank in a figure.

We call α_0 an *initial configuration* if it satisfies the following condition.

$$\exists (i_0, j_0) \in \mathbb{Z}^2 \, [\, \alpha_0(i_0, j_0) = (\#, \#, S, \#, \#) \; \wedge$$
$$\forall (i, j) \in \mathbb{Z}^2 \, [\, (i, j) \neq (i_0, j_0) \; \rightarrow \; \alpha_0(i, j) = (\#, \#, \#, \#, \#)] \,]$$

We call α_f a *final configuration* if it satisfies the following condition.

$$\exists (i_f, j_f) \in \mathbb{Z}^2$$
$$[\; \exists x \in A \, [\, \alpha_f(i_f, j_f) = (x, \#, \$, \#, \#)] \; \wedge$$
$$\forall (i, j) \in \mathbb{Z}^2$$
$$[\, [\, j > j_f \vee (j = j_f \wedge i < i_f) \; \rightarrow$$
$$\exists y \in A \cup \{\#\} \, [\, \alpha_f(i, j) = (y, \#, \#, \#, \#)] \,] \; \wedge$$
$$[\, j < j_f \vee (j = j_f \wedge i > i_f) \; \rightarrow \; \alpha_f(i, j) = (\#, \#, \#, \#, \#)] \,] \,]$$

Let C_{ini} and C_{fin} denote the sets of all initial configurations and final configurations of P, respectively.

Intuitively, an initial configuration is a one such that there is just one cell whose right part contains the start symbol S and all the other parts contain $\#$s, and all other cells are in quiescent states. A final configuration is a one such that the cell whose position is at the rightmost column of the lowermost row among the cells that contain array symbols in their center parts contains the end symbol $\$$ in its right part and contains $\#$s in its up, down, and left parts, and all the other cells contain $\#$s in their up, right, down, and left parts.

A word $w \in A^{2+}$ is said to be *generated* by a PCAAG G, if there exist an initial configuration α_0 and a final configuration α_f such that $\alpha_0 \overset{*}{\Rightarrow} \alpha_f$ and $w_\# = \text{CENTER}(\alpha_f)$. The set of all words generated by G is called a *language* generated by G, and denoted by $L(G)$, i.e.,

$$L(G) = \{w \in A^{2+} \mid \exists \alpha_0 \in C_{\text{ini}}, \; \exists \alpha_f \in C_{\text{fin}} \, [\, \alpha_0 \overset{*}{\Rightarrow} \alpha_f \wedge w_\# = \text{CENTER}(\alpha_f)] \}.$$

Let $\mathcal{L}[\mathcal{C}]$ denote the class of languages generated by a class $\mathcal{C}$ of PCAAGs, i.e.,

$$\mathcal{L}[\mathcal{C}] = \{L \mid \exists G \in \mathcal{C} \, [\, L = L(G)] \}.$$

We now define three subclasses of PCAAG.

Definition 2.4 $G = (A, T, P, S, \$, \#)$ is called a *reversible PCAAG* (RPCAAG) if P is an RPCA. ☐

Definition 2.5 An RPCAAG $G = (A, T, P, S, \$, \#)$ is called a *monotonic RP-CAAG* (MRPCAAG), if each rule

$$(c, d, l, u, r) \rightarrow (c_1, u_1, r_1, d_1, l_1), \cdots, (c_n, u_n, r_n, d_n, l_n)$$

of P satisfies the following conditions.

(1) If $c \in A$ then $c_1, \cdots, c_n \in A$.
(2) If $c = \#$ then the following holds for each i $(1 \leq i \leq n)$.

$$c_i \in A \vee [\, [d \in T \leftrightarrow u_i \in T] \wedge [l \in T \leftrightarrow r_i \in T] \wedge$$
$$[u \in T \leftrightarrow d_i \in T] \wedge [r \in T \leftrightarrow l_i \in T] \,]$$

☐

Intuitively, G is an MRPCAAG, if, in each rule of P, an array symbol is not rewritten into a blank symbol, and when the symbol of the center part remains blank, transmission signals always return to the direction from which they came.

Definition 2.6 An RPCAAG $G = (A, T, P, S, \$, \#)$ is called a *bounded transmission signal RPCAAG* (BRPCAAG), if G satisfies

$$\exists k \in \mathbf{N}, \forall \alpha \in \text{Conf}(C \times U \times R \times D \times L), \forall \alpha_0 \in C_{ini}$$
$$[\, \text{if } \alpha_0 \overset{*}{\underset{P}{\Rightarrow}} \alpha \text{ then } |T_U| + |T_R| + |T_D| + |T_L| \leq k \,],$$

where $T_U = \{(i,j) \mid \text{UP}(\alpha(i,j)) \neq \#\}$, $T_R = \{(i,j) \mid \text{RIGHT}(\alpha(i,j)) \neq \#\}$,
$T_D = \{(i,j) \mid \text{DOWN}(\alpha(i,j)) \neq \#\}$, $T_L = \{(i,j) \mid \text{LEFT}(\alpha(i,j)) \neq \#\}$,

and each rule

$$(c, d, l, u, r) \rightarrow (c_1, u_1, r_1, d_1, l_1), \cdots, (c_n, u_n, r_n, d_n, l_n)$$

of P satisfies the following conditions.

(1) If $c \in A$ then $c_1 = c_2 = \cdots = c_n = c$.
(2) If $c = \#$ then the following holds for each i $(1 \leq i \leq n)$.

$$c_i \in A \vee [\, [d \in T \leftrightarrow u_i \in T] \wedge [l \in T \leftrightarrow r_i \in T] \wedge$$
$$[u \in T \leftrightarrow d_i \in T] \wedge [r \in T \leftrightarrow l_i \in T] \,]$$

☐

Intuitively, G is an BRPCAAG, if there exists a constant k that bounds the total number of transmission signals in any configuration derived from an initial configuration, and, in each rule of P, an array symbol is not rewritten into other symbol, and when the symbol of the center part remains blank, transmission signals always return to the direction from which they came. It should be noted that we cannot decide whether a given RPCAAG satisfies the above condition (1) only from the outer form of the rules of P (it requires some proof).

From the above definitions, it is easy to see that RPCAAG is a subclass of PCAAG, MRPCAAG is a subclass of RPCAAG, and BRPCAAG is a subclass of MRPCAAG.

3   Examples of PCAAGs

Example 3.1 The first example is a simple PCAAG G_1 that generates "¬"-shaped figures consisting of 'a's of all sizes.

$$G_1 = (\{a\}, \{S, A, \$\}, P_1, S, \$, \#),$$

where

$$P_1 = (\mathbf{Z}^2, C_1, U_1, R_1, D_1, L_1, f_{P_1}),$$
$$C_1 = \{a, \#\},$$
$$U_1 = R_1 = D_1 = L_1 = \{S, A, \$, \#\},$$

and f_{P_1} is shown in Fig.1. A derivation example is shown in Fig.2. G_1 is not reversible since the righthand sides of the rules (2) and (3), for example, have a common element.

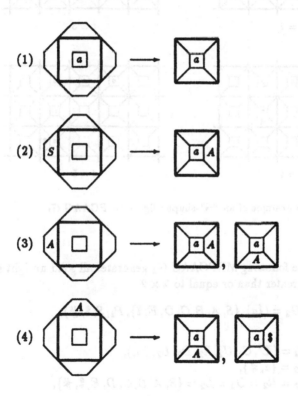

Fig. 1. Rules of PCAAG G_1 that generates all "¬"-shaped figures.

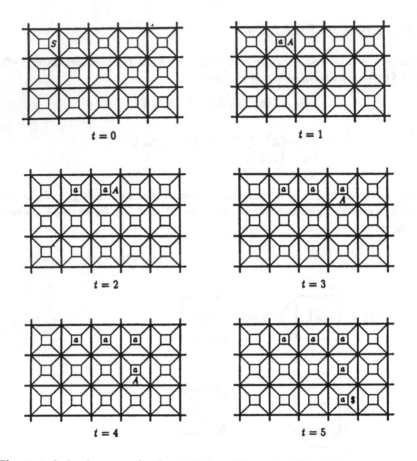

Fig. 2. A derivation example of an "¬"-shaped figure by PCAAG G_1.

Example 3.2 The following MRPCAAG G_2 generates all solid upright squares of 'a's with sizes greater than or equal to 2 × 2.

$$G_2 = (\{a\}, \{S, A, B, C, D, E, \$\}, P_2, S, \$, \#),$$

where

$$P_2 = (\mathbf{Z}^2, C_2, U_2, R_2, D_2, L_2, f_{P_2}),$$
$$C_2 = \{a, \#\},$$
$$U_2 = R_2 = D_2 = L_2 = \{S, A, B, C, D, E, \$, \#\},$$

and f_{P_2} is shown in Fig.3. It is easy to see that G_2 is reversible and monotonic. Fig.4 shows a derivation example.

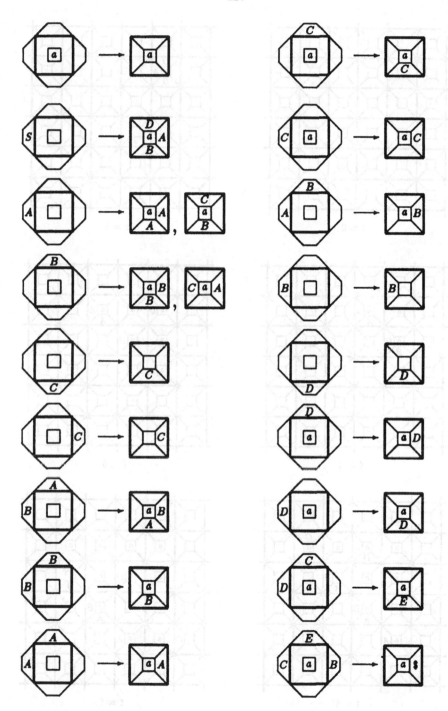

Fig. 3. Rules of MRPCAAG G_2 that generates all squares.

222

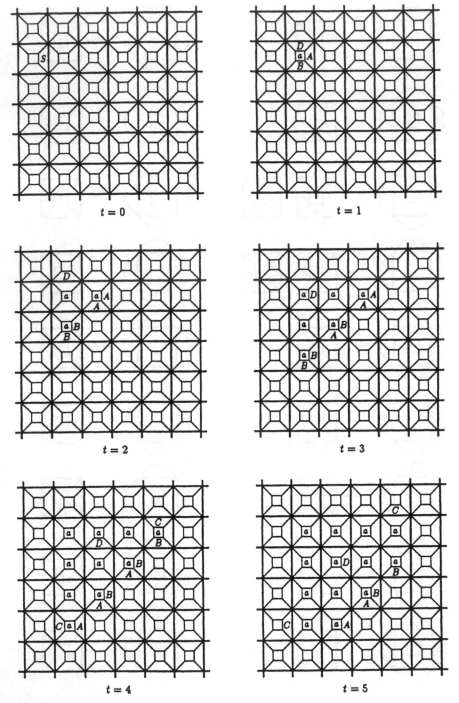

Fig. 4. A derivation example of a square by MRPCAAG G_2.

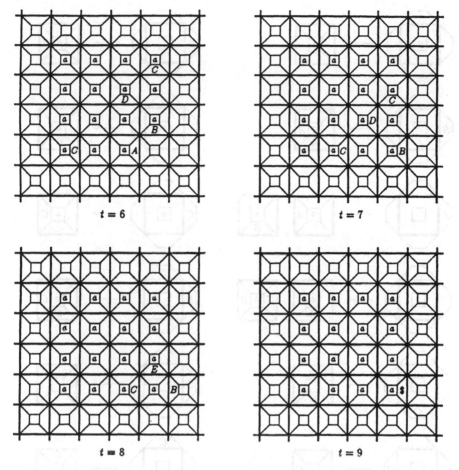

Fig.4. (Continued)

Example 3.3 The following BRPCAAG G_3 generates all hollow thin upright rectangles of 'a's with sizes greater than or equal to 2 × 2. (Note that when the width or the height is 2, they degenerate to solid rectangles.)

$$G_3 = (\{a\}, \{S, A, B, C, \$\}, P_3, S, \$, \#),$$

where

$$P_3 = (\mathbb{Z}^2, C_3, U_3, R_3, D_3, L_3, f_{P_3}),$$
$$C_3 = \{a, \#\},$$
$$U_3 = R_3 = D_3 = L_3 = \{S, A, B, C, \$, \#\},$$

and f_{P_3} is shown in Fig.5. Fig.6 shows a derivation example. We can prove that every configuration of G_3 derived from an initial configuration contains at most 4 transmission symbols.

224

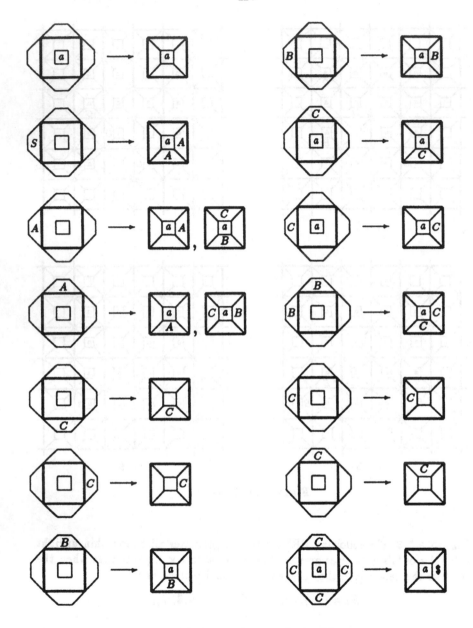

Fig. 5. Rules of BRPCAAG G_3 that generates all hollow rectangles.

225

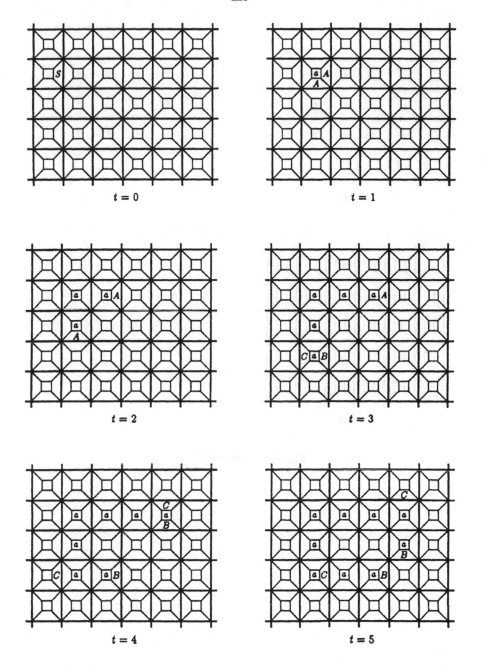

Fig. 6. A derivation example of a hollow rectangle by BRPCAAG G_3.

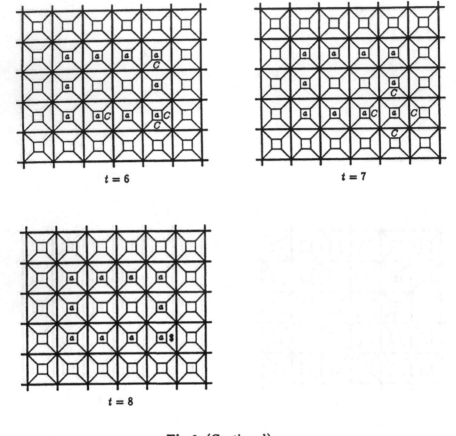

Fig.6. (Continued)

4 Parsing Array Languages by Inverse PCAs

Definition 4.1 A nondeterministic *inverse PCA* (IPCA) is a system defined by

$$Q = (\mathbf{Z}^2, C, U, R, D, L, g_Q)$$

where $\mathbf{Z}, C, U, R, D, L$ are the same as in PCA, and $g_Q : C \times U \times R \times D \times L \rightarrow 2^{C \times D \times L \times U \times R}$ is a local function. □

As in the case of PCA,

$$g_Q(c, u, r, d, l) = \{(c_1, d_1, l_1, u_1, r_1), \cdots, (c_n, d_n, l_n, u_n, r_n)\}$$

is represented by the following figure, which will be helpful to understand the operation of the local function of Q intuitively.

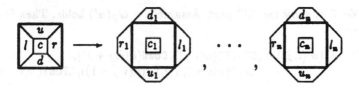

The global function $G_Q : \mathrm{Conf}(C \times U \times R \times D \times L) \rightarrow 2^{\mathrm{Conf}(C \times U \times R \times D \times L)}$ is defined by

$$G_Q(\alpha) = \{\alpha' \mid \forall i, j \in \mathbf{Z}[\; g_{\bar{P}}(\alpha(i, j)) \ni (\mathtt{CENTER}(\alpha'(i, j)), \mathtt{DOWN}(\alpha'(i, j + 1)),$$
$$\mathtt{LEFT}(\alpha'(i + 1, j)), \mathtt{UP}(\alpha'(i, j - 1)),$$
$$\mathtt{RIGHT}(\alpha'(i - 1, j)))]\;\}.$$

Definition 4.2 An IPCA $Q = (\mathbf{Z}^2, C, U, R, D, L, g_Q)$ is called *locally deterministic* if

$$\forall s \in C \times U \times R \times D \times L \; [\; |g_Q(s)| \leq 1\;],$$

and is called *globally deterministic* if

$$\forall \alpha \in \mathrm{Conf}(C \times U \times R \times D \times L) \; [\; |G_Q(\alpha)| \leq 1\;].$$

□

Lemma 4.1 An IPCA Q is globally deterministic iff it is locally deterministic.

Proof. Omitted. □

From now on, globally or locally deterministic IPCA is called simply "deterministic" and denoted by DIPCA.

Let α and α' be configurations of Q. If $\alpha' \in G_Q(\alpha)$, we say that α' is derived from α, and denote it by $\alpha \underset{Q}{\Rightarrow} \alpha'$. The notations $\underset{Q}{\overset{*}{\Rightarrow}}, \underset{Q}{\overset{+}{\Rightarrow}}$ are also defined similarly.

Definition 4.3 Let $P = (\mathbb{Z}^2, C, U, R, D, L, f_P)$ be a PCA. The following $\tilde{P}$ is called an IPCA *corresponding to* P.

$$\tilde{P} = (\mathbb{Z}^2, C, U, R, D, L, g_{\tilde{P}}),$$

where $g_{\tilde{P}}$ is defined from the local function f_P of P in the following manner.

$$\forall (c, u, r, d, l) \in C \times U \times R \times D \times L :$$
$$g_{\tilde{P}}(c, u, r, d, l) = \{ (c', d', l', u', r') \mid (c, u, r, d, l) \in f_P(c', d', l', u', r') \}$$

$\square$

Lemma 4.2 Let P be a PCA, $\tilde{P}$ be the IPCA corresponding to P, and α and α' be any configurations of P. Then the following relation holds.

$$\alpha \underset{P}{\Rightarrow} \alpha' \quad \text{iff} \quad \alpha' \underset{\tilde{P}}{\Rightarrow} \alpha$$

Proof. We first show the "if" part. Assume $\alpha \in G_{\tilde{P}}(\alpha')$ holds. Then from the definitions,

$$\forall i, j \in \mathbb{Z} : g_{\tilde{P}}(\alpha'(i,j)) \ni (\text{CENTER}(\alpha(i,j)), \text{DOWN}(\alpha(i,j+1)),$$
$$\text{LEFT}(\alpha(i+1,j)), \text{UP}(\alpha(i,j-1)), \text{RIGHT}(\alpha(i-1,j))),$$

and thus

$$\forall i, j \in Z : f_P(\text{CENTER}(\alpha(i,j)), \text{DOWN}(\alpha(i,j+1)),$$
$$\text{LEFT}(\alpha(i+1,j)), \text{UP}(\alpha(i,j-1)), \text{RIGHT}(\alpha(i-1,j))) \ni \alpha'(i,j).$$

Therefore $\alpha' \in F_P(\alpha)$.

The "only if" part is proved by tracing above inversely. $\square$

Lemma 4.3 Let P be a PCA, and $\tilde{P}$ be the IPCA corresponding to P. Then the following holds.

$$P \text{ is reversible iff } \tilde{P} \text{ is deterministic.}$$

Proof. If $\tilde{P}$ is deterministic, $\forall s \in C \times U \times R \times D \times L \ [\ |g_{\tilde{P}}(s)| \leq 1 \ ]$. Now suppose $s_1 \neq s_2 \ \wedge \ f_P(s_1) \cap f_P(s_2) \neq \emptyset$ for some $s_1, s_2 \in C \times D \times L \times U \times R$. Then there exists $s' \in f_P(s_1) \cap f_P(s_2)$ such that $g_{\tilde{P}}(s') \supseteq \{s_1, s_2\}$, and this contradicts $\forall s \ [\ |g_{\tilde{P}}(s)| \leq 1 \ ]$. Therefore $\forall s_1, s_2 \in C \times D \times L \times U \times R \ [\ s_1 \neq s_2 \rightarrow f_P(s_1) \cap f_P(s_2) = \emptyset \ ]$, and thus P is reversible.

Conversely, if P is reversible, $\forall s_1, s_2 \in C \times D \times L \times U \times R \ [\ s_1 \neq s_2 \rightarrow f_P(s_1) \cap f_P(s_2) = \emptyset \ ]$ holds. Suppose $|g_{\tilde{P}}(s')| > 1$ for some $s' \in C \times U \times R \times D \times L$. Then there exist $s_1, s_2 \in C \times D \times L \times U \times R$ such that $g_{\tilde{P}}(s') \supseteq \{s_1, s_2\} \ \wedge \ s_1 \neq s_2$. Thus $s_1 \neq s_2 \ \wedge \ f_P(s_1) \cap f_P(s_2) \ni s'$ holds, and this contradicts the assumption. Therefore $\forall s \in C \times U \times R \times D \times L \ [\ |g_{\tilde{P}}(s)| \leq 1 \ ]$, and $\tilde{P}$ is deterministic. $\square$

The next theorem shows that any array language generated by an RPCAAG $G = (A, T, P, S, \$, \#)$ can be parsed by a DIPCA $\tilde{P}$ corresponding to P in the same steps as in the generation of the word.

Theorem 4.1 Let P be an RPCA, $\tilde{P}$ be the IPCA corresponding to P, and α_1 and α_2 be configurations of P such that $\alpha_1 \overset{a}{\underset{P}{\Rightarrow}} \alpha_2$. Then the following holds.

$$\exists \alpha' \left[\alpha_2 \overset{a}{\underset{\tilde{P}}{\Rightarrow}} \alpha' \right] \wedge \forall \alpha'' \left[\text{if } \alpha_2 \overset{a}{\underset{\tilde{P}}{\Rightarrow}} \alpha'' \text{ then } \alpha'' = \alpha_1 \right]$$

Proof. By Lemma 4.2 it is easy to see that $\alpha_2 \overset{a}{\underset{\tilde{P}}{\Rightarrow}} \alpha_1$ holds. On the other hand, by Lemma 4.3 $\tilde{P}$ is deterministic since P is reversible. Therefore, if $\alpha_2 \overset{a}{\underset{\tilde{P}}{\Rightarrow}} \alpha''$ then $\alpha'' = \alpha_1$. $\qquad\qquad\square$

5 Concluding Remarks

In this paper, we introduced PCAAG, and its three subclasses, i.e., RPCAAG, MRPCAAG, and BRPCAAG. We have already obtained results that the generating abilities of RPCAAG, MRPCAAG, and BRPCAAG are precisely characterized by a deterministic two-dimensional Turing machine, a deterministic two-dimensional linear-bounded automaton, and a deterministic two-dimensional erasing marker automaton, respectively. These results will appear in the forthcoming paper.

Acknowledgement   The authors are grateful to Yasunori Yamamoto of National Museum of Ethnology for his helpful discussion.

References

1. Cook, C.R., and Wang, P.S.P., "A Chomsky hierarchy of isotonic array grammars and languages," *Computer Graphics and Image Processing*, 8, 144–152 (1978).
2. Milgram, D.L., and Rosenfeld, A., "Array automata and array grammars," *Information Processing* 71, 69–74 (1972).
3. Morita, K., Yamamoto, Y., and Sugata, K., "The complexity of some decision problems about two-dimensional array grammars," *Information Sciences*, 30, 241–262 (1983).
4. Morita, K., and Harao, M., "Computation universality of one-dimensional reversible (injective) cellular automata," *Trans. IEICE*, E-72, 758–762 (1989).
5. Morita, K., and Ueno, S., "Computation-universal models of two-dimensional 16-state reversible cellular automata," *IEICE Trans. Inf. & Syst.*, E75-D, 141–147 (1992).
6. Morita, K., "Computation-universality of one-dimensional one-way reversible cellular automata," *Information Processing Letters*, 42, 325–329 (1992).
7. Nivat, M., Saoudi, A., and Dare, V.R., "Parallel generation of finite images," *Int. J. Pattern Recognition and Artificial Intelligence*, 3 279–294 (1989).
8. Rosenfeld, A., *Picture Languages*, Academic Press, New York (1979).
9. Rosenfeld, A., "Coordinate grammars revisited: generalized isometric grammars," *Int. J. Pattern Recognition and Artificial Intelligence*, 3 435–444 (1989).

10. Saoudi, A., Rangarajan, K., and Dare, V.R., "Finite images generated by GL-systems," *Int. J. Pattern Recognition and Artificial Intelligence*, 3 459–467 (1989).
11. Siromoney, G., Siromoney, R., and Krithivasan, K., "Abstract families of picture languages," *Computer Graphics and Image Processing*, 1, 284–307 (1972).
12. Siromoney, G., Siromoney, R., and Krithivasan, K., "Picture languages with array rewriting rules," *Information and Control*, 22, 447–470 (1973).
13. Wang, P.S.P., "Hierarchical structures and complexities of parallel isotonic array languages," *IEEE Trans.*, PAMI-5, 92–99 (1983).
14. Yamamoto, Y., Morita, K., and Sugata, K., "Context-sensitivity of two-dimensional regular array grammars," *Int. J. Pattern Recognition and Artificial Intelligence*, 3 295–319 (1989).
15. Yamamoto, Y., and Morita, K., "Two-dimensional uniquely parsable isometric array grammars," *Proc. Int. Colloq. Parallel Image Processing*, Paris, 271–287 (1991).
16. Yamamoto, Y., and Morita, K., "Isometric array grammars which are equivalent to two-dimensional deterministic tape acceptors (in Japanese)," *Research Report of Institute of Mathematical Analysis, Kyoto University*, No.754, 115–124 (1991).

Parallel Recognition of Multidimensional Images
using
Regular Tree Grammars

A. Saoudi
L.I.P.N.

Universite Paris XIII

Centre Scientifique et Polytechnique

Av J. B. Clément

93430 Villetaneuse, France

E-mail:as@litp.ibp.fr

Abstract

In this paper, we use tree grammars, and tree automata for representing a set of Multidimensional images. We show that the set of all full 2^d-trees (Quadtrees, Octrees,...etc) is not a regular set. But every finite set of full 2^d-trees can be represented by a regular tree grammar. We give optimal algorithms for solving the S-equivalence problem of two 2^d-trees and the reduction (i.e. compression) problem of a full 2^d-trees. We give some characterizations of regular trees sets. We present a parallel algorithm for recognizing a multidimensional image of size N in $O(log(N))$ time with $O(N) = N/d$ processors on EREW-PRAM model.

Keywords: Multidimensional images, Parallel algorithms, 2^d-trees, Tree automata, and Tree Grammars.

* This research was supported by CEFIPRA and PRC Maths-Info, France

Introduction

Digital image processing has had tremendous growth in the past twenty years. Its applications range from telecommunications(ISDN and HDTV) to medical imaging and remote sensing([Br 82], [DR 81], [RK 82], and [Sa 90]). For image processing the fundamental data structures are quadtrees and octrees ([DRS 79], [RRP 80], [JT 80] [SR 79], [Sa 80] and [Sa 90]). Quadtrees and octrees data structures have been widely used to represent spatial information in computer vision, computer graphics, and geographic information systems. These data structures are efficient for image processing. Also octrees are useful representation of three-dimensional objects for space planning, computer animation, and navigation.

The first aim of this work is to define a device for representing a set of images (or objects) as a collection of trees. The second aim is to design parallel algorithms for testing the S-equivalence of two 2^d-ary trees, for condensing the full 2^d-trees, and for recognizing multidimensional images represented by 2^d-trees. The S-equivalence problem for two 2^d-trees is the problem to check whether or not two trees can be reduced (i.e. condensed or compressed) to the same tree.

In this paper, the Parallel Random Access Machine (PRAM) is used as the basic model of parallel processing. A PRAM consists of a collection of synchronous deterministic random access machines (RAM) with shared memory. Every processor can read and write from a global random shared memory. There are various types of PRAMs. The Concurrent Read Concurrent Write (CRCW) PRAM allows all the processors to read and write from the same memory location at the same time. However, when more processors attempt to write into the same memory, only one processor succeds nondeterministically to write into the location. In the Concurrent Read Exclusive Write (CREW) PRAM, all processors are allowed to read from the same location at the same time. The EREW-PRAM allows only one access per memory during one step.

In the first section, we give some basic definitions of quadtrees, octrees, and 2^d-trees. We give optimal algorithms for solving S-equivalence problem and the reduction problem for 2^d-trees. In the third section, we use regular tree grammar for representing a set of multidimensional images. We prove that the set of full 2^d-trees is not regular. In the third section, we introduce two new models of finite state automata(M-automata and controlled automata) and then prove that they accept exactly the class of tree sets generated by regular tree grammars. We also prove that the set of reduced (compressed) 2^d-trees is regular (i.e. generated by a regular tree grammar). The last section deals with serial and parallel algorithms for recognizing multidimensional images.

1. Multidimensional Images

We first define the quadtree representation of two-dimensional objects. Consider a $2^n \times 2^n = \sqrt{N} \times \sqrt{N}$ image (i.e array) of unit cells called the universe space array, where n corresponds to the sise of space or resolution if space array is fixed. Only objects within the universe space are represented by quadtrees. The two-dimensional object representation can be obtained by associating to each unit cell a color, say black and white, according to whether that cell is inside or outside the object. In the quadtree (resp. full quadtree) data structure, the root node corresponds to the universe space (or image). The space is divided into 4 quadrants of size $2^{n-1} \times 2^{n-1}$. These quadrants correspond to 4 children nodes of the root. Each of quadrant is recursively subdivided further into quadrants, until each quadrant contains unit cells (resp. cells of a single color).

Let us give an example to make clear how one can obtain a quadtree (rep. full quadtree) from an image.

Example 1: ($\sqrt{N} = 4 = 2^2$ and $d = 2$)

$$I = \begin{array}{|c|c|c|c|}
\hline
0 & 0 & 1 & 1 \\
\hline
0 & 0 & 1 & 0 \\
\hline
1 & 0 & 0 & 0 \\
\hline
1 & 1 & 0 & 0 \\
\hline
\end{array}$$

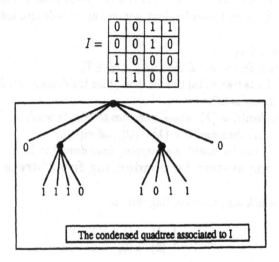

The condensed quadtree associated to I

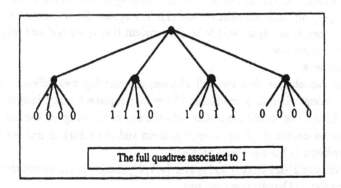

The full quadtree associated to I

We use the symbol • to label the internel node of trees representing images.

We now define the octree representation of three-dimensional objects. Consider a $2^n \times 2^n \times 2^n$ image (i.e array) of unit cubes called the universe space array, where n corresponds to the size of space or resolution if space array is fixed. Only objects within the universe space are represented by octrees. The three-dimensional object representation can be obtained by associating to each unit cube a color, say black and white, according to wether that cube is inside or outside the object. In the octree (resp. full octree) data structure, the root node corresponds to the universe space (or image). The space is divided into 8 octants of size $2^{n-1} \times 2^{n-1} \times 2^{n-1}$. These octants correspond to 8 children nodes of the root. Each of octant is recursively subdivided further into octants, until each octant contains unit cubes (resp. cubes of a single color).

The quadtree and octree stuctures can be generalized to 2^d-tree structure in order to represent multidimensional objects such images. Observe that a quadtree (resp. 2^d-tree) is a 4-ary tree (resp. 2^d-ary tee).

Let us define inductively the set of k-ary trees. Since finite tree can be represented by a term, we will denote trees as terms. Let Σ be a finite alphabet. Then the set of k-ary finite trees over Σ, called k-trees for short, denoted by T_Σ is defined inductively as follows :

(i) If $a \in \Sigma$ then $a \in T_\Sigma$.

(ii) If $t_1, t_2, ..., t_k \in T_\Sigma$ and $a \in \Sigma$ then $a(t_1, ..., t_k) \in T_\Sigma$.

Let t be a k-ary finite tree, called tree for short, then the domain of t, denoted by $dom(t)$, is defined inductively as follows :

(i) If $t \in \Sigma$ then $dom(t) = \{\lambda\}$, where λ denotes the empty word.

(ii) If $t = a(t_1, ..., t_k)$ then $dom(t) = \{\lambda\} \cup (\cup_{i=1}^{k} i.dom(t_i))$.

Note that a tree t can be viewed as a mapping from $dom(t)$ to Σ.

Tree rewriting system for condensing full 2^d-trees

Consider the following tree rewriting system:

$$S = \begin{cases} (1) & \bullet(1, \cdots, 1) \to 1 \\ (2) & \bullet(0, \cdots, 0) \to 0 \end{cases}$$

This reduction system can be used for obtaining the condensed 2^d-trees from full 2^d-trees by applying reduction rules (1) and (2). The system S induceses an equivalence relation. Two trees t_1 and t_2 are said to be S-equivalent (i.e. $t_1 \equiv_S t_2$)if and only if can be reduced to the same tree.

Remarks

One can observe that two full 2^d-trees, representing two different images, can be S-equivalent. Hence, a condensed 2^d-tree can represent many images. For representing an image, one should use 2^d-tree with the size of the image or the full 2^d-tree.

Now we discuss the S-equivalence problem and the reduction problem for 2^d-trees.

Problem (1)(S-eqivalence problem)

Given two 2^d-ary trees t_1 and t_2 over $\{\bullet, 1, 0\}$, check whether or not they are S-equivalent.

Problem (2)(reduction problem)

Given a full 2^d-ary tree t over $\{\bullet, 1, 0\}$, compute the reduced tree S-eqivalent to t.

Proposition 1.1

The S-equivalence problem can be solved in $O(P)$ time and $O(P)$ space on RAM (Random Access Machine), where $P = Max(size(t_1), size(t_2))$.

To prove proposition (1), one can reduce t_1 and t_2 and then test the equality of their reduced trees.

Proposition 1.2

The S-equivalence problem can be solved in $O(log(P))$ time with $O(P/log(P))$ processors on EREW-PRAM model.

Proposition 1.3

The reduction problem can be solved in $O(P)$ time and $O(P)$ space on RAM (Random Access Machine), where $N = size(t)$.

Proposition 1.4

The reduction problem can be solved in $O(log(P))$ time with $O(P/log(P))$ processors on EREW-PRAM model.

2. Regular Tree Grammars and Multidimensional Images

In this section, we will define regular tree grammars for representing sets of 2^d-trees. Then we show that the set of full 2^d-trees, over a fixed alphabet, cannot be generated by a regular tree grammar.

Let us define the class of regular tree grammars that defines the family of regular tree sets.

Definition 2.1

A regular k-ary tree grammar is a structure $G =< V, \Sigma, S, R >$, where:

(i) V is the set of nonterminals,

(ii) Σ is the alphabet,

(iii) $S \in V$ is the start symbol(i.e. axiom), and

(iv) R is the set of rules of the following forms:

$X \rightarrow a(X_1, \cdots, X_k)$ or $X \rightarrow a$, where $a \in \Sigma$ and $X, X_i \in V$.

This what is called the normal form for regular tree grammars. Since each regular tree grammar is equivalent to a regular tree grammar in normal form, we adopt the normal form as a definition.

To prove that the set of full 2^d-trees cannot be generated by regular tree grammars, we need the following lemma.

Lemma 2.1(Pumpping lemma)

Let x be a symbol not in Σ. For each set T of trees, over Σ, generated by regular tree grammar we can find an integer h such that for all trees $t \in T$, if $height(t) \geq h$, then there are trees $t_1, t_2, t_3 \in T_{\Sigma \cup \{x\}}$ such that:

(i) t_1 and t_2 contain exactly one occurence of x as the label of a terminal nodes, $t_3 \in T_\Sigma$;

(ii) $t = t_1 \cdot_x t_2 \cdot_x t_3$;

(iii) $height(t_1 \cdot_x t_3) \leq h$;

(iv) $height(t_2) \geq 1$;

(v) for all $n \geq 0$, $t = t_1 \cdot_x t_2^{nx} \cdot_x t_3 \in T$, where $t_2^{nx} = t_2 \cdot_x \cdots \cdot_x t_2$ (n times).

Note that $t._x t'$ is the tree obtained by substituting x in t by t'.

Proposition 1.1

The set of the full 2^d-trees cannot be generated by a regular tree grammar.

This can be proved by using the above lemma.

Since the family of sets generated by regular tree grammars is closed by union an each set containing only one tree is regular, we can use tree grammar for representing a finite set of images.

In the next section, we will prove that the set of all condensed 2^d-trees is regular.

3. Finite State Tree Automata and Regular Tree Grammars

In this section, we introduce two new finite state tree automata (M-auaomata and C-automata) and give some caracterizations of tree sets generated by regular tree grammars.

Definition 3.1

A **top-down k-ary tree automaton** , a top-down tree automaton for short, is a structure $M =< Q, \Sigma, q_0, \delta, F >$, where Q is a set of finite states, Σ is a finite set of input symbols, q_0 is the set of initial states, $\delta : Q \times \Sigma \rightarrow 2^{Q^k}$ is the transition function and F is a set of terminal states.

A computation of M on the tree t is a tree C on Q such that $dom(C) = dom(t) \cup dom^+(t)$, $C(\lambda) = q_0$ and for each node u we have $(C(u1), ..., C(uk)) \in \delta(C(u), t(u))$. The tree t is accepted by the top-down tree automaton M, if for some computation of M on t, and for node $v \in dom^+(t)$, $C(v)$ is a terminal state (i.e. $C(v) \in F$).

M is said to be deterministic iff for each $q \in Q$ and $f \in \Sigma$, $Card(\delta(q, f) \leq 1$.

The set $dom^+(t)$ is a set of nodes that are immediate successors of terminal nodes. For example, $dom^+(\bullet(1, 0, 1, 1)) = (1 + 2 + 3 + 4)(1 + 2 + 3 + 4)$.

Definition 3.2

A **bottom-up k-ary tree automaton**, called bottom-up tree automaton for short, is a structure $M =< Q, \Sigma, \delta, Q_0, F >$, where Q, Σ, F are defined as for a top-down tree automaton, $\delta : Q^k \times \Sigma \rightarrow 2^Q$ is the transition function, and $Q_0 \subseteq Q$ is the set of initial states.

A computation of the bottom-up tree automaton M on the tree t is a tree C over Q with for each node $u \in dom^+(t)$, $C(u) \in Q_0$ and for each node u, $C(u) \in \delta(< C(u1), ..., C(uk) > , t(u))$. M accepts t iff there is a computation C of M on t such that $C(\lambda) \in F$.

M is said to be deterministic iff for each $< q_1, ..., q_k > \in Q^k$ and $f \in \Sigma$, $Card(\delta(< q_1, ..., q_k >, f)) \leq 1$.

It is well known [Bra 68] and [Th 73] that for each set T of finite trees, the following conditions are equivalent :

(i) T is accepted by a top-down tree automaton.

(ii) T is is generated by a regular tree grammar.

(iii) T is accepted by a deterministic bottom-up tree automaton.

Definition 3.3

A **top-down k-ary tree M-automaton**, a Muller tree automaton for short, is a structure $M =< Q, \Sigma, q_0, \delta, F >$, where Q, Σ, q_0 , δ are defined as before, and $F \subseteq 2^Q$ is a collection of designated sets of states.

The tree t is accepted by the Muller tree automaton M iff there is a computation of M on t such that the set of states occuring in $dom^+(t)$ of this computation belongs to F.

Definition 3.4

A controlled tree automaton is a structure $M =< Q, \Sigma, \delta, q_0, L >$, where Q, Σ, q_0 and δ are defined as for a top-down tree automaton, and L is string language over Q accepted by a classical finite state automaton.

The tree t is accepted by the controlled tree automaton M iff there is a computation C of M on t such that for each path of C, the string lying in this path belongs to L.

Theorem 3.1

For each set of finite trees T, the following conditions are equivalent :

(i) T is accepted by a deterministic Muller tree automaton, and

(ii) T belongs to the boolean closure of tree sets accepted by deterministic top-down tree automata.

This is due to the fact that the acceptance condition for Muller tree automata is a boolean combination of the acceptance condition for top-down tree automata.

Theorem 3.2

For each set of finite trees T, the following conditions are equivalent :

(i) T is accepted by a tree automaton (Those define above), and

(ii) T is a projection of tree set accepted by a deterministic tree automaton.

Let T be a tree set accepted by a non-deterministic tree automaton. By coding trees together with their computations, we obtain a deterministic set of trees over $\Sigma \times Q$. The set of coded trees becomes deterministic and the first projection of the obtained tree set is exactly T. The last part can be proved by observing that the class of tree sets accepted by non-detrministic tree automata are closed under projection.

Theorem 3.3

For each set of finite trees T, the following conditions are equivalent :

(i) T is accepted by a Muller tree automaton,

(ii) T is accepted by a controlled tree automaton, and

(iii) T is accepted by a top-down tree automaton.

The equivalence between (i) and (iii) is consequence of theorem (3.2) and the fact that the class of tree sets accepted by non-deterministic tree automata is closed under union and projection.

(iii) $\Rightarrow$ (ii) is trivial. (ii) $\Rightarrow$ (iii) : this can be proved by simulating, a detrministic finite state automaton accepting the control language L and the given top-down tree automaton, by a top-down tree automaton.

Theorem 3.4

The set of condensed 2^d-trees is accepted by a bottom-up tree automaton.

This because bottom-up tree automata are able to check whether or not we can apply a reduction rule of the system S to an input 2^d-trees. If for an input tree a reduction rule can be applyied to one of its terminal node, the automaton should reach the rejecting state, say "reject". Now if the automaton is in the state "reject", by reading any symbol thae autmaton stay in the state "reject".

4. Parallel Recognition of Multidimensional Images

In this section, we use tree regular grammar and design serial and parallel algorithm for recognizing trees and multidimensional images represented by 2^d-trees.

Theorem 4.1

The recognition problem of a tree of size P can be solved in $O(P)$ time and $O(P)$ space on RAM (i.e. Random Access Machine) model.

Theorem 4.2

The recognition problem of a tree of size P can be solved in $O(log(P))$ time with $O(P/log(P))$ processors on EREW-PRAM model.

Proposition 4.3

The recognition problem of a multidimensional image, represented by a 2^d-tree of size N, can be solved in $O(log(N))$ time and N/d processors on EREW-PRAM model.

Acknowledgement. We thank Prof. M. Nivat for many fruitful discussions.

References :

[Ak 89] S. Akl, The Design of Parallel Algorithms, Prentice-Hall (1989).

[Br 82]M. Brady, *Computational Approaches to Image Understanding*, ACM Computing Surveys, vol. 14, no. 1(1982)3-72.

[Bra 69]M. Brainerd, *Tree Generatin Regular Systems*, Information and Control, vol. 14, (1969)217-231.

[DR 81]L.S. Davis and A. Rosenfeld, *Cooperating processes for low-level vision: A survey*, Artif. Intell. 17(1981)245-265.

[DRS 79] R.D. Dyer, A. Rosenfeld and H. Samet, Region Representation: Bouday code from Quadtrees, Communications of the ACM, vol. 23(1979)171-179.

[IN 82]K.Inoue and A. Nakamura, *Some Topological Properties of Σ-structure automata*, S-C-C 7 (1982)19-27.

[JT 80] C. L. Jackins and S. L. Tanimoto, *Oct-trees and their use in representing Three dimensional objects*, Computer Graphics and Image Processing, 14(1980)249-270.

[NS 91] M. Nivat and A. Saoudi, On Parallel Recognition Algorithms for High Dimensional Images, The 1st Int. Conf. on Parallel Image Processing, Paris (1991).

[RRP 80]S. Ranade, A. Rosenfeld and J.M. Prewitt, *Use of Quadtrees for Image Segmentation*, Computer Science Dept., University of Maryland, TR-878 (1980).

[RK 82] A. Rosenfeld and A. C. Kak, Digital Picture Processing, vol. 1 and 2, (The second edition) Academic Press (1982).

[Ro 90] A. Rosenfeld, Array, Tree and Graph Grammars, in "Syntactic and Structural Pattern Recognition: Theory and Applications ",(H. Bunke and A. Sanfeliu, Eds.), World Scientific (1990).

[SR 80] H. Samet and A. Rosenfeld, Quadtreee Representations of Binary Images, Proc. 5th Int. Conf. on Pattern Recognition, Miami(1980)815-818.

[Sa 90]H. Samet, *The Design and Analysis of Spatial Data Structures*, Addison Wesley (1990).

[SR 92] **A. Saoudi and W. Rytter**, Parallel Algorithms for 2D-Image Recognition, Int. Conf. on Pattern Recognition (1992).

[Th 73]**J. W. Thatcher**, *Tree Automata:an informal survey*, in "Currents in Theory of Computing"(ed; A.V. Aho), Prentice-Hall(1973).

Optimal Parallel Algorithms for Multidimensional Image Template Matching and Pattern Matching

A. Saoudi

L.I.P.N

Universite Paris XIII

Institut Galilée

Av. J. B. Clément

93400 Villetaneuse, France

E-mail: as@litp.ibp.fr

M. Nivat

L.I.T.P

Universite Paris VII

2, Place Jusieu

75251 Paris Cedex 05, France

Abstract

This paper presents efficient and optimal parallel algorithms for multidimensional image template matching on CREW PRAM model. For an N^d image and M^d window, we present an optimal (resp. efficient) algorithm which runs in $O(log(M))$ time with $O((M^d \times N^d)/log(M))$ processors (resp. $O(M^d \times N^d)$). We also present efficient and optimal algorithms for solving the multidimensional array and pattern matching.

Keywords: Parallel algorithms, Image template matching, Array Matching, and Parallel Random Access Machine.

* This research was supported by CEFIPRA and PRC Maths-Info, France

Introduction

Template matching is a basic operation in image processing and computer vision. It is used as a method for filtering, edge detection, image registration and object detection [MR 90]. Template matching problem can be described as comparing template (i.e. window) with all possible windows of a given image. Because of the fundamental nature of this problem and because of its high sequential complexity ($O(N^2 \times M^2)$ on a single processor computer), much attention has been devoted to the developpement of efficient fine grain parallel algorithms. The algorithm developed in [KK 87] for fine grained SIMD hypercubes make significant use of the $O(1)$ data broadcast capability available from the host to hypercube processors. The results in [KK 87] include an algorithm for two-dimensional image template matching on a hypercube with N^2 processors in $O(M^2 + log(N))$ time. Ranka et al.[RS 88] extend the techniques used in [KK 87] and derive algorithms for two-dimensional image template matching on MIMD hypercubes. Our aim is to design efficient and optimal parallel CREW-PRAM algorithms for multidimensional image template matching, array matching and pattern matching.

In this paper, the Parallel Random Access Machine (PRAM) is used as the basic model of parallel processing. A PRAM consists of a collection of synchronous deterministic random access machines (RAM) with shared memory. Every processor can read and write from a global random shared memory. There are various types of PRAMs. The Concurrent Read Concurrent Write (CRCW) PRAM allows all the processors to read and write from the same memory location at the same time. However, when more processors attempt to write into the same memory, only one processor succeds nondeterministically to write into the location. In the Concurrent Read Exclusive Write (CREW) PRAM, all processors are allowed to read from the same location at the same time. The EREW PRAM allows only one access per memory during one step.

The paper is organised as follows: in the next section we give parallel algorithms for 2D (two-dimensional) image template matchings. This make clear the passage from two-dimensional case to multidimensional case. In the second section, we present some parallel algorithms for multidimensional image template matching. The third section deals with the multidimensional pattern array matching. In the last section, we applly array matching techniques to multidimensional array-like pattern matching.

1. Two-Dimensional Image Template Matching

Let $I = (I(i,j))$ be an $N \times N$ binary image, where $i, j \in [1, N]$. Let $W = (W(s,t))_{1 \leq s,t \leq M}$ represent the window of size $M \times M$. Then the two dimensional image template matching problem consists of computing the matrix $T = (T(i,j))_{1 \leq i,j \leq N}$ defined by the following equation:

$$T(i,j) = \sum_{i=1}^{M} \sum_{j=1}^{M} I((i+s-1) \bmod N, (j+t-1) \bmod N) \times W(s,t)$$

Example 1: (N= 7 and M= 3)

$$
I =
\begin{array}{|c|c|c|c|c|c|c|c|}
\hline
1 & 1 & 1 & 0 & 1 & 1 & 1 & 1 \\
\hline
0 & 1 & 1 & 1 & 1 & 0 & 0 \\
\hline
1 & 1 & 1 & 1 & 1 & 1 & 1 \\
\hline
0 & 1 & 0 & 0 & 1 & 0 & 0 \\
\hline
1 & 1 & 1 & 1 & 1 & 1 & 0 \\
\hline
0 & 1 & 0 & 0 & 1 & 0 & 0 \\
\hline
1 & 1 & 1 & 1 & 1 & 1 & 0 \\
\hline
\end{array}
\qquad
T =
\begin{array}{|c|c|c|c|c|c|c|}
\hline
7 & 6 & 6 & 6 & 6 & 6 & 6 \\
\hline
4 & 5 & 5 & 4 & 3 & 1 & 3 \\
\hline
7 & 6 & 6 & 7 & 5 & 5 & 5 \\
\hline
3 & 3 & 3 & 3 & 3 & 0 & 3 \\
\hline
7 & 6 & 6 & 7 & 4 & 4 & 4 \\
\hline
5 & 4 & 4 & 4 & 5 & 3 & 5 \\
\hline
6 & 7 & 6 & 6 & 4 & 3 & 4 \\
\hline
\end{array}
$$

$$
W =
\begin{array}{|c|c|c|}
\hline
1 & 1 & 1 \\
\hline
0 & 1 & 0 \\
\hline
1 & 1 & 1 \\
\hline
\end{array}
$$

Algorithm 1. (Serial algorithm)
(1) **for** $i:= 1$ **to** M **do**
(2) **for** $j:= 1$ **to** M **do**
begin
(3) $T[i,j]:= 0$
(4) **for** $s:= 1$ **to** N **do**
(5) **for** $t:= 1$ **to** N **do**
(7) $T[i,j] := T[i,j] + I[(i + s - 1)modN, (j + t - 1)modN] * W[s,t]$
end

Proposition 1.
The 2D-image template matching problem can be solved in $O(M^2 \times N^2)$ time and $O(M^2)$ space on a RAM (Random Access Machine).

We will use the parallel random machine for designing a parallel algorithm for two-dimensional image template matching problem.

Algorithm 2. (Parallel-Sum)
(1) **for** $i:= log(n) - 1$ **downto** 0 **do**
(2)   **forall** $j:= 2^i$ **to** 2^{i+1} **pardo**
(3) $S[j]:= S[2j] + S[2j + 1]$

This Parallel-Sum algorithm runs in $O(log(n))$ time with $n/2 = O(n)$ processors on EREW-PRAM model.

Using the last parallel algorithm for computing the sum of n values in $O(log(n))$ time on EREW-PRAM with $O(n)$ processors, we can prove the following result.

Theorem 1.
The image template matching problem can be solved in $O(log(M))$ time on CREW-PRAM with $O((M^2 \times N^2))$ processors.

An alternative implementation that uses only $O(n/log(n))$ processors and also runs in $O(log(n))$ time is possible according to the Brent Theorem.

Theorm 2.(Brent 1974)

Any synchronous parallel algorithm of time t that consists of a total of x elementary operations can be implemented by p processors within a time of $(\lceil n/p \rceil + t)$.

Lemma 1.

The sum of n values can be computed in $O(log(n))$ time on EREW-PRAM with $O(n/log(n))$ processors.

This gives an optimal algorithm for parallel sum of n value. Using lemma 1, one can prove the following theorem.

Theorem 3.

The image template matching problem can be solved in $O(log(M))$ time on CREW-PRAM with $O((M^2 \times N^2/log(M)))$ processors.

2. Multidimensional Image Template Matching

In this section, we will generalize the image template matching to the multidimensional case. For this, we first define what we will call the multidimensional image template matching. Let $I = (I(i_1, \ldots, i_d))$ be an N^d binary image, where $i_1, \ldots, i_d \in [1, N]$. Let $W = (W(s_1, \ldots, i_d))_{1 \le i_1, \ldots, i_d \le M}$ represents the window of size M^d. Then the d-dimensional image template matching problem consists of computing the matrix $T = (T(i_1, \ldots, i_d))_{1 \le i_1, \ldots, i_d \le N}$ defined by the following equation:

$$T(i_1, \ldots, i_d) = \sum_{s_1=1}^{M} \cdots \sum_{s_d=1}^{M} I((i_1 + s_1 - 1) \bmod N, \ldots, (i_d + s_d - 1) \bmod N) \times W(s_1, \ldots, s_d)$$

Proposition 2.

The multidimensional image template matching problem can be solveded in $O(M^d \times N^d)$ time and $O(M^d)$ space on a RAM (Random Access Machine).

We will use the parallel random machine for designing a parallel algorithm for multidimensional image template matching problem.

Theorem 4.

The multidimensional image template matching problem can be solved in $O(log(M))$ time on CREW-PRAM with $O((M^d \times N^d))$ processors.

Theorem 5.

The multidimensional image template matching problem can be computed in $O(log(M))$ time on CREW-PRAM with $O((M^d \times N^d/log(M)))$ processors.

3. Multidimensional Array (Image) Matching

In this section, we will use some techniques from the image template matching to solve the high dimensional string (array) matching. For this, we first define what we will call the multidimensional array matching problem. Let $I = (I(i_1, \ldots, i_d))$ be an N^d binary image, where $i_1, \ldots, i_d \in [1, N]$. Let $W = (W(s_1, \ldots, i_d))_{1 \le i_1, \ldots, i_d \le M}$ represents the window of size M^d. Then the multidimensional image template matching problem consists of computing the matrix $T = (T(i_1, \ldots, i_d))_{1 \le i_1, \ldots, i_d \le N}$ defined by the following equation:

$$T(i_1,\ldots,i_d) = \sum_{s_1=1}^{M} \cdots \sum_{s_d=1}^{M} [I((i_1+s_1-1) \bmod N,\ldots,(i_d+s_d-1) \bmod N) = W(s_1,\ldots,s_d)]$$

where $[x=y]=1$ if $x=y$ and $[x=y]=0$ otherwise.

Example 2: (N= 7 and M= 3)

$$I = \begin{array}{|c|c|c|c|c|c|c|} \hline 1 & 1 & 1 & 0 & 1 & 1 & 1 \\ \hline 0 & 1 & 1 & 1 & 1 & 0 & 0 \\ \hline 1 & 1 & 1 & 1 & 1 & 1 & 1 \\ \hline 0 & 1 & 0 & 0 & 1 & 0 & 0 \\ \hline 1 & 1 & 1 & 1 & 1 & 1 & 0 \\ \hline 0 & 1 & 0 & 0 & 1 & 0 & 0 \\ \hline 1 & 1 & 1 & 1 & 1 & 1 & 0 \\ \hline \end{array} \qquad T = \begin{array}{|c|c|c|c|c|c|c|} \hline 8 & 6 & 6 & 6 & 7 & 6 & 6 \\ \hline 4 & 5 & 5 & 4 & 3 & 1 & 3 \\ \hline 9 & 7 & 7 & 9 & 7 & 7 & 6 \\ \hline 3 & 3 & 3 & 3 & 4 & 0 & 4 \\ \hline 9 & 7 & 7 & 9 & 5 & 6 & 5 \\ \hline 5 & 4 & 4 & 4 & 6 & 3 & 6 \\ \hline 6 & 8 & 6 & 7 & 4 & 3 & 4 \\ \hline \end{array}$$

$$W = \begin{array}{|c|c|c|} \hline 1 & 1 & 1 \\ \hline 0 & 1 & 0 \\ \hline 1 & 1 & 1 \\ \hline \end{array}$$

$Match(I,W) = \{(3,1),(3,4),(5,1),(5,4)\}$, where $Match(I,W)$ is the set of positions where W match I.

Note that W match I at the position (i,j) if and only if $T(i,j) = M^d$

Proposition 3.

The multidimensional array matching problem can be solved in $O(M^d \times N^d)$ time and $O(M^d)$ space on a RAM (Random Access Machine).

By using the parallel random machine for designing a parallel algorithm for multidimensional array matching problem, we obtain the following result.

Theorem 6.

The multidimensional array matching problem can be solved in $O(log(M))$ time on CREW-PRAM using $O((M^d \times N^d))$ processors.

Theorem 7.

The multidimensional array matching problem can be solved in $O(log(M))$ time on CREW-PRAM with $O((M^2 \times N^2/log(M)))$ processors.

4. Matching with an Array Represented Pattern

In this section we will use some techniques from the image template matching to solve the high dimensional pattern matching. For this, we first define what we will call the multidimensional pattern matching. Let $I = (I(i_1,\ldots,i_d))$ be an N^d binary image, where $i_1,\ldots,i_d \in [1,N]$. Let P be a d-dimensional pattern. Intuitively, P can be vieuwed as a collection of hypercube of size one(1^d). Let $W = (W(s_1,\ldots,i_d))_{1 \le i_1,\ldots,i_d \le M}$ represents the Then the d-dimensional pattern matching problem consists of computing the set of positions in I where the pattern P occurs. For this, we will reduce the pattern matching problem to the array matching problem. Let $W = (W(s_1, \ldots, i_d))_{1 \le i_1, \ldots, i_d \le M}$ represents the smallest "box" that can contain the pattern P. Then the pattern matching problem can be reduced to the array matching problem. Let us give an exxample to make clear how the pattern matching problem can be reduced to the array matching problem.

Example 3: ($N= 7$ and $M= 3$)

$$
I = \begin{array}{|c|c|c|c|c|c|c|}
\hline
1 & 1 & 1 & 0 & 1 & 1 & 1 \\
\hline
0 & 1 & 1 & 1 & 1 & 0 & 0 \\
\hline
1 & 1 & 1 & 1 & 1 & 1 & 1 \\
\hline
0 & 1 & 0 & 0 & 1 & 0 & 0 \\
\hline
1 & 1 & 1 & 1 & 1 & 1 & 0 \\
\hline
0 & 1 & 0 & 0 & 1 & 0 & 0 \\
\hline
1 & 1 & 1 & 1 & 1 & 1 & 0 \\
\hline
\end{array}
\qquad
T = \begin{array}{|c|c|c|c|c|c|c|}
\hline
4 & 6 & 4 & 5 & 3 & 2 & 3 \\
\hline
8 & 5 & 5 & 8 & 5 & 7 & 5 \\
\hline
3 & 3 & 3 & 3 & 4 & 1 & 4 \\
\hline
9 & 5 & 5 & 9 & 4 & 6 & 4 \\
\hline
3 & 5 & 5 & 3 & 5 & 0 & 4 \\
\hline
7 & 5 & 4 & 8 & 4 & 5 & 4 \\
\hline
6 & 4 & 4 & 5 & 6 & 5 & 6 \\
\hline
\end{array}
$$

$$
P = \quad
\qquad
W = \begin{array}{|c|c|c|}
\hline
0 & 1 & 0 \\
\hline
1 & 1 & 1 \\
\hline
0 & 1 & 0 \\
\hline
\end{array}
$$

$Match(I, W) = \{(4,1), (4,4)\}$, where $Match(I, W)$ is the set of positions where W match I, $\bullet$ is a fixed marked position of the pattern P (i.e. $Mark\_Pos(P) = (1,2)$).
$Match(I, P) = \{(4,2), (4,5)\}$, where $Match(I, W)$ is the set of positions where P match I.
It's easy to see the relation between $Match(I, W)$ and $Match(I, P)$ that we express by the following.

$$\boxed{Match(I, P) = Match(I, W) + Mark\_Pos(P) - (1,1)}$$

Proposition 4.
The multidimensional pattern matching problem can be computed in $O(M^d \times N^d)$ time and $O(M^d)$ space on a RAM (Random Access Machine).
By using the parallel random machine for designing a parallel algorithm for two-dimensional image template matching problem, we obtain the following result.

Theorem 8.
The multidimensional pattern matching problem can be computed in $O(log(M))$ time on CREW-PRAM with $O((M^d \times N^d))$ processors.

Theorem 9.
The multidimensional pattern matching problem can be computed in $O(log(M))$ time on CREW-PRAM with $O((M^d \times N^d)/log(M))$ processors.

As a variation of a two-dimensional pattern matching problem, one can consider that the pattern match the image at the position (i, j) if there exists a rotation of this pattern such that the rotated pattern match the image at position (i, j). Also, one can extend this variation to the multidimensional case. Observe that we can derive algorithms for Pattern Matching with rotation with the same complexity as algorithms for the pattern matching.

Let us mention that in [AL 92], the authors propose a fast parallel algorithm which runs in $O(dlog(N))$ time using on CRCW-PRAM $O(N^d)$ processors . By assuming that the size of the image I is constant, one can notice that our algorithm is faster than the one proposed in [AL 92].

Conclusion

We presented efficient and optimal parallel algorithms for multidimensional image template matching, multidimensional array matching and multidimensional pattern matching on CREW PRAM model. For an N^d image and M^d window, we proved that there is an optimal (resp. efficient) algorithm which runs in $O(log(M))$ time using $O((M^d \times N^d)/log(M))$ processors (resp.$O(M^d \times N^d)$). We conclude by listing our results in the following table.

Time complexity	Processor complexity	Model
$O(log(M))$	$O((M^d \times N^d))$	CREW
$O(log(M))$	$O((M^d \times N^d)/log(M))$	CREW

Table 1.

Acknowledgement. We thank the anonymous referee for pointing us the work done by Amir and Landau [AL 92].

References

[Ak 89] S. Akl, The Design of Parallel Algorithms, Prentice-Hall (1989).

[AL 87] A. Amir and G. M. Landau, Fast Parallel and Serial Multidimensional Approximate Array Matching, Theoretical Computer Science 81(1992)97-114.

[KK 87] P. Kumar and V. K. Krishnan, Efficient Image Template Matching on SIMD Hypercube Machines, Proceedings of the International Conference on Parallel Processing(1987)765-771.

[Ro 82] A. Margalit and A. Rosenfeld, Reducing the expected Computational cost of Template Matching using Run Length Representation, Pattern Recognition Letters(1990)255-265.

[NS 91] M. Nivat and A. Saoudi, On Parallel Recognition Algorithms for High Dimensional Images, The 1st Int. Conf. on Parallel Image Processing, Paris (1991).

[RS 88] S. Ranka and S. Sahni, Image Template Matching on SIMD Hypercube Multicomputer, Proceedings of the International Conference on Parallel Processing(1988)84-91.

[Ro 82] A. Rosenfeld and A. C. Kak, Digital Picture Processing , vol. 1 and 2, Academic Press 1982.

[RK 79] A. Rosenfeld, Picture Languages: Formal Models for Picture Recognition, Academic Press 1979.

[SR 92] A. Saoudi and W. Rytter, On Parallel Algorithms for 2D-Image Recognition, Int. Conf. on Pattern Recognition (1992).

Learning of Recognizable Picture Languages

Rani Siromoney, Lisa Mathew, K.G. Subramanian and V. Rajkumar Dare

Department of Mathematics, Madras Christian College,
Tambaram, Madras 600 059, India

Abstract. Learning of certain classes of two-dimensional picture languages is considered. Linear time algorithms that learn in the limit, from positive data the classes of local picture languages and locally testable picture languages are presented. A crucial step for obtaining the learning algorithm for local picture languages is an explicit construction of a two-dimensional on-line tessellation acceptor for a given local picture language. An efficient algorithm that learns the class of recognizable picture languages from positve data and restricted subset queries, is presented in contrast to the fact that this class is not learnable in the limit from positive data alone.

1 Introduction

Extension to two dimensions of the Chomskian hierarchy of grammars and the corresponding automata characterizations were of great interest during the seventies [4]. Identification of deterministic finite automata recognizing two-dimensional tapes to correspond exactly to the two-dimensional generalization of grammars seemed to be difficult. Recently, there has been renewed interest in such problems related to two-dimensional picture languages [2,3]. A new notion of recognizability for picture languages is introduced in [2] extending the definition of one- dimensional recognizable 'languages in terms of local languages and alphabetic mappings. Some of the open problems cited in [2] are solved in [3] and a characterization of recognizable picture languages given in terms of two-dimensional on-line tessellation automata.

Machine learning has been of great interest and a lot of study has centered around the inductive inference of deterministic finite automata recognizing one-dimensional tapes [6]. For string languages, Yokomori [7] has given learning algorithms for strictly locally testable languages, locally testable languages and regular languages. So far, not much work [5] has been done on the learning of two-dimensional grammars or automata acting on two-dimensional tapes in spite of the fact that these are of great significance in syntactic pattern recognition and also for problems in tiling. In this paper, we examine learning of two-dimensional picture languages. We present linear time algorithms that identify in the limit, from positve data, the classes of local picture languages and locally testable picture languages. A crucial step in obtaining learning algorithms for local picture languages is our explicit construction of a 2-ota [3] from a given local picture language. The existence of such a 2-ota corresponding to a given local picture language is known [1,2]. But construction of such a 2-ota is of significance as it provides a direct method of obtaining the 2-ota from a given local picture language. Thus the learning of local picture languages is aided by the construction of 2-ota coresponding to a local picture language, besides the techniques presented for string languages in [7]. On the other hand, the learning of locally testable picture languages makes use of the extension to picture languages of the condition proposed by Angluin in [1].

It is clear that the class REC of recognizable picture languages cannot be learnt in the limit from positive data alone, since the class of regular languages can be considered as a subclass of REC, by considering a string of length n as a picture of size $(n,1)$. However, we present an algorithm for learning recognizable picture languages from positive data and restricted subset queries.

2 Preliminaries

Let A be a given alphabet. A picture over A is an $m \times n$ rectangular array of elements of A. A^{**} denotes the set of all pictures over A. $A^{m \times n}$ denotes the set of all pictures of size (m,n) with m rows and n columns.

We denote by $B_{k,r}(p)$, the set of all subpictures of size (k,r). For any picture $p \in A^{**}$, we denote by $b(p)$, the picture of size $(m+2, n+2)$ obtained by surrounding p by a special boundary symbol $\# \notin A$

Let p be a picure of size (m,n), $m \geq 1$, $n \geq 1$. If $1 \leq i \leq m$, $1 \leq j \leq n$, $p(i,j)$ denotes the symbol in p with coordinates (i,j). If $1 \leq i \leq i' \leq m$, $1 \leq j \leq j' \leq n$,, $p[(i,j), (i',j')]$ denotes the subpicture z of p such that (i) the size of z is $(i'-i+1, j'-j+1)$ and (ii) for each k,r $(1 \leq k \leq i'-i+1, 1 \leq r \leq j'-j+1)$, $z(k,r) = p(k+i-1, r+j-1)$.

We recall the definitions of local picture languages, locally testable picture languages and recognizable picture languages [2]

Local Picture Languages. A picture language $L \subseteq A^{**}$ is local, if there exists a subset Q of $(A \cup \#)^{2 \times 2}$ such that

$$L = \{p \in A^{**} \, / \, B_{2,2}\,(b(p)) \subset Q\}$$

We write $L = R(Q)$. The family of local picture languages is denoted by LOC.

We give an example of a local picture language.

Example. Let $A = \{a,b,c\}$ and let

$$Q = \left\{ \begin{matrix} \# \; \# & \# \; \# & \# \; b & a \; \# & \# \; \# & \# \; \# & \# \; \# & \# \; a & \# \; c \\ \# \; a & b \; \# & \# \; \# & \# \; \# & a \; c & c \; c & c \; b & \# \; c & \# \; c \end{matrix} \right.$$

$$\begin{matrix} c \; \# & \# \; c & b \; c & c \; c & c \; a & b \; \# & c \; \# & a \; c & c \; a & c \; c \\ a \; \# & \# \; b & \# \; \# & \# \; \# & \# \; \# & c \; \# & c \; \# & c \; a & c \; c & a \; c \end{matrix}$$

$$\left. \begin{matrix} c \; b & c \; c & b \; c & c \; c & a \; b & c \; a & c \; c & b \; c & b \; a \\ b \; c & c \; b & c \; c & c \; c & b \; a & c \; b & a \; b & a \; c & c \; c \end{matrix} \right\}$$

It can be seen that

$R(Q)$ = {pictures of size $2n \times 2n$ ($n \geq 1$) such that the left diagonal is composed of a's, the other diagonal is composed of b's and the remaining cells are filled by c's}.

Locally Testable Picture Languages. An equivalence relation, denoted by $=_{k,r}$ ($k,r \geq 1$) over A^{**}, is defined as follows: Two pictures p_1 and p_2 are equivalent i.e. $p_1 =_{k,r} p_2$, if $B_{k,r}(p_1) = B_{k,r}(p_2)$.

A picture language $L \subset A^{**}$ is (k,r) testable if it is the union of some classes of $=_{k,r}$. $L \subseteq A^{**}$ is locally testable if it is (k,r) testable for some k and r.

Recognizable Picture Languages. Let A and B be two finite alphabets and $h:A \rightarrow B$ be an alphabetic mapping. If L is a picture language over A, $h(L)$ denotes the picture language over B, obtained by replacing every letter $a \in A$ by the corresponding letter $h(a)$ in every picture p of L.

A picture language L over A is recognizable if there exists a finite alphabet B, a local picture language $R(Q)$ over B and an alphabetic mapping $h:B \rightarrow A$ such that $L = h(R(Q))$. The triple (B,Q,h) is called a representation of L. The family of recognizable picure languages is denoted by REC.

3 Learning of Local Picture Languages

Learning of local picture languages is considered in this section. The main idea is to construct a characteristic sample for a local picture language, as done in the string case [7]. In order to do this, first we give a method of constructing a 2-ota for a given local picture language. Informally, the two- dimensional on-line tesselation acceptor (2-ota) [3] M is an infinite array of identical finite state machines in two dimensional space. The input of a picture p to M means that each symbol p(i,j) is placed on the finite state machine situated at coordinates (i,j) This machine is called the (i,j) - cell and the boundary symbol # is placed on any other cell in M.

Definition 3.1 [3] A two-dimensional on-line tessellation acceptor (2-ota) is defined as $M = (K, A \cup \{\#\}, \delta, q_e, q_o, F)$ where K is a finite set of states; A is a finite set of input symbols; $\# \notin A$ is the boundary symbol; $\delta : K^3 \times (A \cup \{\#\}) \to 2^K \cup \{\{q_o\}\}$ is the cell-state transition function; $q_e \in K$ is the motive state; $q_o \in K$ is the quiescent state and $F \subset K$ is a set of accepting states where K corresponds to the state set of any one of the finite state machines in the array.

The state of each (i,j) cell at time $t+1$ depends on the states at time t of cells in its neighbourhood and on the symbol read by it, where the neighbourhood of (i,j) - cell is the set consisting of the (i,j), (i–1,j) and (i,j–1) cells i.e. if $q(i,j)(t)$ is the state of the (i,j)-cell at time t, then

$$q(i,j)(t+1) \in \delta (q(i,j)(t), q(i-1,j)(t), q(i,j-1)(t), a)$$

where a is the symbol in the (i,j)-cell.

The motive state q_e gives M the motivation to begin to read an input picture. Each element in $K' = K - \{q_e, q_o\}$ is called a stable state. δ must satisfy the foll owing conditions:

For any $a \in A \cup \{\#\}$ and for any $p_i \in K$, $i = 1,2,3$

1) $\delta (p_1, p_2, p_3, a) = \{q_o\}$, if and only if $(a = \#)$ or ($p_1 = q_o$ and for each i $(i = 2,3)$, $p_i \in \{q_e, q_o\}$)

2) if $p_1 \in K - \{q_e, q_o\}$ and $a \neq \#$, then $\delta (p_1, p_2, p_3, a) = \{p_1\}$.

Condition (2) implies that each cell stays in the stable state regardless of the state of each cell in its neighbourhood and of the symbol (except the symbol #) in it, once it enters a stable state.

A state configuration c of M is a mapping from I^2 into K, where I denotes the set of all integers. For each $(i,j) \in I^2$, $c(i,j)$ represents the state of the (i,j) cell. The state configuration c of M such that $c(1,1) = q_e$ and $c(v) = q_o$, for each $v \neq (1,1)$ in I^2, is called the primitive state configuration of M.

M accepts a picture p of size (m,n) if and only if when M begins to read the picture p in its primitive state configuration, the (m,n) cell can enter a stable state in F.

It is clear that if a picture p of size (m,n) is presented to M, then at time t $(t \geq 1)$, all cells that read symbols at a distance t–1 from the symbol p(1,1) (i.e. all (i,j) cells such that $i-1+j-1 = t-1$) have attained their ultimate states and it is decided exactly at time $m+n-1$ whether the picture p is accepted by M.

Let p be a picture of size (m,n) over A. A run of M on p is a picture z of size (m,n) over K − {q_e, q_o} such that

i) $z(1,1) = \delta (q_e, q_o, q_o, p(1,1))$ and

ii) for each $(i,j) \neq (1,1)$, $1 \leq i \leq m$, $1 \leq j \leq n$

$z(i,j) \in \delta (q_o, z(i-1,j), z(i,j-1), p(i,j))$

where $z(0,j) = z(i,0) = q_o$. An accepting run on p is a run z of M on p such that $z(m,n) \in F$.

The picture language accepted by M is defined as $L(M) = \{p \in A^{**} /$ there is an accepting run of M on p}

The family of picture languages accepted by 2-ota's is denoted by L(2-ota)

The next lemma gives the explicit construction of a 2-ota for a given local picture language. This is crucial for obtaining the characteristic sample and the learning algorithm.

Lemma 3.1 LOC ⊂ L(2-ota)

Proof. It is proved in [3] that REC = L(2-ota) and in [2] that LOC ⊂ REC. Hence the result is clear but we give a direct, explicit construction of a 2-ota for a given local picture language.

Let L be a local picture language over an alphabet and Q be a subset of $(A \cup \{\#\})^{2 \times 2}$ such that $L = R(Q)$. Q can be written as the union of disjoint sets Q_{ul}, Q_{dl}, Q_{ur}, Q_{dr}, Q_u, Q_d, Q_l, Q_r, Q_m where the sets in order consist of arrays of the form

$$
\begin{array}{ccccccccc}
\# \# & \# \, a & \# \# & a \, \# & \# \# & a \, b & \# \, a & a \, \# & a \, b \\
\# \, a \, ' & \# \, \# \, ' & a \, \# \, ' & \# \, \# \, ' & a \, b \, ' & \# \, \# \, ' & \# \, b \, ' & b \, \# \, ' & c \, d
\end{array}
$$

For a 2 × 2 array $X = \begin{array}{cc} a & b \\ c & d \end{array}$, write $X(1)=a$, $X(2)=b$, $X(3)=c$, $X(4)=d$.

A 2-ota to accept L, is constructed as follows:

$$M = (K, A \cup \{\#\}, \delta, q_e, q_o, F), \text{ where}$$

K, the finite set of states consists of q_o (the quioscent state), q_e (the motive state) and states of the form xyz, $x,y,z \in A \cup \{\#\}$. (The elements of K are obtained when δ is defined); A is the finite set of input symbols and # is the boundary symbol not in A;

The cell-state transition function $\delta : K^3 \times (A \cup \{\#\}) \rightarrow 2^{K'} \cup \{\{q_0\}\}$, where $K' = K - \{q_e, q_0\}$, is defined in the following steps:

1) $\delta (q_0, r, s, \#) = \{q_0\}$, for any $r, s \in K$

2) $\delta (q, r, s, a) = \{q\}$, for $q \in K - \{q_e, q_0\}$, $r, s \in K$ $a \in A$ and $a \neq \#$

3) For every X in Q_{ul}, $\#X(4)\# \in \delta (q_e, q_0, q_0 X(4))$

4) For every X in Q_l, Y in Q_{ul} or Q_l,
$\#X(4)\ Y(4) \in \delta (q_0,\ Y(3)Y(4)Y(2),\ q_0,\ X(4))$

5) For every X in Q_u, Y in Q_{ul} or Q_u,
$Y(4)X(4)\# \in \delta (q_0,\ q_0,\ Y(3)\ Y(4)\ Y(2),\ X(4))$

6) For every X in Q_m, Y in Q_l or Q_m, Z in Q_u or Q_m with $Y(2) = Z(3)$,
$Y(4)X(4)Z(4) \in \delta (q_0,\ Z(3)Z(4)Z(2),\ Y(3)Y(4)Y(2),\ X(4))$

7) For every X in Q_{dl}, Y in Q_{ul} or Q_l
$\#X(2)\ Y(4) \in \delta (q_0,\ Y(3)\ (4)Y(2),\ q_0,\ X(2))$

8) For every X in Q_{ur}, Y in Q_{ul} or Q_u,
$Y(4)X(3)\ \# \in \delta (q_0,\ q_0\ Y(3)Y(4)Y(2), X(3))$

9) For every X in Q_d, Y in Q_l or Q_m, Z in Q_u or Q_m with $Y(2) = Z(3)$,
$Y(4)X(1)Z(4) \in \delta (q_0,\ Z(3)Z(4)Z(2),\ Y(3)Y(4)Y(2),\ X(1))$

10) For every X in Q_r, Y in Q_l or Q_m, Z in Q_u with $Y(2) = Z(3)$,
$Y(4)X(3)Z(4) \in \delta (q_0,\ Z(3)Z(4)Z(2),\ Y(3)Y(4)Y(2),\ X(3))$

11) For every X in Q_{dr}, Y in Q_l or Q_m, Z in Q_m with $Y(2) = Z(3)$,
$Y(4)X(1)Z(4) \in \delta (q_0,\ Z(3)Z(4)Z(2),\ Y(3)Y(4)Y(2),\ X(1))$

Finally, F consists of all the states $Y(4)X(1)Z(4)$ obtained in step (11)

It can be seen that $L(M) = L = R(Q)$.

We illustrate with an example.

Example : Let $A = \{a, b\}$.

$L = \{$pictures with $n \geq 2$ rows, whose uppermost row is composed of a's and other rows of b's$\}$

L is a local picture language [2] with

$$Q_{ul} = \left\{ \begin{matrix} \# & \# \\ \# & a \end{matrix} \right\}, \qquad Q_{dl} = \left\{ \begin{matrix} \# & b \\ \# & \# \end{matrix} \right\}, \qquad Q_{ur} = \left\{ \begin{matrix} \# & \# \\ a & \# \end{matrix} \right\},$$

$$Q_{dr} = \left\{ \begin{matrix} b & \# \\ \# & \# \end{matrix} \right\}, \qquad Q_u = \left\{ \begin{matrix} \# & \# \\ a & a \end{matrix} \right\}, \qquad Q_d = \left\{ \begin{matrix} b & b \\ \# & \# \end{matrix} \right\},$$

$$Q_l = \left\{ \begin{matrix} \# & a \\ \# & b \end{matrix} , \begin{matrix} \# & b \\ \# & b \end{matrix} \right\}, \qquad Q_r = \left\{ \begin{matrix} a & \# \\ b & \# \end{matrix} , \begin{matrix} b & \# \\ b & \# \end{matrix} \right\}, \qquad Q_m = \left\{ \begin{matrix} a & a \\ b & b \end{matrix} , \begin{matrix} b & b \\ b & b \end{matrix} \right\}$$

The 2-ota $M = (K, A \cup \{\#\}, \delta, q_e, q_o, F)$ with δ defined below, accepts L.

1) $\delta\,(q_o, r, s, \#) = \{q_o\}$, for any $r, s \in K$

2) $\delta\,(q, r, s, a) = \{q\}$, for $q \in K - \{q_e, q_o\}$, $r, s \in K$, $a \in A$ $(a \ne \#)$

3) $\delta\,(q_e, q_o, q_o, a) = \{\#a\#\}$

4) $\delta\,(q_o, \#a\#, q_o, b) = \{\#ba\}$

   $\delta\,(q_o, \#ba, q_o, b) = \{\#bb\}$

   $\delta\,(q_o, \#bb, q_o, b) = \{\#bb\}$

5) $\delta\,(q_o, q_o, \#a\#, a) = \{aa\#\}$

   $\delta\,(q_o, q_o, aa\#, a) = \{aa\#\}$

6) $\delta\,(q_o, aa\#, \#ba, b) = \{bba\}$

   $\delta\,(q_o, aa\#, bba, b) = \{bba\}$

   $\delta\,(q_o, bba, bbb, b) = \{bbb\}$

   $\delta\,(q_o, bbb, bbb, b) = \{bbb\}$

9) $\delta\,(q_o, bbb, \#bb, b) = \{bbb\}$

11) $\delta\,(q_o, bba, \#bb, b) = \{bbb\}$

Steps 7,8 and 10 do not yield anything new.

Finally, $F = \{bbb\}$

Characteristic Sample. We now introduce the notion of a characteristic sample for a local picture language and show the existence of a characteristic sample for every local picture language, using the 2-ota constructed in lemma 3.1. This is a crucial step in the learning of local picture languages.

Definition 3.2 Let L be a local picture language over A and $L = R(Q)$ for some finite $Q \subseteq (A \cup \{\#\})^{2\times2}$. Q is said to be minimal if $L = R(Q')$, for some finite $Q' \subseteq (A \cup \{\#\})^{2\times2}$, implies $Q \subset Q'$.

Lemma 3.2 Let L be a local picture language. Then there is a minimal Q for L such that $L = R(Q)$.

Proof. Let Q_i $(i=1,...,n)$ be subsets of $(A\cup\{\#\})^{2\times2}$, incomparable with respect to the inclusion relation and $L = R(Q_i)$, $i=1,...,n$. Let $Q = Q_1 \cap ... \cap Q_n$. Then $B_{2,2}(b(p)) \subset Q_i$, for $i = 1,...,n$. so that $B_{2,2}(b(p)) \subset Q$. Thus $L = R(Q)$ and $Q \subset Q_i$, for $i=1,...,n$.

Remark. We assume hereafter that Q is minimal for any local picture language $L = R(Q)$.

Definition 3.3 Let T be a finite subset of A^{**}. Let $Q_T = \{B_{2,2}(b(p))/p \in T\}$. The set $L = R(Q_T)$ is called the local picture language associated with T.

The following lemma is easy to prove.

Lemma 3.3 Let T, T' be finite subsets of A^{**}. Then

(i) $T \subset R(Q_T)$

(ii) If $T \subset T'$, then $R(Q_T) \subset R(Q_{T'})$ and

(iii) If L is an arbitrary local picture language such that $T \subset L$ then $R(Q_T) \subset L$.

Definition 3.4 Let L be a local picture language. A finite subset U is called a characteristic sample for L if and only if L is the smallest local picture language containing U.

The following lemma is clear as in the case of strings [7].

Lemma 3.4 If U is the characteristic sample for a local picture language L, then (i) $L = R(Q_U)$; (ii) If $U \subset T \subset L$ for a finite T, then $L = R(Q_T)$.

Theorem 3.1 There exists a characteristic sample for any local picture language.

Proof. Let A be the alphabet of the local picture language L with $|A| = m$. Applying Lemma 3.1 we construct a 2-ota with state set K, where $|K| \leq m(m+1)^2 + 2$.

Consider the set

$$U = \{p \in A^{**} / 2 < n_1, n_2 < |K|; \text{ size } (p) = (n_1, n_2), p \in L\}$$

Then, as in [7], it is easy to prove that U is a characteristic sample for L.

Learning Algorithm for Local Picture Languages. We now present an algorithm that learns an unknown local picture language, in the limit from positive data.

Algorithm LOC

Input : A sequence of positive presentations of L.
Output : An increasing sequence Q_{E_i} such that $R(Q_{E_i})$ are local picture
 languages.

Procedure:

> Initialize E_0 to Φ
> Construct the initial $Q_{E_0} = \Phi$
> repeat (for ever)
> > let Q_{E_i} be the current conjecture
> > read the next positive example
> > scan p to obtain $B_{2,2}(b(p))$
> > $Q_{E_{i+1}} = Q_{E_i} \cup B_{2,2}(b(p))$
> > $E_{i+1} = E_i \cup \{p\}$
> > Output $Q_{E_{i+1}}$ as new conjecture

Lemma 3.5 Let $Q_{E_0}, Q_{E_1} \ldots Q_{E_i}, \ldots$ be the sequence of conjectures by the algorithm LOC. Then (i) for all $i \geq 0$, $R(Q_{E_i}) \subset R(Q_{E_{i+1}}) \subset L$ and (ii) there exists $r \geq 0$ such that for all $i \geq 0$, $R(Q_{E_r}) = R(Q_{E_{r+i}}) = L$

Proof. Since $E_i \subset E_{i+1}$, $R(Q_{E_i}) \subset R(Q_{E_{i+1}})$ (Lemma 3.3)

Consider a sufficiently large $n_0 > 0$ such that $U_L \subset E_{n_0}$ where U_L is a characteristic sample for L. Since U_L is finite, this is always possible. Then, from Lemma 3.4, $L = R(U_L) = R(E_{n_0})$, nothing that $E_{n_0} \subset L$. Also, since $E_i \subset L$, for all $i > n_0$, we again obtain by Lemma 3.4, that $R(Q_{E_i}) = L$

Summarizing all the lemmas, we obtain the following theorem.

Theorem 3.2 Given an unknown local picture language L, the algorithm LOC learns in the limit a set Q_E such that $R(Q_E) = L$.

Time Analysis. The time complexity of the algorithm LOC depends on the size of the positive data provided. The size of an example p, where p is an $m \times n$ picture is $A(p) = mn$. Hence the running time of the algorithm is a function of N, the sum of the sizes of all the positive data provided. Each time a new example is provided, $B_{2,2}(b(p))$ and hence the new conjecture Q_E is computed in time $O(A(p))$. Hence total time required is $\sum\limits_{p \in E_r} O(A(p)) = O(N)$ where E_r is the set of positive data with which the algorithm converges to a correct conjecture.

4 Locally Testable Picture Languages

We now present a learning algorithm for locally testable picture languages.

A test set $T_{k,r}$ over A is any finite subset of $\{B_{k,r} (b(p))/p \in A^{**}\}$. Clearly, for each locally testable picture language L, we can find a test-set $T_{k,r}$ such that L is defined by $T_{k,r}$ and corresponding to each $T_{k,r}$, there exists a unique locally testable language defined by it.

We note that the cardinality of $\{B_{k,r} (b(p)) / p \in A^{**}\}$ is finite.

The following condition for learnability of picture languages follows from that given by Angluin in [1] for string languages.

Condition A. Let $F = \{L_1, L_2 ...\}$ be an indexed class of nonempty recursive picture languages. Then F satisfies Condition A, if for each nonempty finite $S \subseteq \Sigma^*$, the set

$$C(S) = \{L/S \subset L, L = L_i \text{ for some i}\} \text{ is of finite cardinality.}$$

Theorem 4.1 [1] If F satisfies Condition A, then it is learnable in the limit, from positive data.

Theorem 4.2 The class of (k,r) testable picture languages is learnable in the limit from positive data.

For a fixed (k,r), a (k,r) testable picture language L is obtained by giving a test set $T_{k,r}$ such that L is defined by $T_{k,r}$. Since the number of distinct $T_{k,r}$ is finite, when k and r are fixed there exist only a finite number of (k,r) testable picture languages. Hence this class of languages satisfies Condition A.

Theorem 4.3 There is an algorithm which, given any unknown (k,r) testable picture language U, learns in the limit a test set T_E, such that $L(T_E) = U$. Further this algorithm runs in time $O(N)$ where N is the sum of the areas $A(p)$ of the given pictures.

Algorithm LT

Input : A sequence of positive presentations of L
Output : A sequence of test sets for (k,r) testable languages.

Procedure:

    Initialize $E_0 = \Phi$
    Construct the initial test set $T_\Phi = \Phi$
    repeat (forever)
        Let T_{E_i} be the current test set
        read the next positive example p
        scan p to compute $B_{k,r}(b(p))$
        $E_{i+1} = E_i \cup \{p\}$
        $T_{E_{i+1}} = T_{E_i} \cup \{B_{k,r}(b(p))\}$
        Output $T_{E_{i+1}}$ as a conjecture

5 Recognizable Picture Languages

We now present a learning algorithm for Recognizable Picture Languages.

Lemma 5.1 [3] Given any language $L \in REC$, there exists a 2-ota M such that $L(M) = L$.

Let $M = (K, A \cup \{\#\}, \delta, q_e, q_0, F)$ be a 2-ota such that $L(M) = L \in REC$. Let $\Gamma = A \times (K - \{q_e, q_0\})$ and h_1, h_2 alphabetic mappings on Γ given by $h_1(a,q) = a$, $h_2(a,q) = q$. A picture X over Γ is called a computation description picture if $h_2(X)$ is a run of M on $h_1(X)$ and is called an accepting computation description picture if $h_2(X)$ is an accepting run.

The following lemma can be proved as in the case of strings [6].

Lemma 5.2

i) The alphabet Γ contains atmost $O(m(n-2))$ elements where n = number of states of the minimal 2-ota M_L for L and m = $|A|$.

ii) For $p \in L$, let $d(p)$ be a picture over Γ representing an accepting computation description for p. $X(p) = h_2(d(p))$ is called a valid picture for p. Let $Val(p) = \{X(p) / X(p)$ is a valid picture for p$\}$.

    Then $|Val(p)| \subset n^{A(p)}$.

iii) Let S be a local picture language over A such that $L = h(S)$ and R_S be a characteristic sample for L. Then there is a finite subset S_L of L such that $R_S \subseteq Val(S_L)$.

    From Lemma 5.2, we obtain a learning algorithm for REC.

Algorithm REC

Input : A positive presentation of L, $n = |K|$ for the minimal 2-ota for L.
Output : A sequence of conjectures of the form $h(R(Q))$
Query : Restricted subset query

Procedure:

> Initialize E_0 to Φ
> Construct initial $Q_0 = \Phi$
> repeat (forever)
> > let Q_i be the current conjecture
> > read next positive example p
> > compute $Val(p) = \{\alpha_1, \alpha_2,..., \alpha_t\}$
> > for each j scan α_j to compute $Q_{i,j} = B_{2,2} (b(\alpha_j))$
> > ask if $h(R(Q_{i,j})) \subset L$ or not
> > $Val(p) = Val(p) - \{\alpha_j/$ the answer is no$\}$
> > $E_{i+1} = E_i \cup Val(p)$
> > $Q_{i+1} = Q_i \cup \{B_{2,2} (b(\alpha)) / \alpha \in Val(p)\}$
> > Output Q_{i+1} as new conjecture

Lemma 5.3 Let n be the number of states of the minimal 2-ota accepting the recognizable picture language L. After atmost $t(n)$ subset queries, the algorithm REC produces a conjecture Q_{E_i} such that E_i includes a characteristic sample for a local picture language U such that $L = h(U)$ where $t(n)$ is a polynomial in n, which depends on U.

This is a consequence of Lemma 5.2 and the fact that the maximum size of pictures in L is bounded by a polynomial in n by the proof of Theorem 3.1. Summarizing, we obtain the following theorem.

Theorem 5.1 Given an unknown recognizable picture language L, the algorithm REC efficiently learns in the limit from positive data and restricted subset queries, a subset Q of $(A \cup \{\#\})^{2\times2}$ such that $L = h(R(Q))$.

References

1.   D. Angluin, Inductive inference of formal languages from positive data, Information and control 45:117-135, 1980.

2.   D. Giammarressi and A. Restivo, Recognizable picure languages, Proc. of the International Colloquium on Parallel Image Processing (Eds. M. Nivat, A. Saoudi and P.S.P. Wang), Paris (1991), 3-16.

3.   K. Inoue and I. Takanami, A characterization of recogniable picture languages, Tech. Report, Yamaguchi University, Ube, Japan, 1991.

4. R. Siromoney, Array Languages and Lindenmayer systems - a survey in the Book of *L*, eds. G. Rozenberg and A. Salomaa, Springer - Verlag, Berlin, 1985.

5. R. Siromoney, K.G. Subramanian and Lisa Mathew, Learning of Pattern and picture languages, Proc. of the International Colloquium on Parallel Image Processing (Eds. M. Nivat, A. Saoudi and P.S.P. Wang), Paris (1991).

6. L. Pitt, Inductive inference, DFAs and computational complexity, in Analogical and Inductive Inference, Lecture notes in Artificial Intelligence 397, Springer Verlag, 1989, 18-44.

7. Yokomori, Learning local languages from positive data, in Proc. Fujitsu IIAS - SIS Workshop on Computational Learning Theory '89, Numazu (1989).

Circular DNA and Splicing Systems

Rani Siromoney, K.G. Subramanian and V. Rajkumar Dare

Department of Mathematics, Madras Christian College,
Tambaram, Madras 600 059, India.

Abstract. Circular strings representing DNA molecules and certain recombinant behaviour are formalized. Various actions of splicing schemes on linear and circular DNA molecules are examined. It is shown that there is a difference in the regularity result of Culik and Harju [1] between the linear and circular strings. A consequence of this result is that a conjecture of Head [4] that the circular string language of a splicing system under an action on circular strings is regular, when the set of initial circular strings is regular, is disproved.

1 Introduction

Splicing systems [3] acting on sets of linear strings, is a new generative formalism that enables a mathematical analysis of the action of restriction enzymes and a ligase that allow DNA molecules to be cleaved and reassociated to produce further molecules. This gives rise to a new study of informational macro molecules, making use of the wealth of knowledge from formal language theory.

DNA exist not only as linear molecules but also as circular molecules. Head [4] proposes an extension of the splicing operation on linear strings to other actions on circular strings as well. Conventional formal language theory deals mainly with linear strings. Rosenfeld [6] has considered grammars whose languages are cycles which are cyclically ordered sequences of symbols. Siromoney et al [8,9] have considered cycle grammars for description of kolam patterns. In this paper, circular strings are treated in a formal manner. Various actions on circular as well as circular and linear strings that correspond to behaviours of circular DNA molecules are examined. Regularity of circular string languages is related to that of string languages.

An interesting property is that the generative capacity of the action of splicing schemes on circular strings or on linear and circular strings with a finite number of initial linear strings or initial circular strings is higher in contrast to the regularity property for the splicing of linear strings established in Culik and Harju [1]. Head [4] has conjectured that for every splicing system $G = (A,T,P,I,J)$ with $I \subseteq A$, $J \subset A^\wedge$ regular, both $A \cap L(G)$ and $A^\wedge \cap L(G)$ are regular. This conjecture is disproved in the case of splicing systems (A,T,P,J) without initial linear strings (i.e. $I = \Phi$), under action SA1 (formalized in definition 2.3) as a consequence of our result (Theorem 3.2). Another feature is that certain splicing languages that cannot be obtained by splicing operation of linear strings [2] can be generated by splicing schemes acting on circular strings. In particular, we examine splicing schemes with a finite number of circular linear strings that use sites with null-context and crossings of length one.

2 Splicing Systems

Let A be an alphabet and A^* be the set of all linear strings over A. The empty string is denoted by λ. A^* is the free monoid generated by A. A circular string c of symbols $a_1,...,a_n$ in this order is denoted by

$$c = {}^\wedge a_1 ... a_n$$

where $a_1...a_n$, $a_i \in A$ is one of the linearized forms of c. In fact, c is an equivalence class in A^* with respect to the binary relation ~ defined as follows: For x, y $\in A^*$, x ~ y if y is a cyclic permutation of x. We denote the set of all circular strings over A by $A^\wedge$.

Example. c = $^\wedge$abbac is a circular string over A = {a,b,c} and abbac, bbaca, bacab, acabb, cabba are the linearized forms of c in A^*.

We define regular and context-free circular languages as suggested by [4].

Definition 2.1 A set S $\subset A^\wedge$ is regular (respectively context-free) if there exists a regular (respectively context-free) string language L in A^* such that $^\wedge a_1$ ... a_n is in S if and only if at least one cyclic permutation of a_1 ... a_n is in L.

Remark. We note that the language {a^n b a^n/n $\geq$ 1} is a CF string language in {a,b}* whereas {$^\wedge a^n b a^n$/n $\geq$ 1} = {$^\wedge a^{2n} b$/n $\geq$ 1} is a regular circular language.

Next, we define splicing schemes and their actions on linear strings and circular strings as proposed in Head [4] for modeling the behaviours of circular and linear DNA molecules under certain recombinant processes.

Definition 2.2 A splicing scheme is a triple S = (A,T,P) where A the alphabet, is a finite set of symbols; T the set of triples, is a finite subset of $A^* \times A^* \times A^*$ and P the pairing relation of the scheme, is a binary relation on T satisfying the following condition : If p,x,q,u,y,v are in A^*, (p,x,q), (u,y,v) in T and (p,x,q) P (u,y,v), then x = y.

Definition 2.3 Let S = (A,T,P) be a splicing scheme.

a) Let $^\wedge$hpxq and $^\wedge$wuxv be two circular strings in $A^\wedge$ with (p,x,q) P (u,x,v). Then the action SA1 of S on these two circular strings is to produce the single longer circular string $^\wedge$hpxvwuxq. We write

$$\{\,^\wedge hpxq,\ ^\wedge wuxv\}\ \Rightarrow\ \{\,^\wedge hpxvwuxq\}$$
$$SA1$$

b) Let $^\wedge hpxqwuxv$ be a circular string with (p,x,q) P (u,x,v). The action SA2 of S on this circular string is to produce the pair of circular strings $^\wedge hpxv$ and $^\wedge wuxq$. We write

$$\{\,^\wedge hpxqwuxv\}\ \Rightarrow\ \{\,^\wedge hpxv,\ ^\wedge wuxq\}$$
$$SA2$$

c) Let hpxqk be a linear string and $^\wedge wuxv$ be a circular string with (p,x,q) P (u,x,v). The action SA3 of S on this pair is to produce the single linear string hpxvwuxqk. We write

$$\{hpxqk,\ ^\wedge wuxv\}\ \Rightarrow\ hpxvwuxqk$$
$$SA3$$

d) Let hpxqkuxvz be a linear string. The action SA4 of S on this linear string is to produce the pair of strings hpxvz and $^\wedge kuxq$. We write

$$\{hpxqkuxvz\}\ \Rightarrow\ \{hpxvz,\ ^\wedge kuxq\}$$
$$SA4$$

Definition 2.4 A splicing system is $G = (A,T,P,I,J)$ where (A,T,P) is a splicing scheme, $I \subseteq A^*$, $J \subseteq A^\wedge$. The language generated by G with respect to SAi $(i = 1,2,3,4)$ is

$$L^i(G)\ =\ \{s/s \in X \text{ and } Y \Rightarrow_* s\}$$
$$SAi$$

where $X = \begin{cases} A^\wedge, & i=1,2 \text{ and} \\ A^* \cup A^\wedge, & i=3,4 \end{cases}$
$Y = \begin{cases} J, & i=1,2; \ I=\Phi \\ I \cup J, & i=3 \\ I, & i=4, \ J=PHI \end{cases}$

and $\Rightarrow_*$ is the transitive closure of $\Rightarrow$
$\ \ \ \ SAi$ $\ SAi$

Remark. While studying the complexity of kolam patterns, Siromoney [7] introduced two operations "cut and connect" and "cut and join". It is interesting to note that these two operations depict all the four actions SA1 to SA4 just defined. In figure 2, action SA1 on two circular DNA molecules producing a single circular DNA is shown as the "cut and connect" operation. This reflects the recombinant process given in ([5], Fig. 32.5). In figure 3, action SA2 on a single circular DNA molecule producing a pair of circular strings is shown as the "cut and join" operation. This shows the behaviour of a single circular DNA molecule haing a so called "direct repeat" ([5], Fig 32.6). Action SA3 involves a linear and a circular DNA to produce a single linear string, which reflects the behaviour of linear and circular DNA molecules under certain recombination processes ([5], Fig.32.18, 32.19). This is captured by circularising the linear string and applying "cut and connect" operation to yield a single circular string which is linearized. Action SA4 is on a single linear string to produce a pair, one of which is linear and the other circular. This is obtained by first circularizing the single linear string, applying the "cut and join" operation to produce two circular strings one of which is linearized and the other kept as a circular string. This reflects the behaviour of a single linear DNA molecule having a "direct repeat" ([5], Fig.32.18, 33.9) In all four actions, the operation (cut and join and cut and connect) is performed at the "site" i.e. when the crossings are same in the pair. This is similar to the recombinant behaviour represented as operations on pair of linear strings.

We now relate the regularity in $A^{\wedge}$ in the sense of Head [4] to the regularity in $A^{\cdot}$ via the recognizability of a monoid.

For the alphabet A, define $\overline{A} = \{\overline{a} : a \in A \cup \{\lambda\}\}$, $A \cap \overline{A} = \Phi$

Let $C_1 = \overline{A}A^{\cdot}$ and O be a binary operation on C_1 which is defined by $\overline{a}x \ O \ \overline{b}y = \overline{a}xby$ for $a,b \in A \cup \{\lambda\}$ and $x,y \in A^{\cdot}$. Let $\square$ be a new symbol not in C_1. Let $C = C_1 \cup \{\square\}$. Then (C, O) is a monoid, on defining $\overline{a}x \ O \ \square = \square \ O \ \overline{a}x = \overline{a}x$ for $a \in A \cup \{\lambda\}$ and $x \in A^{\cdot}$.

Let $h : C \to A^{\cdot}$ be a morphism given by $h(\overline{a}x) = ax$, $h(\square) = \lambda$ for $a \in A \cup \{\lambda\}$ and $x \in A^{\cdot}$.

The action SA1 in Definition 2.3(a) can now be reworded in order to introduce the notion of splicing semigroup on circular strings.

Let $P = \{A_1, A_2, ..., A_n\}$ where $A_i \subset A^{\cdot} \times \{a_i\} \times A^{\cdot}$ and $a_i \in A \cup \{\lambda\}$. Let $\alpha_1, \alpha_2 \in C$. Then

$\alpha_1 \mid_p \alpha_2 = \beta$,   if $\alpha_1 = \overline{a}vwu$, $\alpha_2 = \overline{a}qhp$ and $(u,a,v),(p,a,q) \in A_i$ for some $A_i \in P$ and $\alpha_1 \ O \ \alpha_2 = \beta$

$\alpha_1 \mid_p \alpha_2 = \square$,   otherwise.

$|_p$ is the action SA1 of Definition 2.3(a), on noting that any triple (p,x,q) with crossing x having length > 1, is equivalently replaced by a triple (p,a,q$_1$) where a is a single letter and aq$_1$ = xq

Let J $\subset$ C. M(J) is the minimum subset of C containing J and closed under $|_p$.

M(J) is called the splicing semigroup of C generated by J.

Let H : C $\rightarrow$ A$^\wedge$ be such that H(α) = $^\wedge$h(α), $\alpha \in$ C.

For a splicing system G = (A,T,P,J) we have L^1(G) = H(M(J)), where L^1(G) is as in Definition 2.4.

Definition 2.5 Let M be a monoid and L $\subset$ M, then L is recognizable in M if and only if there exists a morphism h : M $\rightarrow$ M′, where M′ is a finite monoid, such that L = h^{-1}(M$_1$), where M$_1$ $\subset$ M′.

Theorem 2.1 A subset L of C is recognizable if and only if h(L) is regular in A*.

Proof. Let L be recognizable in C. Then there exists a finite monoid M and a morphism f : C $\rightarrow$ M such that L = f^{-1}(M$_1$) where M$_1$ $\subset$ M.

Define a morphism $\overline{f}$: A* $\rightarrow$ M such that $\overline{f}$(a) = f($\overline{a}$). Then we have h(L) = $\overline{f}^{-1}$(M$_1$). Hence h(L) is regular in A*. Since L = h^{-1}(h(L)), L is recognizable, if h(L) is regular in A*.

Theorem 2.2 A subset L of C is recognizable if and only if H(L) is recognizable in $^\wedge$A in the sense of Head [4].

The proof follows directly from Theorem 2.1.

3 Generative Power and Comparisons

In this section, we assume that I and J are finite, in a splicing system G = (A,T,P,I,J). First we consider action SA1 of Definition 2.3(a) and language L^1(G) of Definition 2.4. We call L^1(G) as a circular splicing language under action SA1.

Theorem 3.1 There exist circular splicing languages under action SA1, which cannot be generated by any linear splicing system [3], when the circular string language is considered in any of its linearized forms.

Proof. It is known that L = (aa)* is not a linear splicing language [2]. But the corresponding circular language L$_1$ = { $^\wedge$(aa)^n/n $\geq$ 0} can be generated under action SA1 by the splicing system $G_1 = (A_1,T_1,P_1,\Phi,J_1)$ where $A_1 = \{a\}$, $T_1 = \{(\lambda,a,\lambda)\}$, $J_1 = \{^\wedge\lambda, ^\wedge aa\}$ and (λ,a,λ) P_1 (λ,a,λ).

Remark. Rosenfeld [6] has shown that the family of cycle languages obtained by rewriting cycles with CF rules is the same as the family of cycle languages obtained by circularizing linear strings of CF languages. But we note that there is a difference between this result and the one for splicing, as noted in Theorem 3.1.

Theorem 3.2 There exist circular splicing languages under action SA1 that are strictly CF circular string languages.

Proof. The language $\{^\wedge(ca)^n(cb)^n/n \geq 1\}$ is a strictly CF circular string language since it does not have any linearized string form which is regular. But it is generated under action SA1 by the splicing system

$$G_2 = (A_2,T_2,P_2,\Phi,J_2), \text{ where } A_2 = \{a,b,c\}$$

$$T_2 = \{(ca,c,b), (cb,c,a)\}, J_2 = \{^\wedge cacb\} \text{ and } (ca,c,b) \text{ P } (cb,c,a)$$

Remark. Head [4] has conjectured that the class of languages generated under action SA1 by splicing systems (A,T,P,J) with $J \subset A^\wedge$ is regular. But we note that as a consequence of Theorem 3.2, this is not true, thus disproving Head's conjecture.

Theorem 3.3 There exist regular circular string languages that are not circular splicing languages under action SA1.

Proof. The language $L_3 = \{^\wedge a^n b/n \geq 1\}$ is a regular circular string language. Suppose it is generated under action SA1 by a splicing system. Then the action SA1 on circular strings $^\wedge a^n b$, $^\wedge a^m b$ in L_3 produces $^\wedge a^n ba^m b$ which is not in L_3.

Definition 3.1 Given the splicing system (A,T,P,I,J) a triple in T of the form (λ,a,λ), $a \in A$ is said to have null context and crossing of length one.

Theorem 3.4 The circular splicing language under action SA1 of a splicing system $G = (A,T,P,\Phi,J)$ with J finite and triples in T having null context and crossings of length one, is regular. If the context is non-null in the triples in T, the language need not be regular (even when the crossings are of length one).

Proof. We first write every initial circular string in J in the linear form aw where $(\lambda,a,\lambda) \in T$. We then construct a FSA M to accept all such linear strings. Corresponding to every two paths in M of the forms

$$\text{Path} \quad 1: \quad q_0 \xrightarrow{a} q_r \longrightarrow \qquad \cdots \qquad \longrightarrow q_m$$

$$\text{Path} \quad 2: \quad q_0 \xrightarrow{a} q_s \longrightarrow \qquad \cdots \qquad \longrightarrow q_k$$

where q_0 is the initial state of M and q_m and q_k are the final states of M, we add two edges in M with label a, one from q_k to q_r and the other from q_m to q_s. Also, corresponding to every path of the form as in path 1, we add an edge from q_m to q_r with label a. The resulting automaton clearly accepts the language $L^1(G)$ in a linearized form. Hence $L^1(G)$ is regular. The second statement in the theorem is a consequence of the proof of Theorem 3.2.

We now consider splicing systems under action SA3. I and J are finite as before.

Theorem 3.5 There exist context-free string languages that are generated by splicing systems under action SA3.

Proof. The proof is similar to the proof of Theorem 3.2 and follows from the fact that the context-free language $\{(ca)^n(cb)^n / n \geq 1\}$ is generated by the splicing system (A,T,P,I,J) where $A = \{a,b,c\}$, $T = \{(ca,c,b), (cb,c,a)\}$,

$$I = \{cacb\}, \quad J = \{{}^\wedge cacb\} \text{ and } (ca,c,b) \ P \ (cb,c,a).$$

Theorem 3.6 The string language L of a splicing system G under action SA3, having triples with null context and crossings of length one, is regular.

Proof. The idea of the proof is similar to the proof of Theorem 3.4. Let $G = (A,T,P,I,J)$ be a splicing system with triples in T of the form (λ,a,λ), $a \in A$. We construct a finite state automaton to accept the strings of I. Corresponding to every triple (λ,a,λ) in T with (λ,a,λ) P (λ,a,λ) and every circular string ${}^\wedge aw$ in J, we add a loop with label a at every state q which has an edge of the form $q \xrightarrow{a} q'$ emerging out of it. The resulting automaton accepts L.

Remark. Starting with finite sets I and J of (linear) strings and circular strings in a splicing system, it is clear that the resulting language of (linear/circular) strings can only be finite, if action SA2 alone or action SA4 alone are considered.

We now examine splicing systems (A,T,P,I) with $I \subset A^*$, an equal matrix language [10], under splicing action on linear strings introduced in Head [4]. The resulting language need not be regular unlike the case when I is taken as regular [1].

Theorem 3.7 There exist splicing systems $G = (A,T,P,I)$ with $I \subset A^*$, an equal matrix language, under splicing action on pairs of linear strings, such that L(G) is (i) regular (ii) context-free.

Proof. The languages $L_1 = \{a^n b^n \mid n \geq 1\}$, $L_2 = \{a^n b^n c^n \mid n \geq 1\}$ are equal matrix languages. Consider the splicing systems $G_1 = (\{a,b\}, \{(a,b,\lambda)\}, P_1, I_1)$ where (a,b,λ) P_1 (a,b,λ) and $I_1 = L_1$ and $G_2 = (\{a,b,c\}, \{(b,c,\lambda)\}, P_2, I_2)$ where (b,c,λ) P_2 (b,c,λ) and $I_2 = L_2$. Then $L(G_1) = \{a^n b^m /n, m \geq 1\}$, which is regular and $L(G_2) = \{a^n b^n c^m \mid n,m \geq 1\}$ which is context-free.

Conclusion. Languages of circular strings have been treated as regular or context-free according as they admit linearized forms of languages that are regular or context-free under well known notions formal language theory. The study of circular string languages gives rise to different aspects as well. For instance, for finite state automaton to accept a language of circular strings, the position where the FSA is allowed to start reading the circular string matters. In reading the circular strings in Figure 6 from the 'middle a' (marked $\bar{a}$) a FSA has no way of knowing whether there are an equal number of a's between the marked a and the letter b in the 'upper' and in the 'lower' portions of the circular string and so cannot accept such circular strings, whereas if the marked a's are as in Figure 7 a FSA in the usual sense can be constructed. Another aspect that is of interest is the notion of product of circular strings. Here the action of splicing scheme on circular strings produces circular strings and there is increase in generative power as well, even if we start with a finite number of circular strings in the splicing scheme.

References

1. K. Culik II and T. Harju (1991), The regularity of splicing systems and DNA, Discrete Applied Mathematics 31, 261-277.

2. R.W. Gatterdam (1989), Splicing systems and regularity, Int. J. Comp. Math., 31, 63-67.

3. T. Head (1987), Formal language theory and DNA : Analysis of the generative capacity of specific recombinant behaviours, Bull. Math. Bio., 49, 737-759.

4. T. Head (1991), Splicing schemes and DNA, manuscript.

5. B. Lewin (1990), Genes IV, New York, Oxford Univ. Press.

6. A. Rosenfeld (1975), A note on cycle grammars, Inform. Contr. 27, 374-377.

7. G. Siromoney (1985), Studies on the traditional art of Kolam, Working paper, May 1985.

8. G. Siromoney and R. Siromoney, (1987), Rosenfeld's cycle grammars and Kolam in Graph-grammars and application to Computer Science, Lecture Notes in Computer Science, 291, Springer-Verlag.

9.   G. Siromoney, R. Sirmoney and T. Robinson (1989), Kambi Kolam and Cycle grammars, In A Perspective in Theoretical Computer Science - Commemorative Volume for Gift Siromoney, Ed. R. Narasimhan, World Scientific, 267-300.

10.   R. Siromoney (1969), On Equal Matrix Languages, Inform. Contr. 14, 135-151.

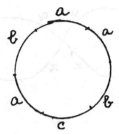

Fig. 1. Circular string ⌃abacba

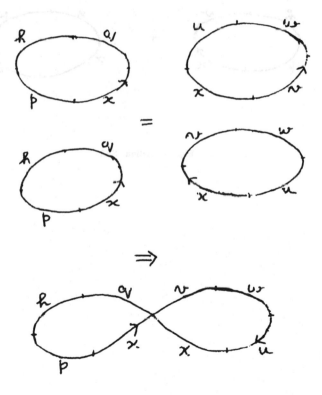

Fig. 2. Action SA1

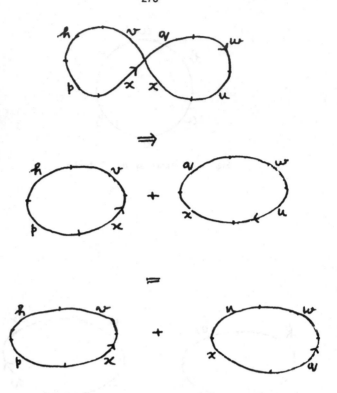

Fig. 3. Action SA2

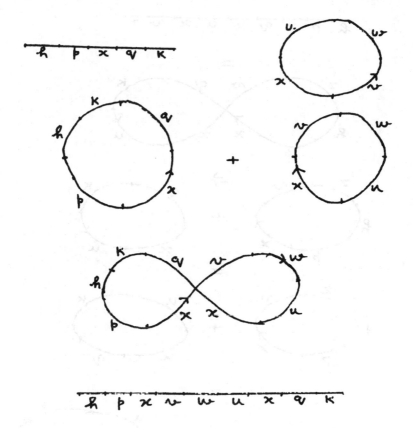

Fig. 4. Action SA3

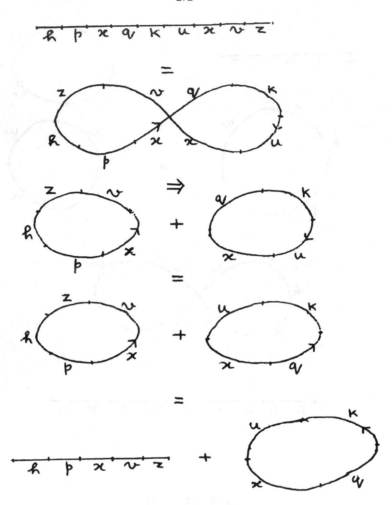

Fig. 5. Action SA4

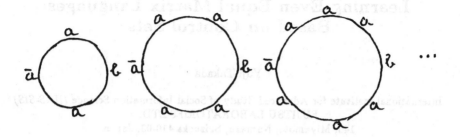

Fig. 6. Circular strings with marked 'middle a'

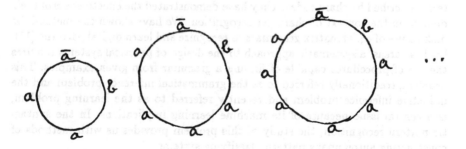

Fig. 7.

Learning Even Equal Matrix Languages Based on Control Sets

Yuji Takada

International Institute for Advanced Study of Social Information Science *(IIAS-SIS)*
FUJITSU LABORATORIES LTD.
140, Miyamoto, Numazu, Shizuoka 410-03, Japan
email : yuji@iias.flab.fujitsu.co.jp

Abstract. An equal matrix grammar is a parallel rewriting system. In this paper, we shall show a subclass of equal matrix languages, called *even equal matrix languages*, for which the learning problem is reduced to the problem of learning regular sets.

1 Introduction

An equal matrix grammar introduced by Siromoney [7] is a parallel rewriting system. It is a general extension of a regular grammar and closely related to a multitape automaton, which is a generalization of a finite automaton. This type of parallel rewriting systems has been investigated in several areas such as L-systems (n-parallel right linear grammars [12]) and Petri nets [2]. In particular, in the syntactic pattern recognition, Siromoney, Huq, Chandrasekaran, and Subramanian [8] have made use of equal matrix grammars to recognize patterns described by chain codes. They have demonstrated the effectiveness of their method in the context of character recognition. We have shown the method for making use of equal matrix grammars to recognize and learn digital patterns [11]. In these areas, a systematic approach to the design of practical systems requires the use of procedures capable of finding a grammar from given examples. This problem, traditionally referred to as the grammatical inference problem and the inductive inference problem and recently referred to as the learning problem, is a central issue because of its machine learning implications. In the syntactic pattern recognition, the study of this problem provides us with methods of constructing autonomous pattern classifying systems.

These observations lead us to the study of the learning method for equal matrix grammars. In [9], we have given some result on the learning problem for patterns described by chain codes with equal matrix grammars proposed by Siromoney, Huq, Chandrasekaran, and Subramanian [8]. In [11], we have also shown a learning method for equal matrix grammars with structural information based on control sets. In this paper, we shall show a subclass of equal matrix languages, called *even equal matrix languages*, for which the learning problem is reduced to the problem of learning regular sets. This is based on our former result for even linear languages [10].

2 Preliminaries

Let Σ denote an alphabet, Σ^* denote the set of all strings over Σ including the null string λ, and Σ^+ denote $\Sigma^* - \{\lambda\}$. $lg(w)$ denotes the length of a string w. A *language* over Σ is a subset of Σ^*.

We denote a *nondeterministic finite automaton*, abbreviated *NFA*, (over an alphabet Σ) M by a 5-tuple $(K, \Sigma, \delta, q_0, F)$, where K is a finite nonempty set of states, Σ is a finite input alphabet, δ is a transition function from $K \times \Sigma$ to 2^K, q_0 in K is the initial state, and $F \subseteq K$ is the set of final states. The transition function δ can be extended to a function from $2^K \times \Sigma^*$ to 2^K in the usual way. The set accepted by M, denoted $L(M)$, is the set $L(M) = \{u \in \Sigma^* \mid \delta(q_0, u) \cap F \neq \emptyset\}$. A *deterministic finite automaton*, abbreviated *DFA*, M is a 5-tuple $(K, \Sigma, \delta', q_0, F)$, where K, Σ, q_0, and F have the same meaning as for *NFA*s, but δ' is a function from $K \times \Sigma$ to K.

A subset R of Σ^* is called a *regular set* (over Σ) if and only if R is accepted by an *NFA*.

Let k be a positive integer. An *equal matrix grammar of order k*, abbreviated *k-EMG* (over an alphabet Σ), is a $(k+3)$-tuple $G = (N_1, \ldots, N_k, \Sigma, \Pi, S)$, where

1. $N_1, \ldots, N_k$ are finite nonempty pairwise disjoint sets of *nonterminals*,
2. S is not in $N_1 \cup \cdots \cup N_k \cup \Sigma$ and is called the *start symbol*.
3. Π is a finite nonempty set of the following three types of *matrix rules*.
   (a) *initial matrix rules* of the form $[S \rightarrow u_1 A_1 \cdots u_k A_k]$,
   (b) *nonterminal matrix rules* of the form $[A_1 \rightarrow u_1 B_1, \ldots, A_k \rightarrow u_k B_k]$,
   (c) *terminal matrix rules* of the form $[A_1 \rightarrow w_1, \ldots, A_k \rightarrow w_k]$,
   where for each i $(1 \leq i \leq k)$, A_i, B_i are in N_i and u_i, w_i are in Σ^*.

In what follows, we may assume that any matrix rule has a unique label and denoted with its label such as $\pi : [S \rightarrow x]$ or $\pi' : [A_1 \rightarrow x_1, \ldots, A_k \rightarrow x_k]$, where π and π' are labels. Also, we may refer any matrix rule by its label, so $\pi \in \Pi$ means that the matrix rule with label π is in Π.

Let $G = (N_1, \ldots, N_k, \Sigma, \Pi, S)$ be a k-EMG. We denote $N_1 \cup \cdots \cup N_k \cup \Sigma \cup \{S\}$ by V. For any $x, y \in V^*$, $x \Longrightarrow_G^\pi y$ if and only if either

1. $x = S$ and $\pi : [S \rightarrow y]$ is in Π or
2. there exist $u_1, \ldots, u_k$ in Σ^*, $z_1, \ldots, z_k$ each z_i in $(N_i \cup \Sigma)^*$, and $A_1, \ldots, A_k$ each A_i in N_i such that $x = u_1 A_1 \cdots u_k A_k$, $y = u_1 z_1 \cdots u_k z_k$, and $\pi : [A_1 \rightarrow z_1, \ldots, A_k \rightarrow z_k]$ is in Π.

We write $x \Longrightarrow_G^\alpha y$ if and only if either

1. $x = y$ and $\alpha = \lambda$, or
2. there exist $x_0, \ldots, x_n \in V^*$ such that $x = x_0$, $y = x_n$, and $x_i \Longrightarrow_G^{\pi_i} x_{i+1}$ for each i and $\alpha = \pi_1 \cdots \pi_n$.

$x \Longrightarrow_G^\alpha y$ is called a *derivation from x to y with an associate word α* in G.

A k-EMG G is said to be *unambiguous* if and only if for any string $w \in \Sigma^*$, $S \Longrightarrow_G^\alpha w$ and $S \Longrightarrow_G^{\alpha'} w$ imply $\alpha = \alpha'$.

The *language generated by* a k-EMG G, denoted $L(G)$, is the set

$$L(G) = \{w \in \Sigma^* \mid S \underset{G}{\overset{\alpha}{\Longrightarrow}} w\}$$

and is called the *equal matrix language of order k* (abbreviated k-EML) generated by G. A language L is a k-EML if and only if there exists a k-EMG G such that $L = L(G)$ holds.

Clearly, a 1-*EML* is regular and for $k \geq 2$, the class of k-*EMLs* contains some context-sensitive languages; the context-sensitive language $L_{ww} = \{ww \mid w \in \Sigma^*\}$ is a 2-*EML* and for any $k \geq 3$, the context-sensitive language $L_k = \{a_1^n \cdots a_k^n \mid n \geq 1,$ each distinct $a_i \in \Sigma\}$ is a k-*EML*. Also, there exists a context-free language which is not a k-*EML* for any $k \geq 1$ (Ibarra [4]); the context-free language $\bigcup_{i \geq 0} \{a^n b^n \mid n \geq 1\}^i$ is not a k-*EML* for any $k \geq 1$.

Let G be a k-EMG. Then $A(G)$ denotes a set

$$A(G) = \{\alpha \mid S \underset{G}{\overset{\alpha}{\Longrightarrow}} w, \; w \in L(G)\}.$$

A subset C of $A(G)$ is called a *control set* for G and

$$L_C(G) = \{w \in \Sigma^* \mid S \underset{G}{\overset{\alpha}{\Longrightarrow}} w, \; \alpha \in C\}$$

is called the *language generated by G with the control set C*. From the definition, for any control set C for G, $L_C(G) \subseteq L(G)$ always holds. We note that the definition of control sets is slightly different from the usual one such as in [6], where a control set is any language over the set of labels of rules.

3 Even Equal Matrix Languages

In this section, we introduce a subclass of k-EMGs, called *even k-EMGs*, and show that there is a normal form, called *even normal form*, for even k-EMGs.

Definition 1. An *even k-EMG* $G = (N_1, \ldots, N_k, \Sigma, \Pi, S)$ is a k-EMG such that each matrix rule in Π is of the form

$$\pi_S : [S \rightarrow A_1 \cdots A_k],$$
$$\pi_N : [A_1 \rightarrow u_1 B_1, \ldots, A_k \rightarrow u_k B_k], \text{ or}$$
$$\pi_T : [A_1 \rightarrow w_1, \ldots, A_k \rightarrow w_k],$$

where for each i, $A_i, B_i \in N_i$, $u_i \in \Sigma^+$, $w_i \in \Sigma^*$, and $lg(u_1) = \cdots = lg(u_k)$.

A language L is an *even k-EML* if and only if there exists an even k-EMG G such that $L = L(G)$.

Again, an even 1-*EML* is regular and for $k \geq 2$, the class of even k-*EMLs* contains some context-sensitive languages (two context-sensitive languages L_{ww} and L_k described in the previous section are an even 2-*EML* and an even k-*EML*, respectively).

Definition 2. An even k-EMG G is said to be *in even normal form* if and only if every nonterminal matrix rule of G is of the form $\pi_N : [A_1 \rightarrow a_1 B_1, \ldots, A_k \rightarrow a_k B_k]$ and every terminal matrix rule is of the form $\pi_T : [A_1 \rightarrow \lambda, \ldots, A_{k-1} \rightarrow \lambda, A_k \rightarrow w_k]$ where each $A_i, B_i \in N_i$, each $a_i \in \Sigma$, and w_k is a string such that $lg(w_k) < k$.

Now we shall show that for any even k-EML L there effectively exists an even k-EMG G in even normal form such that $L = L(G)$ holds.

Lemma 3. *For any even k-EMG G, there effectively exists an even k-EMG G' such that $L(G) = L(G')$ and all nonterminal matrix rules of G' are of the form $\pi'_N : [A'_1 \rightarrow a_1 B'_1, \ldots, A'_k \rightarrow a_k B'_k]$, where each A'_i, B'_i are nonterminals of G' and each a_i is in Σ.*

Proof. We construct G' from G in the following way. For any nonterminal matrix rule $\pi_N : [A_1 \rightarrow u_1 B_1, \ldots, A_k \rightarrow u_k B_k]$ with each $u_i = a_{i,1} \cdots a_{i,m}$ $(m > 1)$, we introduce new nonterminals $C_{i,1}, \ldots, C_{i,m-1}$ into N_i for each i and replace π_N by new nonterminal matrix rules

$$\pi_1 : [A_1 \rightarrow a_{1,1} C_{1,1}, \ldots, A_k \rightarrow a_{k,1} C_{k,1}],$$
$$\pi_2 : [C_{1,1} \rightarrow a_{1,2} C_{1,2}, \ldots, C_{k,1} \rightarrow a_{k,2} C_{k,2}],$$
$$\vdots$$
$$\pi_{m-1} : [C_{1,m-2} \rightarrow a_{1,m-1} C_{1,m-1}, \ldots, C_{k,m-2} \rightarrow a_{k,m-1} C_{k,m-1}],$$
$$\pi_m : [C_{1,m-1} \rightarrow a_{1,m} B_1, \ldots, C_{k,m-1} \rightarrow a_{k,m} B_k].$$

It is easy to verify that $L(G) = L(G')$. $\qquad\qquad\square$

In what follows, by Lemma 3, we assume that an even k-EMG G has nonterminal matrix rules of the form $\pi_N : [A_1 \rightarrow a_1 B_1, \ldots, A_k \rightarrow a_k B_k]$ only.

Definition 4. An even k-EMG G is said to be *primary* if and only if G has exactly one initial matrix rule and exactly one terminal matrix rule.

Lemma 5. *A language L is an even k-EML if and only if there are finite number of primary even k-EMGs $G_1, \ldots, G_n$ such that $L = L(G_1) \cup \cdots \cup L(G_n)$ holds.*

Proof. Since $G_1, \ldots, G_n$ are all primary even k-EMGs, $L(G_1), \ldots, L(G_n)$ are even k-EMLs. It is easy to verify that the class of even k-EMLs is closed under union. Hence, $L(G_1) \cup \cdots \cup L(G_n)$ is an even k-EML.

Let $G = (N_1, \ldots, N_k, \Sigma, \Pi, S)$ be an even k-EMG which has n number of initial matrix rules and m number of terminal matrix rules. Then we define primary even k-EMGs $G_{(1,1)}, \ldots, G_{(1,m)}, \ldots, G_{(n,1)}, \ldots, G_{(n,m)}$ in the following way. For each i, j $(1 \leq i \leq n, 1 \leq j \leq m)$, $G_{(i,j)} = (N_1, \ldots, N_k, \Sigma, \Pi_{(i,j)}, S)$ where $\Pi_{(i,j)}$ consists of the i-th initial matrix rule, all nonterminal matrix rules, and the j-th terminal matrix rule of G. Then it is easy to verify that $L(G) = \bigcup_{i=1}^{n} \bigcup_{j=1}^{m} L(G_{(i,j)})$. $\qquad\qquad\square$

Note that a primary even k-EMG $G_{(i,j)}$ constructed from G in the proof of Lemma 5 may have useless nonterminal matrix rules. We can effectively remove these useless nonterminal matrix rules and useless nonterminals in the way similar to context-free grammars (cf. [3]). In the sequel, for any even k-EMG G, we can effectively have finite number of primary even k-EMG $G_1, \ldots, G_n$ such that each G_i is reduced (that is, G_i has no useless matrix rule and no useless nonterminal), each $L(G_i)$ is not empty, and $L(G) = L(G_1) \cup \cdots \cup L(G_n)$.

Lemma 6. *Let G be a primary even k-EMG whose terminal matrix rule is π_T : $[A_1 \to w_1, \ldots, A_k \to w_k]$. For any i ($1 \le i \le k-1$), if $w_i = v_i d$ where $v_i \in \Sigma^*$ and $d \in \Sigma$, then there effectively exists an even k-EMG G' such that $L(G) = L(G')$ and all terminal matrix rules are of the form*

$$\pi_T' : [A_1' \to w_1, \ldots, A_{i-1}' \to w_{i-1},$$
$$A_i' \to v_i, A_{i+1}' \to w_{i+1}',$$
$$A_{i+2}' \to w_{i+2}, \ldots, A_k' \to w_k]$$

where $lg(w_{i+1}') = lg(w_{i+1}) + 1$.

Proof. We define $G' = (N_1', \ldots, N_k', \Sigma, \Pi', S)$ in the following way.

1. N_i' has a nonterminal S_i and for each j ($1 \le j \le k$ and $j \ne i$), $N_j' = \{S_j\}$.
2. The initial matrix rule $\pi_S' : [S \to S_1 \cdots S_{i-1} S_i S_{i+1} \cdots S_k]$ is in Π'.
3. Let $\pi_S : [S \to A_1 \cdots A_k]$ be the initial matrix rule of G. For any nonterminal matrix rule $\pi_N : [A_1 \to a_1 B_1, \ldots, A_k \to a_k B_k]$ of G, we introduce a new nonterminal $(B_1, \ldots, B_{i-1}, A_i, B_{i+1}, \ldots, B_k)$ into N_i' and introduce the nonterminal matrix rule

$$\pi_N' : [S_1 \to a_1 S_1, \ldots, S_{i-1} \to a_{i-1} S_{i-1},$$
$$S_i \to a_i(B_1, \ldots, B_{i-1}, A_i, B_{i+1}, \ldots, B_k), S_{i+1} \to d S_{i+1},$$
$$S_{i+2} \to a_{i+2} S_{i+2}, \ldots, S_k \to a_k S_k]$$

into Π'. Moreover, if G has the terminal matrix rule $\pi_T : [A_1 \to w_1, \ldots, A_k \to w_k]$, then we also introduce the terminal matrix rule

$$\pi_T' : [S_1 \to w_1, \ldots, S_{i-1} \to w_{i-1},$$
$$S_i \to v_i, S_{i+1} \to d w_{i+1},$$
$$S_{i+2} \to w_{i+2}, \ldots, S_k \to w_k]$$

into Π'.

4. Whenever any new nonterminal $(B_1, \ldots, B_{i-1}, A_i, B_{i+1}, \ldots, B_k)$ is introduced, we apply the following procedure. For any nonterminal matrix rules $\pi_{N1} : [A_1 \to a_1 B_1, \ldots, A_k \to a_k B_k]$ and $\pi_{N2} : [B_1 \to b_1 C_1, \ldots, B_k \to b_k C_k]$ of G, we introduce the new nonterminal $(C_1, \ldots, C_{i-1}, B_i, C_{i+1}, \ldots, C_k)$ into N_i' if not in N_i', and introduce the nonterminal matrix rule

$$\pi_N' : [S_1 \to b_1 S_1, \ldots, S_{i-1} \to b_{i-1} S_{i-1},$$
$$(B_1, \ldots, B_{i-1}, A_i, B_{i+1}, \ldots, B_k) \to b_i(C_1, \ldots, C_{i-1}, B_i, C_{i+1}, \ldots, C_k),$$
$$S_{i+1} \to a_{i+1} S_{i+1}, S_{i+2} \to b_{i+2} S_{i+2}, \ldots, S_k \to b_k S_k]$$

into Π'. Moreover, if G has the terminal matrix rule $\pi_T : [B_1 \rightarrow w_1, \ldots, B_k \rightarrow w_k]$, then we introduce the terminal matrix rule

$$\pi'_T : [S_1 \rightarrow w_1, \ldots, S_{i-1} \rightarrow w_{i-1},$$
$$(B_1, \ldots, B_{i-1}, A_i, B_{i+1}, \ldots, B_k) \rightarrow v_i, S_{i+1} \rightarrow a_{i+1}w_{i+1},$$
$$S_{i+2} \rightarrow w_{i+2}, \ldots, S_k \rightarrow w_k]$$

into Π'.

Since G has a finite number of matrix rules, we introduce a finite number of new nonterminals and, in the sequel, this procedure is applied finitely.

Claim 1 $A_1 \cdots A_k \Longrightarrow_G^\alpha w_1 \cdots w_{i-1} v_i d w_{i+1} \cdots w_k$ *and G has a nonterminal matrix rule $\pi_N : [B_1 \rightarrow b_1 A_1, \ldots, B_k \rightarrow b_k A_k]$ if and only if*

$$S_1 \cdots S_{i-1}(A_1, \ldots, A_{i-1}, B_i, A_{i+1}, \ldots, A_k)S_{i+1} \cdots S_k$$
$$\xrightarrow[G']{\alpha'} w_1 \cdots w_{i-1} v_i b_{i+1} w_{i+1} \cdots w_k.$$

Proof. We prove this claim by an induction on the length of associate words. For the terminal matrix rule $\pi_T : [A_1 \rightarrow w_1, \ldots, A_{i-1} \rightarrow w_{i-1}, A_i \rightarrow v_i d, A_{i+1} \rightarrow w_{i+1}, \ldots, A_k \rightarrow w_k]$ of G, there is a nonterminal matrix rule $\pi_N : [B_1 \rightarrow b_1 A_1, \ldots, B_k \rightarrow b_k A_k]$ of G if and only if G' has

$$\pi'_T : [S_1 \rightarrow w_1, \ldots, S_{i-1} \rightarrow w_{i-1},$$
$$(A_1, \ldots, A_{i-1}, B_i, A_{i+1}, \ldots, A_k) \rightarrow v_i, S_{i+1} \rightarrow b_{i+1}w_{i+1},$$
$$S_{i+2} \rightarrow w_{i+2}, \ldots, S_k \rightarrow w_k].$$

Then $A_1 \cdots A_k \Longrightarrow_G^{\pi_T} w_1 \cdots w_{i-1} v_i d w_{i+1} \cdots w_k$ if and only if

$$S_1 \cdots S_{i-1}(A_1, \ldots, A_{i-1}, B_i, A_{i+1}, \ldots, A_k)S_{i+1} \cdots S_k$$
$$\xrightarrow[G']{\pi'_T} w_1 \cdots w_{i-1} v_i b_{i+1} w_{i+1} \cdots w_k.$$

Suppose that the claim holds for any associate word of length n or less. The construction of G' ensures that G' has the nonterminal matrix rule

$$\pi'_N : [S_1 \rightarrow b_1 S_1, \ldots, S_{i-1} \rightarrow b_{i-1}S_{i-1},$$
$$(B_1, \ldots, B_{i-1}, C_i, B_{i+1}, \ldots, B_k) \rightarrow b_i(A_1, \ldots, A_{i-1}, B_i, A_{i+1}, \ldots, A_k),$$
$$S_{i+1} \rightarrow c_{i+1}S_{i+1}, S_{i+2} \rightarrow b_{i+2}S_{i+2}, \ldots, S_k \rightarrow b_k S_k]$$

if and only if there are two nonterminal matrix rules $\pi_{N1} : [C_1 \rightarrow c_1 B_1, \ldots, C_k \rightarrow c_k B_k]$ and $\pi_{N2} : [B_1 \rightarrow b_1 A_1, \ldots, B_k \rightarrow b_k A_k]$ of G. Then

$$B_1 \cdots B_{i-1}B_iB_{i+1} \cdots B_k$$
$$\xrightarrow[G]{\pi_{Nj}} b_1 A_1 \cdots b_{i-1}A_{i-1}b_i A_i b_{i+1}A_{i+1} \cdots b_k A_k$$
$$\xrightarrow[G]{\alpha} b_1 w_1 \cdots b_{i-1}w_{i-1}b_i v_i d b_{i+1}w_{i+1} \cdots b_k w_k$$

if and only if

$$S_1 \cdots S_{i-1}(B_1, \ldots, B_{i-1}, C_i, B_{i+1}, \ldots, B_k)S_{i+1}S_{i+2} \cdots S_k$$

$$\overset{\pi'_N}{\underset{G'}{\Longrightarrow}} b_1 S_1 \cdots b_{i-1}S_{i-1}b_i(A_1, \ldots, A_{i-1}, B_i, A_{i+1}, \ldots, A_k)c_{i+1}S_{i+1}b_{i+2}S_{i+2} \cdots b_k S_k$$

$$\overset{\alpha'}{\underset{G'}{\Longrightarrow}} b_1 w_1 \cdots b_{i-1}w_{i-1}b_i v_i c_{i+1}b_{i+1}w_{i+1}b_{i+2}w_{i+2} \cdots b_k w_k.$$

This completes the proof of the claim. □

The construction of G' ensures the followings.

1. G has a nonterminal matrix rule $\pi_N : [B_1 \to b_1 A_1, \ldots, B_i \to b_i A_i, \ldots, B_k \to b_k A_k]$ and $S \overset{\pi_S \pi_N}{\underset{G}{\Longrightarrow}} b_1 A_1 \cdots b_i A_i \cdots b_k A_k$ if and only if

$$S \overset{\pi'_S \pi'_N}{\underset{G'}{\Longrightarrow}}$$
$$b_1 S_1 \cdots b_{i-1}S_{i-1}b_i(A_1, \ldots, A_{i-1}, B_i, A_{i+1}, \ldots, A_k)dS_{i+1}b_{i+1}S_{i+1} \cdots b_k S_k.$$

2. $S \overset{\pi_S \pi_T}{\underset{G}{\Longrightarrow}} w_1 \cdots w_{i-1}v_i d w_{i+1} \cdots w_k$ if and only if

$$S \overset{\pi'_S \pi'_T}{\underset{G'}{\Longrightarrow}} w_1 \cdots w_{i-1}v_i d w_{i+1} \cdots w_k.$$

Hence, by the claim, $L(G) = L(G')$. This completes the proof of Lemma 6. □

Lemma 7. *Let G be a primary even k-EMG whose terminal matrix rule is $\pi_T :$ $[A_1 \to w_1, \ldots, A_k \to w_k]$. For any i $(2 \le i \le k)$, if $w_i = dv_i$ where $d \in \Sigma$ and $v_i \in \Sigma^*$, then there effectively exists an even k-EMG G' such that $L(G) = L(G')$ and all terminal matrix rules are of the form*

$$\pi'_T : [A'_1 \to w_1, \ldots, A'_{i-2} \to w_{i-2},$$
$$A'_{i-1} \to w'_{i-1}, A'_i \to v_i,$$
$$A'_{i+1} \to w_{i+1}, \ldots, A'_k \to w_k]$$

where $lg(w'_{i-1}) = lg(w_{i-1}) + 1$.

Proof. We define $G' = (N'_1, \ldots, N'_k, \Sigma, \Pi', S)$ in the following way.

1. For each j $(1 \le j \le k$ and $j \ne i)$, $N'_j = \{S_j\}$.
2. Let $\pi_S : [S \to A_1 \cdots A_k]$ be the initial matrix rule of G. For any nonterminal matrix rule $\pi_N : [A_1 \to a_1 B_1, \ldots, A_k \to a_k B_k]$ of G, we introduce a new nonterminal $(A_1, \ldots, A_{i-1}, B_i, A_{i+1}, \ldots, A_k)$ into N'_i and introduce the initial matrix rule

$$\pi'_S : [S \to S_1 \cdots S_{i-1}(A_1, \ldots, A_{i-1}, B_i, A_{i+1}, \ldots, A_k)S_{i+1} \cdots S_k]$$

into Π'. Moreover, if G has the terminal matrix rule $\pi_T : [A_1 \to w_1, \ldots, A_k \to w_k]$, then we introduce a new nonterminal $(A_1, \ldots, A_{i-1}, T_i, A_{i+1}, \ldots, A_k)$ into N'_i and introduce the initial matrix rule

$$\pi''_S : [S \to S_1 \cdots S_{i-1}(A_1, \ldots, A_{i-1}, T_i, A_{i+1}, \ldots, A_k)S_{i+1} \cdots S_k]$$

and the terminal matrix rule

$$\pi'_T : [S_1 \rightarrow w_1, \ldots, S_{i-2} \rightarrow w_{i-2},$$
$$S_{i-1} \rightarrow w_{i-1}d, (A_1, \ldots, A_{i-1}, T_i, A_{i+1}, \ldots, A_k) \rightarrow v_i,$$
$$S_{i+1} \rightarrow w_{i+1}, \ldots, S_k \rightarrow w_k]$$

into Π'.

3. Whenever any new nonterminal $(A_1, \ldots, A_{i-1}, B_i, A_{i+1}, \ldots, A_k)$ is introduced, we apply the following procedure. For any nonterminal matrix rules $\pi_{N1} : [A_1 \rightarrow a_1 B_1, \ldots, A_k \rightarrow a_k B_k]$ and $\pi_{N2} : [B_1 \rightarrow b_1 C_1, \ldots, B_k \rightarrow b_k C_k]$ of G, we introduce the new nonterminal $(B_1, \ldots, B_{i-1}, C_i, B_{i+1}, \ldots, B_k)$ into N'_i if not in N'_i, and introduce the nonterminal matrix rule

$$\pi'_N : [S_1 \rightarrow a_1 S_1, \ldots, S_{i-1} \rightarrow a_{i-1} S_{i-1},$$
$$(A_1, \ldots, A_{i-1}, B_i, A_{i+1}, \ldots, A_k) \rightarrow b_i(B_1, \ldots, B_{i-1}, C_i, B_{i+1}, \ldots, B_k),$$
$$S_{i+1} \rightarrow a_{i+1} S_{i+1}, \ldots, S_k \rightarrow a_k S_k]$$

into Π'. Moreover, if G has the terminal matrix rule $\pi_T : [B_1 \rightarrow w_1, \ldots, B_k \rightarrow w_k]$, then we introduce a new nonterminal $(B_1, \ldots, B_{i-1}, T_i, B_{i+1}, \ldots, B_k)$ into N'_i, and introduce the nonterminal matrix rule

$$\pi''_N : [S_1 \rightarrow a_1 S_1, \ldots, S_{i-1} \rightarrow a_{i-1} S_{i-1},$$
$$(A_1, \ldots, A_{i-1}, B_i, A_{i+1}, \ldots, A_k) \rightarrow d(B_1, \ldots, B_{i-1}, T_i, B_{i+1}, \ldots, B_k),$$
$$S_{i+1} \rightarrow a_{i+1} S_{i+1}, \ldots, S_k \rightarrow a_k S_k]$$

into Π', and for any nonterminal matrix rule π_N such that

$$S \overset{\pi_S \pi_N \cdots \pi_T}{\underset{G}{\Longrightarrow}} c_1 u_1 w_1 \cdots c_i u_i w_i \cdots c_k u_k w_k \quad \text{(each } c_j \in \Sigma \text{ and each } u_j \in \Sigma^*\text{)},$$

we introduce the terminal matrix rule

$$\pi'_T : [S_1 \rightarrow w_1, \ldots, S_{i-2} \rightarrow w_{i-2},$$
$$S_{i-1} \rightarrow w_{i-1}c_i, (B_1, \ldots, B_{i-1}, T_i, B_{i+1}, \ldots, B_k) \rightarrow v_i,$$
$$S_{i+1} \rightarrow w_{i+1}, \ldots, S_k \rightarrow w_k]$$

into Π'.

Since G has a finite number of matrix rules, we introduce a finite number of new nonterminals and, in the sequel, this procedure is applied finitely.

We can prove the following claim by the induction on the length of associate words similar to the claim of Lemma 6.

Claim 2 $A_1 \cdots A_k \overset{\alpha}{\Longrightarrow}_G w_1 \cdots w_{i-1} d v_i w_{i+1} \cdots w_k$ and G has a nonterminal matrix rule π_N such that $S \overset{\pi_S \pi_N}{\Longrightarrow}_G c_1 C_1 \cdots c_i C_i \cdots c_k C_k$ if and only if

$$S_1 \cdots S_{i-1}(A_1, \ldots, A_{i-1}, B_i, A_{i+1}, \ldots, A_k) S_{i+1} \cdots S_k$$
$$\overset{\alpha'}{\underset{G'}{\Longrightarrow}} w_1 \cdots w_{i-1} c_i v_i w_{i+1} \cdots w_k.$$

The construction of G' ensures the followings.

1. G has a nonterminal matrix rule $\pi_N : [A_1 \to a_1 B_1, \ldots, A_i \to a_i B_i, \ldots, A_k \to a_k B_k]$ and $S \overset{\pi_S \pi_N}{\Longrightarrow}_G a_1 B_1 \cdots a_i B_i \cdots a_k B_k$ if and only if G' has the initial matrix rule $\pi'_S : [S \to S_1 \cdots S_{i-1}(A_1, \ldots, A_{i-1}, B_i, A_{i+1}, \ldots, A_k) S_{i+1} \cdots S_k]$.

2. $S \overset{\pi_S \pi_T}{\Longrightarrow}_G w_1 \cdots w_{i-1} d v_i w_{i+1} \cdots w_k$ if and only if

$$S \overset{\pi'_S \pi'_T}{\Longrightarrow}_{G'} w_1 \cdots w_{i-1} d v_i w_{i+1} \cdots w_k.$$

Hence, by the claim, $L(G) = L(G')$. This completes the proof of Lemma 7. □

Theorem 8. *For any even k-EML L, there effectively exists an even k-EMG G in even normal form such that $L = L(G)$ holds.*

Proof. Let G' be an even k-EMG such that $L = L(G')$ holds. We construct an even k-EMG G in even normal form in the following way.

By Lemma 5, from G', we have a finite number of primary even k-EMGs $G_1, \ldots, G_n$ such that $L = L(G_1) \cup \cdots \cup L(G_n)$ holds. For each primary even k-EMG $G_j = (N_{j,1}, \ldots, N_{j,k}, \Sigma, \Pi_j, S)$ $(1 \le j \le n)$ which has the terminal matrix rule $\pi_T : [A_1 \to w_1, \ldots, A_k \to v_i d, \ldots, A_k \to w_k]$ for some i, by Lemma 6, we have an even k-EMG G'_j such that $L(G_j) = L(G'_j)$ and all terminal matrix rules are of the form

$$\pi'_T : [A'_1 \to w_1, \ldots, A'_{i-1} \to w_{i-1},$$
$$A'_i \to v_i, A'_{i+1} \to w'_{i+1},$$
$$A'_{i+2} \to w_{i+2}, \ldots, A'_k \to w_k]$$

where $lg(w'_{i+1}) = lg(w_{i+1}) + 1$. Then let G_j be G'_j. We apply this procedure to each G_j until all terminal matrix rules of G_j are of the form $\pi_T : [A_1 \to \lambda, \ldots, A_{k-1} \to \lambda, A_k \to w_k]$.

Next we apply Lemma 5 and then Lemma 7 to each G_j repeatedly so that each G_j has terminal matrix rules of the form $\pi_T : [A_1 \to a_1, \ldots, A_{k-1} \to a_{k-1}, A_k \to a_k w_k]$ where each $a_i \in \Sigma$. Then we introduce a new nonterminal B_i into $N_{j,i}$ for each i and replace π_T by the nonterminal matrix rule $\pi'_N : [A_1 \to a_1 B_1, \ldots, A_k \to a_k B_k]$ and the terminal matrix rule $\pi'_T : [B_1 \to \lambda, \ldots, B_{k-1} \to \lambda, B_k \to w_k]$. We apply this to each G_j until all terminal matrix rules of G_j are of the form $\pi_T : [A_1 \to \lambda, \ldots, A_{k-1} \to \lambda, A_k \to w_k]$ where $lg(w_k) < k$.

By the above process, we have the m number of even k-EMGs in even normal form. Let $G = (\bigcup_{j=1}^m N_{j,1}, \ldots, \bigcup_{j=1}^m N_{j,k}, \Sigma, \bigcup_{j=1}^m \Pi_j, S)$. Then this even k-EMG G is in even normal form and $L = L(G)$ holds. □

4 Representation Theorem

In this section, we shall show a representation theorem for even k-EMLs (Theorem 16), which says that there effectively exists an even k-EMG *fixed for an alphabet* such that any even k-EML over the alphabet is generated by the even k-EMG with a *regular* control set *unique for the even k-EML*.

Definition 9. A *universal* even *k-EMG* over an alphabet Σ is the even *k-EMG* $G^e = (\{S_1\}, \ldots, \{S_k\}, \Sigma, \Pi^e, S)$ such that Π^e consists of the following matrix rules.

$$\Pi^e = \{\pi_S^e : [S \to S_1 \cdots S_k]\}$$
$$\cup \{\pi_N^e : [S_1 \to a_1 S_1, \ldots, S_k \to a_k S_k] \mid a_i \in \Sigma\}$$
$$\cup \{\pi_T^e : [S_1 \to \lambda, \ldots, S_{k-1} \to \lambda, S_k \to w_k] \mid w_k \in \Sigma^* \text{ and } lg(w_k) < k\}.$$

Note that for any alphabet Σ, a universal even *k-EMG* G^e is unique up to renaming of nonterminals.

Remark 10. A universal even *k-EMG* G^e is in even normal form and if the cardinality of Σ is m then G^e has $1 + m^k + \sum_{i=0}^{k-1} m^i$ number of matrix rules.

Remark 11. For any alphabet Σ, a universal even *k-EMG* G^e is unambiguous.

Let $G = (N_1, \ldots, N_k, \Sigma, \Pi, S)$ be an even *k-EMG* in even normal form and $G^e = (\{S_1\}, \ldots, \{S_k\}, \Sigma, \Pi^e, S)$ be a universal even *k-EMG* over an alphabet Σ. Then we define a homomorphism h from Π^* to Π^{e^*} such that

$$h(\pi) = \begin{cases} \pi_S^e : [S \to S_1 \cdots S_k] & \\ \qquad \text{if } \pi : [S \to A_1 \cdots A_k], & \\ \pi_N^e : [S_1 \to a_1 S_1, \ldots, S_k \to a_k S_k] & \\ \qquad \text{if } \pi : [A_1 \to a_1 B_1, \ldots, A_k \to a_k B_k], & \\ \pi_T^e : [S_1 \to \lambda, \ldots, S_{k-1} \to \lambda, S_k \to w_k] & \\ \qquad \text{if } \pi : [A_1 \to \lambda, \ldots, A_{k-1} \to \lambda, A_k \to w_k]. & \end{cases}$$

We also define the *NFA* $M = (\{S, q_f\} \cup (N_1 \times \cdots \times N_k), \Pi^e, \delta, S, \{q_f\})$ corresponding to G, where $q_f \notin N_1 \cup \cdots \cup N_k$ and δ is defined as follows.

$$\delta(S, \pi_S^e) =$$
$$\{(A_1, \ldots, A_k) \mid \pi_S \in h^{-1}(\pi_S^e) \text{ and } \pi_S : [S \to A_1 \cdots A_k]\},$$
$$\delta((A_1, \ldots, A_k), \pi_N^e) =$$
$$\{(B_1, \ldots, B_k) \mid \pi_N \in h^{-1}(\pi_N^e) \text{ and } \pi_N : [A_1 \to a_1 B_1, \ldots, A_k \to a_k B_k]\},$$
$$\delta((A_1, \ldots, A_k), \pi_T^e) = \{q_f\}$$
if $\pi_T \in h^{-1}(\pi_T^e)$ and $\pi_T : [A_1 \to \lambda, \ldots, A_{k-1} \to \lambda, A_k \to w_k]$.

Lemma 12. *For any $w \in \Sigma^*$, any $(A_1, \ldots, A_k) \in N_1 \times \cdots \times N_k$, and any $\alpha \in A(G)$, $A_1 \cdots A_k \Longrightarrow_G^\alpha w$ if and only if $S_1 \cdots S_k \Longrightarrow_{G^e}^{h(\alpha)} w$ and $q_f \in \delta((A_1, \ldots, A_k), h(\alpha))$.*

Proof. We prove this lemma by an induction on the length of associate words. By the definitions of h and δ, $A_1 \cdots A_k \Longrightarrow_G^{\pi_T} w$ if and only if $S_1 \cdots S_k \Longrightarrow_{G^e}^{h(\pi_T)} w$ and $q_f \in \delta((A_1, \ldots, A_k), h(\pi_T))$.

Inductively suppose that for any $\alpha \in \Pi^*$ with $lg(\alpha) \le n$ the assertion holds. If $A_1 \cdots A_k \Longrightarrow_G^{\pi_N} a_1 B_1 \cdots a_k B_k \Longrightarrow_G^\alpha a_1 w_1 \cdots a_k w_k$, then $S_1 \cdots S_k \Longrightarrow_{G^e}^{h(\alpha)}$

$w_1 \cdots w_k$ and $q_f \in \delta((B_1, \ldots, B_k), h(\alpha))$ by the inductive hypothesis. Since $\pi_N : [A_1 \rightarrow a_1 B_1, \ldots, A_k \rightarrow a_k B_k]$ is in Π, by the definition of δ we have $(B_1, \ldots, B_k) \in \delta((A_1, \ldots, A_k), h(\pi_N))$. Hence,

$$S_1 \cdots S_k \stackrel{h(\pi_N)}{\underset{G^e}{\Longrightarrow}} a_1 S_1 \cdots a_k S_k \stackrel{h(\alpha)}{\underset{G^e}{\Longrightarrow}} a_1 w_1 \cdots a_k w_k$$

and q_f is in

$$\delta(\delta((A_1, \ldots, A_k), h(\pi_N)), h(\alpha)) = \delta((A_1, \ldots, A_k), h(\pi_N)h(\alpha))$$
$$= \delta((A_1, \ldots, A_k), h(\pi_N \alpha)).$$

Conversely, suppose that $S_1 \cdots S_k \underset{G^e}{\overset{\pi_N^e}{\Longrightarrow}} a_1 S_1 \cdots a_k S_k \underset{G^e}{\overset{\alpha^e}{\Longrightarrow}} a_1 w_1 \cdots a_k w_k$, $(B_1, \ldots, B_k) \in \delta((A_1, \ldots, A_k), \pi_N^e)$, and $q_f \in \delta((B_1, \ldots, B_k), \alpha^e)$. Then by the inductive hypothesis $B_1 \cdots B_k \underset{G}{\overset{\alpha}{\Longrightarrow}} w$ where $\alpha \in h^{-1}(\alpha^e)$. By the definition of δ, there exists $\pi_N : [A_1 \rightarrow a_1 B_1, \ldots, A_k \rightarrow a_k B_k]$ in Π and $\pi_N \in h^{-1}(\pi_N^e)$, therefore $A_1 \cdots A_k \underset{G}{\overset{\pi_N \alpha}{\Longrightarrow}} a_1 w_1 \cdots a_k w_k$. □

Lemma 13. *For any even k-EML L, there exists a regular control set C for a universal even k-EMG G^e such that $L = L_C(G^e)$ holds.*

Proof. Let $G = (N_1, \ldots, N_k, \Sigma, \Pi, S)$ be an even k-EMG in even normal form which generates L. Let h be a homomorphism from Π^* to Π^{e*} and $M = (\{S, q_f\} \cup (N_1 \times \cdots \times N_k), \Pi^e, \delta, S, \{q_f\})$ be the NFA corresponding to G defined in the above way. From definitions, $S \underset{G}{\overset{\pi_S}{\Longrightarrow}} A_1 \cdots A_k$ if and only if $S \underset{G^e}{\overset{h(\pi_S)}{\Longrightarrow}} S_1 \cdots S_k$ and $(A_1, \ldots, A_k) \in \delta(S, h(\pi_S))$. Therefore, by Lemma 12, for any $w \in \Sigma^*$, $S \underset{G}{\overset{\alpha}{\Longrightarrow}} w$ if and only if $S \underset{G^e}{\overset{h(\alpha)}{\Longrightarrow}} w$ and $q_f \in \delta(S, h(\alpha))$. Hence, for the regular control set $C = L(M)$, $L = L_C(G^e)$ holds. □

Lemma 14. *For any even k-EML L, a control set C for a universal even k-EMG G^e such that $L = L_C(G^e)$ is regular and unique. Moreover,*

$$C = \{\alpha^e \mid S \underset{G^e}{\overset{\alpha^e}{\Longrightarrow}} w, \ w \in L\}.$$

Proof. Let $C = \{\alpha^e \in \Pi^{e*} \mid S \underset{G^e}{\overset{\alpha^e}{\Longrightarrow}} w, \ w \in L\}$. Then, clearly, $L = L_C(G^e)$.

To show that C is unique, assume that C' is another control set such that $L = L_{C'}(G^e)$. Since G^e is unambiguous by Remark 11, for any string $w \in L$ there exists a unique associate word α^e such that $S \underset{G^e}{\overset{\alpha^e}{\Longrightarrow}} w$. Since C and C' are subsets of $A(G^e)$, $w \in L$ if and only if $\alpha^e \in C$ if and only if $\alpha^e \in C'$. Therefore, $C = C'$.

Let G be an even k-EMG in even normal form which generates L. Let C'' be a regular control set which the NFA corresponding to G accepts. Since a control set with which G^e generates L is unique, $C = C''$, hence C is regular. This completes the proof. □

We also have the converse of Lemma 13.

Lemma 15. *Let G^e be a universal even k-EMG and C be a regular control set for G^e. Then $L = L_C(G^e)$ is an even k-EML.*

Proof. Let $G^e = (\{S_1\}, \ldots, \{S_k\}, \Sigma, \Pi^e, S)$ be a universal even k-EMG and $M = (N, \Pi^e, \delta, S, F)$ be an NFA such that $C = L(M)$ holds. From G^e and M, we can define an even k-EMG $G = (\{S_1\}, \ldots, \{S_{k-1}\}, N, \Sigma, \Pi, S)$ and a homomorphism h from Π^* to Π^{e*} as follows.

1. if $\delta(S, \pi_S^e) \ni A$ and $\pi_S^e : [S \to S_1 \cdots S_k]$, then $\pi_S : [S \to S_1 \cdots A]$ is in Π and $h(\pi_S) = \pi_S^e$,
2. if $\delta(A, \pi_N^e) \ni B$ and $\pi_N^e : [S_1 \to a_1 S_1, \ldots, S_k \to a_k S_k]$, then $\pi_N : [S_1 \to a_1 S_1, \ldots, A \to a_k B]$ is in Π and $h(\pi_N) = \pi_N^e$.
3. if $\delta(A, \pi_T^e) \cap F \neq \emptyset$ and $\pi_T^e : [S_1 \to \lambda, \ldots, S_{k-1} \to \lambda, S_k \to w_k]$ then $\pi_T : [S_1 \to \lambda, \ldots, S_{k-1} \to \lambda, A \to w_k]$ is in Π and $h(\pi_T) = \pi_T^e$,

By the similar argument in the proof of Lemma 13, it is easy to prove that for any $w \in \Sigma^*$, $S \Longrightarrow_{G^e}^{\alpha^e} w$ and $\delta(S, \alpha^e) \cap F \neq \emptyset$ if and only if $S \Longrightarrow_G^{\alpha} w$ where $\alpha \in h^{-1}(\alpha^e)$. Therefore, $L = L_C(G^e) = L(G)$ is an even k-EML. $\square$

By combining Lemmas 13, 14, and 15, we have the following representation theorem.

Theorem 16. *A language L is an even k-EML if and only if L is generated by a universal even k-EMG with a regular control set unique for L.*

This theorem implies that, to identify an even k-EML L, we have only to identify a unique regular control set C for a fixed universal even k-EMG.

From Theorem 16 we have the following corollaries, which show that the class of even k-EMLs has similar properties for regular sets.

Corollary 17. *The class of even k-EMLs is closed under Boolean operations.*

Corollary 18. *The following problems are solvable for even k-EMLs L_1 and L_2.*

1. $L_1 \subseteq L_2$,
2. $L_1 = L_2$.

5 Learning Based on Control Sets

In this section we consider the learning problem for even k-EMLs based on control sets. We shall show that the problem of learning even k-EMLs is reduced to the problem of learning regular sets.

We may assume that a learning algorithm (1) gets strings as examples, (2) outputs strings for queries, and (3) outputs representations for languages as conjectures. Although various learning methods for formal languages have been proposed up to now (see [1] and [5], for example), this assumption seems to be general enough to include any former learning situation. Therefore, without loss of generality, we may assume that a learning algorithm for regular sets gets strings, outputs strings, and outputs representations for regular sets such as *DFA*s and regular expressions, and a learning algorithm for even *k-EML*s gets strings, outputs strings, and outputs even *k-EMG*s. Note that a hypothesis space of a learning algorithm for even *k-EML*s depends on a hypothesis space of a learning algorithm for regular sets. For example, if the class of *DFA*s is a hypothesis space of a learning algorithm for regular sets, then the class of all even *k-EMG*s in even normal form to which *DFA*s are corresponding is a hypothesis space of a learning algorithm for even *k-EML*s, which is a proper subclass of all even *k-EMG*s. If the class of *NFA*s is a hypothesis space of a learning algorithm for regular sets, then the class of all even *k-EMG*s in even normal form is a hypothesis space of a learning algorithm for even *k-EML*s. The time complexity of learning algorithms may depend on its hypothesis space. For example, for certain regular set R, the minimum state *DFA* may be exponentially larger than a minimum state *NFA*, therefore, for certain even *k-EML L*, the minimum-size of even *k-EMG*s depends on a class of even *k-EMG*s under the consideration.

The representation theorem (Theorem 16) implies that, to identify an even *k-EML L*, we have only to identify a unique regular control set $C = \{\alpha^e \mid S \Longrightarrow_{G^\bullet}^{\alpha^\bullet} w, \ w \in L\}$ for a universal even *k-EMG* fixed for an alphabet. To construct a learning algorithm for even *k-EML*s, we have only to prepare a front-end processing algorithm and add it to a learning algorithm for regular sets. As we mentioned before, a learning algorithm (1) gets strings as examples, (2) outputs strings for queries, and (3) outputs representations for languages as conjectures. Then the front-end processing algorithm performs the following three auxiliary tasks.

1. It converts an input string w to an associate word α^e of a universal even *k-EMG* G^e by parsing w in G^e and gives α^e to a learning algorithm for regular sets.
   Let m be the cardinality of an alphabet Σ and $lg(w) = n$. To parse w in G^e, the front-end processing algorithm first computes the quotient x and the residue y of $\frac{n}{k}$, which takes time polynomial of n. Then it divides w into k numbers of substrings $w_1, \ldots, w_k$ such that $lg(w_i) = x$ for each i $(1 \leq i \leq k - 1)$ and $lg(w_k) = x + y$, which also takes time polynomial of n. Let $w_i = a_{i,1} \cdots a_{i,x} u_i$ $(a_{i,j} \in \Sigma, u_i \in \Sigma^*)$ for each i $(1 \leq i \leq k)$. If the front-end processing algorithm finds the nonterminal matrix rule $\pi_N : [S_1 \rightarrow a_{1j}S_1, \ldots, S_k \rightarrow a_{kj}S_1]$ for each j $(1 \leq j \leq x)$, and finds the terminal matrix rule $\pi_T : [S_1 \rightarrow \lambda, \ldots, S_{k-1} \rightarrow \lambda, S_k \rightarrow u_k]$ of G^e then it parses w successfully, otherwise, it fails. This takes time polynomial of n and the size of the set Π^e of matrix rules of G^e, which is bounded by a polynomial of m by Remark 10. If w can be parsed successfully then the length of the unique associate word of w is $\frac{n}{k} + 1$ or parsing fails after at most $\frac{n}{k} + 1$ steps otherwise. Therefore, the time complexity of parsing w is bounded by a polynomial of n and m.

2. It gets an associate word α^e of G^e from the learning algorithm for regular sets and converts α^e to an output string w by generating w in G^e. To generate w, for each i $(1 \leq i \leq k)$ the front-end processing algorithm con-

catenates symbols or strings of matrix rules into w_i according to α^e and then concatenates $w_1, \ldots, w_k$ into w. This obviously takes time polynomial of n and m.

3. It gets a representation for a regular set and converts the representation to an even k-EMG in even normal form. The time complexity of this conversion depends on a representation for regular sets. In the proof of Theorem 16, we have shown the conversion of NFAs to even k-EMGs. It is easy to verify that it takes time polynomial of the size of NFAs and the size of G^e. Since any regular expression and any regular grammar can be converted to an NFA in polynomial time in the size of each representation and a DFA is a special case of an NFA, converting a representation for a regular set to an even k-EMG takes time polynomial of the size of the representation for a regular set and the size of G^e, thus the cardinality of the alphabet. We note that in any case, a hypothesis space consists of k-EMGs in even normal form only.

The configuration of a learning algorithm for even k-EMLs is illustrated in Figure 1.

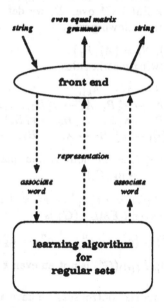

Fig. 1. The configuration of a learning algorithm for even k-EMLs

The front-end processing algorithm reduces the problem of identifying an unknown even k-EML to the problem of identifying an unknown regular control set for a universal even k-EMG G^e in polynomial time of sizes of strings, associate words, and representations for regular sets. Hence, we have the following theorem.

Theorem 19. *The problem of learning even k-EMLs is reduced to the problem of learning regular sets. Moreover, if the time complexity of a learning algorithm for regular sets is polynomial in the number of given examples, the maximum length of any given example, and the size of the representation for a regular set,, then the time complexity of a learning algorithm for even k-EMLs with using the learning algorithm for regular sets is also polynomial of the same parameters.*

This theorem immediately gives the correctness and the time complexity of a learning algorithm for even k-EMLs.

6 Even Equal Matrix Control sets

In this section, we consider a class of languages generated by a universal even k-EMG with even equal matrix control sets.

Let k and l be integers such that $k \geq 1$ and $l \geq 2$. Then let

$$a_{1,1}, \ldots, a_{k,1}, \ldots, a_{1,l}, \ldots, a_{k,l}$$

be $k \times l$ number of distinct symbols and $\Sigma = \{a_{i,j} \mid 1 \leq i \leq k,\ 1 \leq j \leq l\}$.

For a universal even k-EMG G^e over Σ, we define the even l-EMG $G = (N_1, \ldots, N_l, \Pi^e, \Pi, S)$ in the following way.

1. For each j $(1 \leq j \leq l)$, $N_j = \{A_j, B_j\}$.
2. Π consists of the following matrix rules.

$$\Pi = \left\{ \begin{array}{l} \pi_S : [S \rightarrow A_1 \cdots A_k], \\ \pi_1 : [A_1 \rightarrow \pi_S^e B_1, A_2 \rightarrow \pi_2^e B_2, \ldots, A_k \rightarrow \pi_l^e B_k], \\ \pi_2 : [B_1 \rightarrow \pi_l^e B_1, \ldots, B_k \rightarrow \pi_l^e B_k], \\ \pi_T : [B_1 \rightarrow \pi_1^e, B_2 \rightarrow \lambda, \ldots, B_{k-1} \rightarrow \lambda, B_k \rightarrow \pi_T^e] \end{array} \right\}$$

where each π_j^e $(1 \leq j \leq l)$ is the nonterminal matrix rule of G^e such that $\pi_j^e : [S_1 \rightarrow a_{1,j} S_1, \ldots, S_k \rightarrow a_{k,j} S_k]$.

It is easy to verify that $L(G) = \{\pi_S^e \pi_1^{e^n} \cdots \pi_l^{e^n} \pi_T^e \mid n \geq 1\}$ and the language generated by G^e with the even l-EML $L(G)$ is

$$L_{L(G)}(G^e) = \{a_{1,1}^n \cdots a_{k,1}^n \cdots a_{1,l}^n \cdots a_{k,l}^n \mid n \geq 1\}.$$

Then by Ibarra's result [4], $L_{L(G)}(G^e)$ is not an even k-EML but an even $(k \times l)$-EML.

Thus with even equal matrix control sets we have another class of languages, for which if we *a priori* know the order of even equal matrix control sets then we can construct a learning algorithm using the one for regular sets. We, however, do not consider any property of these classes; this is one of our future work.

Acknowledgments

The author would like to thank Dr. Takashi Yokomori, University of Electro-Communications, for his helpful comments and suggestions.

References

1. D. Angluin and C. H. Smith. Inductive inference : Theory and methods. *ACM Computing Surveys*, 15(3):237–269, 1983.

2. S. Crespi-Reghizzi and D. Mandrioli. Petri nets and commutative grammars. Internal Report 74-5, Laboratorio di Calcolatori, Instituto di Electtrotecnica ed Electtronica del Politecnico di Milano, 1974.

3. J. E. Hopcroft and J. D. Ullman. *Introduction to Automata Theory, Languages, and Computation*. Addison-Wesley, Reading, Massachusetts, 1979.

4. O. H. Ibarra. Simple matrix languages. *Information and Control*, 17:359–394, 1970.

5. L. Pitt. Inductive inference, dfas, and computational complexity. In K. P. Jantke, editor, *Proceedings of 2nd Workshop on Analogical and Inductive Inference, Lecture Notes in Artificial Intelligence, 397*, pages 18–44. Springer-Verlag, 1989.

6. A. Salomaa. *Formal Languages*. Academic Press, Inc., New York, 1973.

7. R. Siromoney. On equal matrix languages. *Information and Control*, 14:135–151, 1969.

8. R. Siromoney, A. Huq, M. Chandrasekaran, and K. G. Subramanian. Pattern classification with equal matrix grammars. In *Proceedings of the IAPR Workshop on Syntactic and Structural Pattern Recognition*, 1990.

9. Y. Takada. Learning semilinear sets from examples and via queries. To appear in *Theoretical Computer Science*.

10. Y. Takada. Grammatical inference for even linear languages based on control sets. *Information Processing Letters*, 28(4):193–199, 1988.

11. Y. Takada. Learning equal matrix grammars with structural information. In *LA winter Symposium*, pages 61–72, 1990.

12. D. Wood. Bounded parallelism and regular languages. In G. Rozenberg and A. Salomaa, editors, *L Systems, Lecture Notes in Computer Science, No.15*, pages 292–301. Springer-Verlag, 1974.

Parallel Dynamic Programming Algorithms
for Image Recognition Based on its Contour

Giang Vu Thang
Department for Pattern Recognition
Institute of Computer Science
Lieu giai, Tu liem, Ba dinh, Ha noi, Viet nam

Abstract. The dynamic programming is an important procedure for speech understanding and image recognition. In this paper, the application of dynamic programming to recognise the image basing on its contour is recalled, then the parallelism of this process is analysed and exploited. Some parallel dynamic programming algorithms for evaluating the smallest value of the minimum intersegment distance path in a data table are designed in detail on several kinds of the specialized parallel computer, such as 2D Systolic Array, 1D Systolic Array and an associative computer of the SIMD type. Their computation complexities are respectively estimated as $O(k)$, $O(k)$ and $O(kw)$, while the cost $O(mn)$ is obtained on serial computer, where $k=\max(2n,m)$, m is the number of segments of a contour to be recognised, n gives the number of segments of the sample contour and w is the bit-width of the used associative fields.

1. Introduction.

In this paper, we present some methods to parallelize the dynamic programming process on different computational models. In the second section of the paper, the application of dynamic programming in shape matching based on its contour will be briefly presented. In the third one, a serial searching algorithms for estimating the value of a minimum intersegment distance path in a data table will be analyzed in detail. The computation complexity equals to $O(nm)$, where m is the number of segments of a contour to be recognized and n gives the number of segments of the sample contour. In the fourth section of paper, we describe a parallel version of the previous algorithm for a model of 2D systolic array with k^2 available processors, where $k=\max(2n,m)$. The complexity of the algorithm is $O(k)$. Then an improved version of this algorithm will be presented on the 1D systolic array in the fifth section. In the sixth section, a problem concerning with estimation of the value of a minimal distance path in a table on an associative computer of the SIMD type with $(2k-1)$ processors will be solved. A parallel associative algorithm is developed with complexity $O(kw)$, where w is the bit-width of the used associative fields. The last one is devoted to conclusion and some open problems.

II. The application of dynamic programming in image matching based on its contour.

In [1], one application of Dynamic Programming technique in shape matching by its contour was presented in detail. The main idea of this method consists in extraction of some kinds of the sample image, whereby their contour are extracted in a favour condition. (See fig. 1)

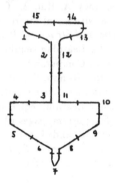

Fig. 1

Fig. 1 is a contour of any kind of airplane. This contour is represented by the chain code. (by a special technique presented in [1]). This contour is separated into some distinguished segments numbered by 1,2,...,n (n=15, for example) (See fig. 2).

Using the Fourier coefficients, we can describe these segments of sample contour in an efficient man-

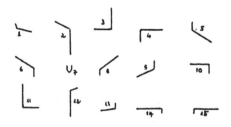

Fig. 2

ner by 6 Fourier coefficients denoted by

$$F^i = (F_1^i, F_2^i, ..., F_6^i, F_{-6}^i, F_{-5}^i, ..., F_{-1}^i), \ i = \overline{1,n}$$

F_j^i gives the j-th coefficient of i-th extracted segment. By the standardization method, we can get the standardized vector SF^i of F^i in a form

$$SF^i = (SF_1^i, ..., SF_6^i, SF_{-6}^i, ..., SF_{-1}^i), \ i = \overline{1,n}$$

In this paper, we don't pay so much attention on the standardization process of the contour segments. Reader can be advised to the reference [1] for a more detailed information. In order to recognize a new image with a new set of m vectors of the standardized Fourier coefficients

$$SA^j = (SA_1^j, SA_2^j, ..., SA_6^j, SA_{-6}^j, SA_{-5}^j, ..., SA_{-1}^j)$$

where $j = \overline{1,m}$, the distance measure function will be formulated by a formula

$$D(SF^i, SA^j) = a_{ij} = \sum_{k=-6}^{6} |SF_k^i - SA_k^j|^2,$$

$$i = \overline{1,n}, \ j = \overline{1,m}$$

Depending on this distance measure function, a intersegment distance table A will be for - med as

$$A = \begin{pmatrix} a_{11} & a_{12} & \cdots & a_{1m} \\ a_{21} & a_{22} & \cdots & a_{2m} \\ \vdots & \vdots & \ddots & \vdots \\ a_{n1} & a_{n2} & \cdots & a_{nm} \end{pmatrix}$$

where m gives the number of segments of the contour to be recognized, n gives the number of segments of the sample contour. The distance between two contours is defined by the measured value of the minimum cost path crossing the intersegment distance-table. This path starts at any certain position in the first column of the table A and terminates at any certain point in the last column of this distance table. (See fig. 3 for a more detailed explanation).

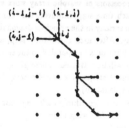

$$(i-1,j-1) \quad (i-1,j)$$
$$(i,j-1) \quad i,j$$

Fig. 3

The desired minimum distance path in table A will be found by applying the dynamic programming technique, which is based on the following principle that " the minimal distance path from the first column of table A to the last column is a concatenation of the minimal distance path from the first column to any certain element a_{ij} in table and the minimal distance path from this point to the last column."

The searching strategy is performed by the formula (1)

$$D(i,j) = min\{a_{i+1,j+1} + D(i+1,j+1), a_{i+1,j}$$

$$+ D(i+1,j), a_{i,j+1} + D(i,j+1)\} \quad (1)$$

where D(i,j) is a value of the minimum cost path, which starts at point a_{ij} and terminates at any certain point of the last column of the distance table A.

Since the beginning point of a desired path is not known, it is necessary to consider all the possibilities and to duplicate the rows of table A (see fig. 4)

$$A = \begin{pmatrix} a_{11} & a_{12} & \cdots & a_{1m} \\ a_{21} & a_{22} & \cdots & a_{2m} \\ \vdots & \vdots & \ddots & \vdots \\ a_{n1} & a_{n2} & \cdots & a_{nm} \\ a_{11} & a_{12} & \cdots & a_{1m} \\ a_{21} & a_{22} & \cdots & a_{2m} \\ \vdots & \vdots & \ddots & \vdots \\ a_{n1} & a_{n2} & \cdots & a_{nm} \end{pmatrix}$$

Fig. 4

Clearly, a desired path will have a form as shown in fig. 3 according to the searching strategy (1) with m=5, 2n=6. A new image, which is considered to be recognized, belongs to any certain sample shape, where the a minimum value of the minimum distance paths is obtained.

III. A sequential searching algorithm for the minimum distance path in a intersegment distance table.

Let us consider an intersegment distance table with m columns and 2n rows. With the searching strategy presented in (1), at each point (i,j) of the table, there exist only three possibilities of the path coming to it. Those are paths through (i,j-1), (i-1,j-1) and (i-1,j) respectively (see fig.3). By comparing three values of the paths coming to point (i,j), it is able to decide which path will be a path coming to (i,j) with a minimum cost and at (i,j) the value of this path with the minimum cost will be estimated.

The dynamic programming problem, which is applied in our case is sequential in nature. The value of the minimal cost path to each point in the first column is precisely the value a_{i1} at that point. In the next column of table (say, j-th column), except a point (1,j), at each point (i,j) i=2,3,...,2n on j-th column, three distance values are needed to be evaluated. Those are the values of the minimal cost path to $(i,j-1)$, to $(i-1,j-1)$ and to $(i-1,j)$ respectively. Comparing three these values we can estimate the value of the minimal cost path starting at the first column of the table to (i,j).

Consequently, the minimal distance value at point (i,j) is attended to calculate the value of a minimal cost path coming to points $(i+1,j)$, $(i+1,j+1)$, $(i,j+1)$. The searching process is carried out from the left to the right, from the second column to the last column of table and in each column, the calculation is performed from top to bottom.

A version of the sequential algorithm can be given in a form as below.

Algorithm 1
    for j=2,3,...,m
        for i=1,2,. . . ,2n
            if (i=1) then $a(1,j) \leftarrow a(1,j) + a(1,j-1)$
            TG $\leftarrow$ min $(a(i,j), a(i-1,j-1), a(i-1,j))$
            $a(i,j) \leftarrow a(i,j) + TG$
        Continue
    Continue

When the algorithm is finished, $a(i,m)$ i=1,2, ..., 2n give the values of the possible minimal cost paths starting from the first column to points (i,m) i=1, ..., 2n in the last column of table a respectively. The smallest value of a desired path is selected by $(2n-1)$ comparisons. In other words, it is fulfilled by a algorithm with a complexity $O(n)$.

The complexity of this sequential algorithm can be estimated by

$$C1 = (n-1)(6n-2)$$
$$= 6nm - 6n - 2m + 2$$
$$C1 = O(mn)$$

IV. Parallel searching algorithm on a 2D systolic array.

In this section of the paper, we shall present a parallel version of the dynamic programming searching algorithm for a 2D systolic array. One among the main reasons to apply our problem on 2D systolic array system is the efficiency of information propagation between the processors. As analyzed before, the old value of one point in the intersegment distance table is updated depending on three values contained in three other points of table A. Then, the new values of this considered points will be propagated to three neighbor points according to the directions specified by the propagation strategy (1) (See fig. 3).

The organization of a systolic array is formulated in a following form. It is assumed that a 2D systolic array with a size k will be designed, where k= max(2n,m) (See fig. 5).

The systolic array consists of k^2 processing elements denoted by P_{ij} $i = \overline{1,k}$, $j = \overline{1,k}$, each processor P_{ij} contains three registers R_1, R_2, R_3, one own local memory keeping the value of a_{ij} i,j=1,k and one state register R_s. The processor P_{ij} with those indices ij,

which don't exist in the intersegment distance table will store a value 0 in its own local memory. Three registers R_1, R_2, R_3 receive the propagated information of the west, west-north and north neighbor processors respectively.

Processor P_{ij} $(i = \overline{1,k}, j = \overline{1,k})$ is called active $(R_s = 1)$ iff the contents of its three registers R_1, R_2, R_3 were already loaded. Firstly all the processors P_{i1}, $i = \overline{1,k}$ of the first column are active with the registers R_1, R_2, R_3 to be loaded by zero. The processors P_{1j}, $j = \overline{2,k}$ of the first row (except P_{11}) are passive with the registers R_2, R_3 to be loaded by a value G (G

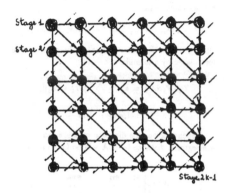

Fig. 5

is a possible largest value). A processor only loads the propagated values of the neighbor processors into its registers R_1, R_2, R_3, iff it is still passive. All the active processors of systolic array perform the same operations at each time step, which contain 2 comparisons, 1 addition and 1 propagating operation of value of its local memory to three neighbor processors, which are the south, south-east and east processors respectively.

The processors of systolic array work simultaneously at each time step, however, the information is propagated serially in the principal diagonal direction of the array. After each calculation step, a group of processors lying on a line parallel to the second diagonal of array will be activated. These groups formulate $(2k-1)$ different stages of a systolic array (See fig. 5 for easier explanation).

The parallel algorithm on a 2D systolic array can be written, in a following form.

Algorithm 2.

While P_{kk} not active

    Do for all the active processors P_{ij}

        Begin

            TG $\longleftarrow$ min (R_1, R_2, R_3)

            $a_{ij} \longleftarrow a_{ij} + TG$

            Propagation of value a_{ij}

        End

It is very easy to find that the algorithm 2 will terminate after 2k-1 iterations and gives the possible minimal cost paths with their values kept in the memories of processors $P_{ik}, i = \overline{1,k}$.

In order to choose a smallest value of the values stored in memories of P_{ik}, $i = \overline{1,k}$, k-1 comparisons are needed according to a certain sequential algorithm.

Briefly, the total complexity of the parallel dynamic programming searching algorithm on an 2D systolic array with k^2 available processors will be estimated by O(k), where k = max(2n,m).

V. An Improved Parallel Dynamic Programming Algorithm for a 1D Systolic Array.

As seen in the section IV of this paper, the problem estimating the value of a minimal cost path can be solved on a 2D systolic array possessing k^2 processors. The parallel algorithm will terminate after 2k-1 iterations and gives the possible smallest values of minimal cost paths starting at the first column of the coefficient table A and reaching the terms of the last column. However, such algorithm brings some disadvantages, which we have to overcome. Firstly, the number k^2 of necessary processors is too large. It means that each processor is only used for one term in the table. Secondly, there are many processors to be idled during the performance of the algorithm. In each calculation step, only a group of processors, which are located on a line parallel to the second diagonal of the array was activated. This event is a serious disadvantage, which we have to prevent.

In this section, we shall present an improved version of the above mentioned algorithm. An 1D systolic array will be designed, which contains only 2k-1 processors denoted by $P_1, P_2, ..., P_{2k-1}$. Each processor $P_i(i = 1, 2, ..., 2k-1)$ has 3 registers denoted by R_r^i, R_w^i, R_l^i and a small private memory M^i in order to store the input data. The register R_r^i receives the propagated data from a processor P_{i-1}, while the register R_l^i receives the propagated data from a neighbor processor P_{i+1}. The register R_w^i keeps the processed data of P_i for its next calculation step. A processor will be called active iff the data from table A is loaded into its memory M. In other cases, it is passive. Each processor P_i (i=1,...,2k-1) has two outputs to propagate its processed data and 3 inputs to load the data in. The configuration of a 1D systolic array with the placement and timing of input data can be seen in fig. 6 for k=6.

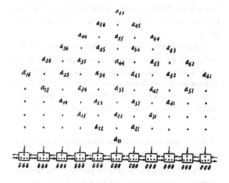

Fig. 6

First of all, the contents of registers R_r^i, R_w^i, R_l^i of the first k-1 processors P_i , i=1,2,...,k-1 are set to G (G

is a largest number, which the register R can accommodate) while the registers R_r^i, R_w^i, R_l^i of k remaining processors P_i (i=k, k+1,..., 2k-1) are initialized by 0. The activity of a processor P_i can be described in the following manner. Firstly, the input data of table A is loaded into a local memory M^i of the processor P_i. The contents of three registers R_r^i, R_w^i, R_l^i are compared, from which the smallest value is chosen. The content of memory M^i is summed with this smallest value. The sum is stored in the register R_w^i. Simultaneously, this sum will be also propagated to two neighbor processors P_{i-1} and P_{i+1} in the left and right directions respectively.

The algorithm consists of 2k-1 iterations, in the first k iterations, at I th calculation step (I=1, ...,k), there are I processors to be active with the indeces K-I+2j+1, J=0,...,I-1. In (k-1) remaining iterations, at I-th calculation step (I=k+1,..., 2k-1), there are 2k-I processors to be active, the indices of these active processors are (I-k)+2j+1, j=0,...,2k-I-1.

The parallel searching algorithm on a 1D systolic array can be written in a following form.

For I=1,2,...,2k-1

    Load data in array from table A

    Do { for all active processors P_i of array }

        Begin

            $TG^i \longleftarrow$ min (R_r^i, R_w^i, R_l^i)

            $M^i \longleftarrow M^i + TG^i$

            { propagation of processed data }

            $R_w^i \longleftarrow M^i$

            $R_r^{i-1} \longleftarrow M^i$

            $R_r^{i+1} \longleftarrow M^i$

End;

Algorithm terminates after 2k-1 iterations and gives k possible smallest values of the minimal cost paths from the first column of table to k terms of the last column in k first registers $R^i_\bullet$ (i=1,...,k) of the array.

The smallest value of minimal cost path can be estimated by any certain algorithm presented in [4] with a computation complexity O(k). Briefly, the computation complexity for estimating the value of a minimal cost path in table A will be obtained with complexity O(k) on 2k-1 processors of 1D systolic array.

VI. Parallel dynamic programming searching algorithm on an associative computer of the SIMD type.

In this section of paper, we present a parallel dynamic programming searching algorithm for a specialized parallel computer of the SIMD type. First of all, we assume that one number with bit-width w in its bit-representation is sufficient to represent the intersegment distance or the value of every possible path in table. We call G the number with 1's in all w bit positions, it means that $G = (111_\bullet...1)$

The algorithm uses (2k-1) associative fields denoted by $A_1, A_2, ..., A_{2k-1}$. Each field A_i; $i = \overline{1,2k-1}$ is subdivided into 4 subfields denoted by A^1_i, A^2_i, A^3_i and A^4_i

A^1_i contains the values of intersegment distance in table, while A^2_i contains the propagated values from field A^1_{i-2} at the corresponding places. A^3_i contains the propagated values from field A^1_{i-1} at the higher one position places and A contains the propagated values from field A^1_{i-1} at the lower-one position places. (See fig. 7). ,

In order to efficiently perform the operations on the associative fields, firstly, the fields $A^j_i, i = \overline{1,2k-1}$, j=2, 3, 4 are assigned by 0 and a placement of the intersegment distance values a_{ij} in the fields of MDA is given as follow with k=4 (Fig. 8)

• denotes those places, where we were not interested in their values. In the associative algorithm, which we present below, the mask register M has a very important role. We call mask(u,v) a mask function, which creates in the mask register M of an associative module u values 0 at u first places of M and v values 1 at v next places. For example M= mask(3,4) implies M=(00011110...0).

The parallel dynamic programming searching algorithm on an associative parallel computer with one associative module and (2k-1) available processors can be given in following form.

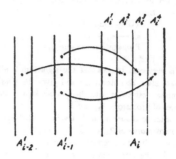

Fig. 7

Algorithm 3
  For i=1,2,...,k
    M ⎯ mask(k-i,1)
    A^3_i ⎯ G(M)
    A^3_i ⎯ G(M)
  Continue

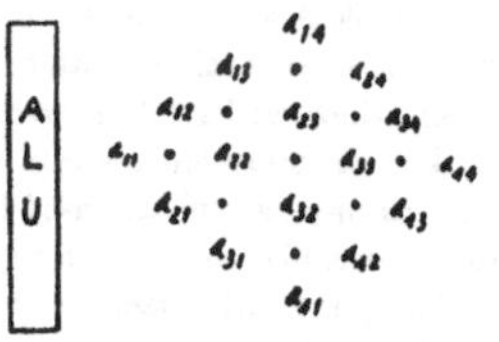

Fig. 8

For i=1,2,...,2k-1
  M ⎯ mask($|k - i|$, 2i-1)
  TG ⎯ min (A^2_i, A^3_i, A^4_i)
  A^1_i ⎯ $(A^1_i + TG)(M)$
  { propagation }
  A^2_{i+2} ⎯ $A^1_i(M)$
  A^3_{i+1} ⎯ $SHIFT^{+1}(A^3_{i+1})$
  λ^3_{i+1} ⎯ $A^1_i(M)$
  A^3_{i+1} ⎯ $SHIFT^{-1}(A^3_{i+1})$
  A^4_{i+1} ⎯ $SHIFT^{-1}(A^4_{i+1})$
  A^4_{i+1} ⎯ $A^1_i(M)$
  A^4_{i+1} ⎯ $SHIFT^{+1}(A^4_{i+1})$
Continue

By counting the computational complexities of mac- ro instructions used in this algorithm, it is very easy to estimate the cost of algorithm in a form O(kw) , where k=max(2n,m) w is the number of the necessary bits, which are sufficient to represent the cost of any certain path crossing the intersegment distance table.

After algorithm 3 terminates, the distance values of the possible, minimal distance paths from the first column to the last column of table A are placed on k last fields in the associative memory. The algorithm estimating the smallest cost of the minimal distance paths can be shown below and finally, the cost of the desired path will be placed at a first element of field A_1^1 from top.

$$M \leftarrow mask(k-1,1)$$
for $i=2k-2,2k-3,...,k$
$$TG \leftarrow A_{i+1}^1$$
$$TG \leftarrow SHIFT^{+1}(TG)$$
$$M \leftarrow SHIFT^{+1}(M)$$
$$A_i^1 \leftarrow min(A_i^1,TG)(M)$$
Continue

Obviously, the computation complexity $O(kw)$ is obtained in this algorithm.

Briefly, the total computation complexity $o(kw)$ of the parallel associative algorithm on an associative computer of the SIMD type with one available associative module and $(2k-1)$ processors is obtained.

VII. Conclusion.

Dynamic programming is an important procedure in speech processing and image recognition field. In this publication, the application of this technique for the shape recognition depending on the extracted segments of its contour is described. Three versions of the dynamic programming searching algorithms for the minimal distance path in an intersegment distance table are presented on the different models of computers : sequential computer, 2D systolic array with k^2 processors, 1D systolic array with 2k-1 processors and parallel associative computer of the SIMD type with one available associative module and (2k-1) processors. The computation complexities $O(mn)$, $O(k)$, $O(k)$, $O(kw)$ are obtained on these different models respectively, where m is the number of the extracted segments of the contour to be recognized, n gives the number of the extracted segments of the sample contour and k= max(2n,m).

References

1. John w. Gorman, O. Robert Mitchell and Frank P. Kuhl, Partial shape recognition using dynamic programming. EEE Transactions on PAMI, Vol. 10, N 2 , March 1988.

2. Basile Iouka and Maurice Tchuente, Dynamic programming on 2D systolic array,Technical report

3. Richter, K.: Parallel Computer System SIMD In proc. AIICSR Conference I. Plander, ed. North-Holland, 1984, pp 309-313.

4. Miklosko J. Vajtersic, Klette R., Vrto I., Fast algorithm and their implementation on specialized parallel computers. Veda- North Holland, Bratislava-Ansterdam 1989.

5. Alberto Martelli, An application of heuristic search methods to edge and contour detection. Computer methods in image analysis, Edited by J. K. Aggarwal, Richard O. Duda, Azriel Rosenfeld. pp 217-227

6. M. Vajtersic, Giang Vu Thang, Threshold and histogram algorithms for a parallel associative computer. Computer and Artificial Intelligence, 5(1986), No 2 , pp 143-161.

7. Giang Vu Thang, Parallel Algorithms of His - togram Equalization on Parallel Computer of the SIMD Type. Computer and Artificial Inte - lligence, Vol. 10, No. 5, pp 465 - 476 (1991)

Neural Model for Pattern Recognition

Jing-xue WANG   Masashi NAKAMURA*
Takashi JIMBO   Masayoshi UMENO

Department of Electrical & Computer Engineering,
Nagoya Institute of Technology,
Gokiso-cho, Showa-ku, Nagoya 466, JAPAN

*Department of Electrical Engineering,
Suzuka College of Technology,
Shiroko-cho, Suzuka 510-02, JAPAN

1   Introduction

Using neural model, many pattern recognition model are proposed such as perceptron[1],Hopfield model[2], neo-cognitoron[3]. Particularly learning algorithms such as back-propagation[4] were proposed for multi-layered perceptron type network with hidden-layers and they have been studied extensively because of their possibility of the application to character and voice recognitions. But when we show a pattern to the input-layer of multi-layers perceptron network, they may have some problems.

1. Larger size of input layer, increasing the number of connecting weights and calculations.

2. If an unit has a very large connecting weight after learning process, the area of the input layer which contains it determines the output of the system. In the extreme case, an input variation of only one unit will make the system output change completely.

To solve these problem, a learning network model with limited connecting weight value has been proposed, the combination of feature detection and recognition which was popular in pattern recognition has also been applied to character recognition. In such system, some geometrical features are detected from an input pattern before the pattern is recognized. But it is difficult for the feature detector to distinguish the noise such as extra dot or line break down from the feature such as point or ends of line. When noise is detected as a feature, the recognition will be failed.

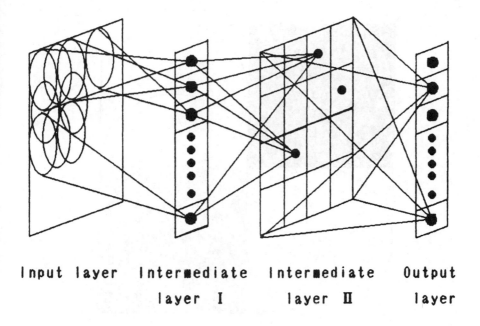

Input layer   Intermediate   Intermediate   Output

layer I   layer Ⅱ   layer

Figure 1: The structure of system

In this paper, we propose a handwritten character recognition system using simple engineering model[5]. Dividing the input layer into the overlapping receptive fields, the number of the connection between units are reduced very much and the recognition rate for distorted pattern is improved very much.

2 Structure of the system

The system is multi-layered network consisted of input layer, intermediate layer I, intermediate layer II and output layer as show in Fig.1.

The unit in each layer are connected to the unit in next layer. There is neither connection between units in the same layer nor feedback from output layer to input layer. Each unit in intermediate layer I are connected to the units in a restricted area in input layer, and receptive field of the unit in intermediate layer I. We drew the circles to show the overlapping of the receptive fields clearly in Fig.2, but the shape of the receptive field in not required to be circle.

In computer simulation, the regular square area circumscribed to the circle is used as a receptive field as Fig.2. By this method, in spite of increasing the number of units in input layer and intermediate layer I, the number of the connection in whole system is not much increased. For example, but the size of input layer is $N \times N$ and the size of receptive field is $m \times m$. Then K, the

Figure 2: Some receptive fields in the dotted area in Fig.1.

number of intermediate layer I is

$$K = \frac{N}{m} \times \frac{N}{m} + 2 \times \frac{N}{m} \times \left(\frac{N}{m} - 1\right)$$
$$= \frac{N}{m} \times \left(3 \times \frac{N}{m} - 2\right) \tag{1}$$

And the number of connections between input layer and intermediate layer I are $K \times m^2$, that is $3 \times N^2 - 2 \times N \times m$.

Table.1: Comparison of the number of the connection
(The size of receptive field is 6 × 6)

Size of input layer	Size of intermediate layer I	Number of connection	
		This model	Conventional model
$N \times N$	K		$K \times N^2$
12 × 12	8	288	1,152
24 × 24	40	1,440	23,040
30 × 30	65	2,340	58,500
36 × 36	96	3,456	124,416
48 × 48	176	6,336	405,504

Table 1 shows the number of connections between input layer and intermediate layer I of this model whose receptive field size is 6 × 6 and conventional multi-layered perceptron model. Clearly the number of connections of this model is much smaller than that of the conventional model.

Each unit in the intermediate layer II and output layer is connected to all units in the intermediate layer I and the intermediate layer II, respectively. In order words, all units in the intermediate layer I form one receptive field for every unit in the intermediate layer II, and all units in the intermediate layer II form one receptive field for every unit in the output layer.

The unit in each layer except in the input layer receive the weighted signals from the unit of its receptive field in the processing layer, sums up them and output a signed to following layer via proper output function. This process is expressed as follows.

$$I_i^k = \sum_{j=1}^{n} W^{k-1}{}_j{}^k{}_i O_j^{k-1} - \theta_i^k, \tag{2}$$

$$O_i^k = f(I_i^k) \tag{3}$$

Where I_k and O_k are the summed up input and the output signals of i-th unit in k-th layer, respectively $W^{k-1}{}_j{}^k{}_i$ is the connecting weight from j-th unit in $(k-1)$-th layer to i-th unit in k-th layer. The units in different layers have different functions depending on the layer but the units in one layer have same function. Following expressions are used as output functions of the unit in each layer. For the function of intermediate layer I,

$$f(X) = \begin{cases} 1 & (X > 0) \\ 0 & (X \le 0) \end{cases} \tag{4}$$

for the intermediate layer II,

$$f(X) = \frac{2}{1 + e^{-x}} - 1, \tag{5}$$

and for the output output layer,

$$f(X) = \frac{1}{1 + e^{-x}}. \tag{6}$$

we used.

There are two kinds of connecting weight, one is fixed and another is variable. All connecting weight between the input layer and the intermediate layer I are fixed to unity. The system realize the learning by changing the variable connecting weight.

3 Detection of line and learning rule

A handwritten character is a combination of lines. If the system can detect correct lines from the input pattern with noise, the input character will be

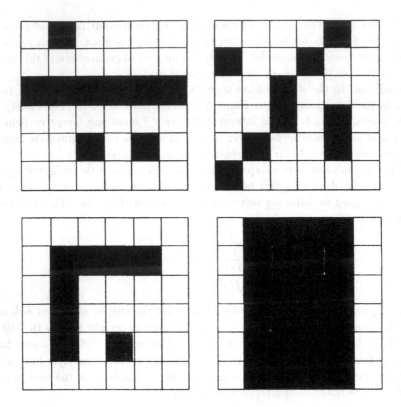

Figure 3: Examples of input pattern in a receptive field that make the unit in the intermediate layer I fire.

recognized correctly. A unit in the intermediate layer I sums up the signals from its receptive field in the input layer and judge whether there exists a line or not by the sum value. In other words, the input layer and the intermediate layer I detect lines. If the threshold value of the unit in the intermediate layer I is adjusted properly, the system can detect correct lines from the input pattern with noise.

For example, let the threshold value is 6 for 6 × 6 receptive field. Then , the system recognizes a part of line in the patterns shown in Fig.3.

In these figures, the black units represents a part of line and the shaded unit represents noise. It means that if more than 6 black units exist in the receptive field of the intermediate layer I, its cell is fireed.

The system which utilizes the feature detectors sometimes requires a pattern thinning process before the features of the pattern are detected. In our system, a part of line with any width can be detected without line thinning process.

The intermediate layer II and the output layer recognize the character by using the output of the intermediate layer II. The connecting weights are adjusted by learning process with error back propagation rule[5]. The connection

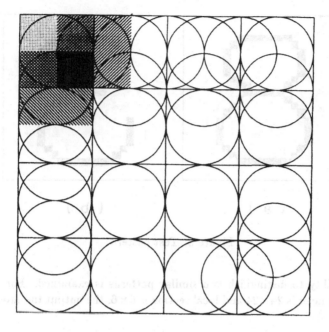

Figure 4: Relative positions of the 40 receptive fields.

$\Delta W^{k-1}_{\ j}\ ^{k}_{\ i}(t)$ of connecting weight $W^{k-1}_{\ j}\ ^{k}_{\ i}(t)$ is as follows.

$$\Delta W^{k-1}_{\ j}\ ^{k}_{\ i}(t+1) = -\varepsilon d^{k}_{i} O^{k-1}_{j} + \alpha \Delta W^{k-1}_{\ j}\ ^{k}_{\ i}(t), \qquad (7)$$

where α is the acceleration of the learning and ε is the learning coefficient. d is the general error.

$$d_k = \sum_l W^k_i\ ^{k+1}_i d^{k+1}_i f'(I^k_i) \qquad (8)$$

for the intermediate layer

$$d^{out}_i = (O^{out}_i - T_i)f'(I^{out}_i) \qquad (9)$$

for the output layer.

4 Computer simulation

The recognition of hand-written numeral characters by this system is simulated with a personal computer. In the simulation, $24 \times 24, 40, 30$ and 10 units are used in the input layer, the intermediate layer I, the intermediate layer II and the output layer, respectively. The receptive field in the input layer consists of 6×6 units and arranged as shown in Fig.4.

The connecting weights between the input layer and the intermediate layer II are fixed to unity as mentioned before, and the threshold value in the intermediate layer I are set to be 4.

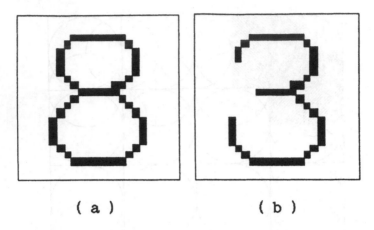

(a)　　　　　　　　　　(b)

Figure 5: Test pattern

The ability to distinguish two similar patterns is examined. For example, when input layer is 24×24 and local section is 6×6, the output the intermediate layer I are

$$01100110100001101111111101111101001101001$$

and

$$01100110000001101110110011111001001101001$$

for the pattern shown in Fig.5(a)and(b), respectively. This result proves that this system can distinguish these patterns if the connecting weights are set properly.

The connecting weights between the intermediate layers I and II and the output layer were adjusted by learning. The initial value of each connecting weight is random value from -0.5 to 0.5, and the threshold is also random value from 0 to 1. $\alpha = 0.7$, $\varepsilon = 0.3$ are also used.

Numeral characters are put into the computer by hand through a digitizer with magnetic pen. The information on the writing order is not used, The size of input pattern is normalized by the square circumscribed to the input pattern as shown in Fig.6. After this normalization, the input data are changed to 24×24 1-bit signals.

After the learning of 300 numeral characters written by 30 persons. the system can recognize the unlearned patterns written by 120 persons with the recognition rate of 92.7%. Some patterns are shown in Fig.7.

5　Conclusion

A model for the numeral character recognition is proposed. Applying the concept of the receptive field to the input layer of the system, the number of connecting

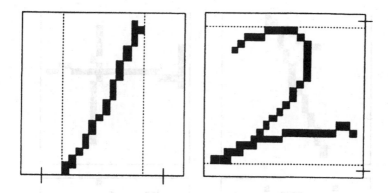

Figure 6: Normalization of input pattern.

weights is decreased very much and the recognition rate for the patterns with noise is improved very much. The computer simulation shows that this system can recognize noisy numeral pattern with the recognition rate of 92.7%.

References

[1] M.Midsky and S.Papert:"Perceptron-An Essay in Computational Geometry ",M.I.T,Press,1969.

[2] J.J.Hopfield and D.W.Tank:"Neural Computation of Decisions in Optimization Problems",Biol.Cybern, pp.141-152,1985.

[3] K,Fukushima:"Neural model of pattern recognition system, neo- cognitron", Trans.IEICE J62-A .pp.658-665,1979.

[4] D.E.Rumelhart, G,E.Hinton and R.J.Williams:"Learning Representation by Back-propagation Errors", Nature, Vol.323, pp.533-536, 1986.

[5] M.Umeno, J.X.Wang and T.Jimbo:"The Mechanism and Simulation of Human Pattern Recognition",Bull.Nagoya Inst. of Tech, pp.173-178,1989.

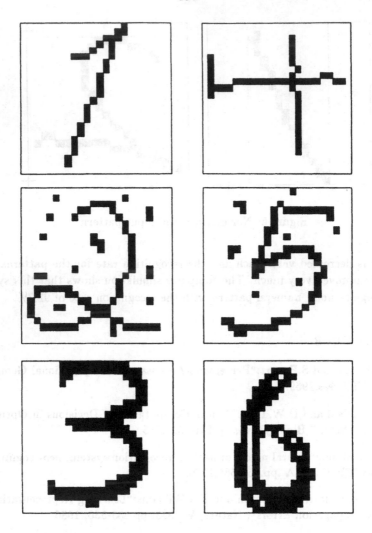

Figure 7: Example of patterns recognized correctly.

Three-Dimensional Sequential/Parallel Universal Array Grammars and Object Pattern Analysis

P. S. P. Wang
College of Computer Science
Northeastern University, Boston, MA 02115

Abstract

We introduce a sequential/parallel parsing algorithm for analyzing 3-dimensional objects represented by 3-d array grammars. The mechanism serves as a compromise between purely sequential methods which take too much time, and purely parallel methods which take too much hardware for large digital arrays.

Keywords: parallel parsing, 3-d universal array grammars, pattern generation, object representation and recognition.

1 Introduction

For the past 3 decades, there has been growing interests in automatic scene analysis by mechanical means such as computers or robots. In dealing with image processing, pattern recognition, computer vision and object recognition problems, pictorial patterns often consist of subpatterns of simple(r) types that are combined in particular ways. The subpatterns in turn may consist of still simpler sub-subpatterns, and so on. As pointed out by Fu[3] and Rosenfeld [7], this method of describing patterns in terms of subpatterns. sub-patterns, etc. is analogous to describing sentences in terms of clauses, phrases, etc. Since one can determine the syntactic structure of a sentence by parsing it in accordance with grammatical rules, this suggests that it should be possible to determine the structure of a pictorial pattern by "parsing" it in accordance with the rules of a "picture grammar".

As pointed out by Roesnfeld, since such an idea was first raised by Minsky [6], there have been many syntactic and structural methods developed for solving scene analysis, image understanding and pattern recognition problems. But most of them are sequential and are limited to two-dimensional space only. Today, in dealing with more and more complicated problems, there is a need to establish a more general model for higher dimensional images and patterns.[15,17]

In this article, we introduce such a model known as "array grammar", which has several advantages over others. It is a powerful pattern generative model generalized from Chomsky's phrase structure grammar[1]; is sufficiently flexible to extend to higher dimensions[9]; has been shown more accurate than some other methods for clustering analysis[13,14]; can be highly parallel and as powerful as tessellation or cellular automata[4]; and can provide a sequential/parallel model that serves as a compromise between a purely sequential model, which takes too much time for large arrays, and a purely parallel one, which normally requires too much hardware for large digital patterns[11]. Besides, it provides a good setting to get inside and indepth views of multi-dimensional parallel computation. automata and language theory[3,11].

This research was motivated by the work done at MKT [2,5,8,10] and is a continuation of [16].

2 Notations, Definitions and Examples

We use the basic definitions and notations of array grammars from [12.15].

Definition 2.1

A 3-d array grammar is $G = (Vn, Vt, P, S, \#)$, where

Vn:	Set of nonterminals,
Vt:	Terminal,
$S \in Vn$	Start symbol,
$\# \notin Vn \cup Vt$	Blank symbol,
$P: \alpha \longrightarrow \beta,$	$\alpha(x,y,z) \longrightarrow \beta(x,y,z)$

During derivation, the locations of each nonterminal that should be applied (replaced) are specified(by their (x,y,z) coordinates).

Definition 2.2

Parallel derivation: when a rule is applied, it is applied to all nonterminals of the α simultaneously (under specifications).

Let $\Longrightarrow$ be a binary relation between two sentential arrays (arrays that are connected and derivable from the initial sentential array with a singleton S surrounded by infinite blank symbols in the 3-d cartesian space) α and β.

We say $\alpha \Longrightarrow \beta$ if α produces (generates, derives) β. Let $\overset{\cdot}{\Longrightarrow}$ be a transitive and reflexive closure of $\Longrightarrow$. Then the language (pattern) generated by G is denoted as follows:

$$L(G) = \{ R \mid S \overset{\cdot}{\Longrightarrow} R \in V_t^{++}$$

according to the coordinates specified and are connected(according to the 6-neighborhood) }

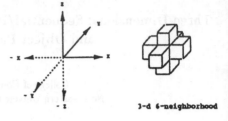

3-d 6-neighborhood

Example 2.1:

$Gu = (Vn, Vt, P, S, \#),$
where $Vn = \{S\}$, $Vt = \{*\}$ and

P: (1) S $\#$ --> S S

        $\#$ S
  (2) S   --> S

  (3) $\#$ S   --> S S

       S   S
  (4) $\#$ --> S

  (5) Sa $\#$ --> Sa S
      [where a means the left symbol
      is above right symbol]
      (along z-axis)

  (6) Sb $\#$ --> Sb S
      [where b means the left symbol
      is below right symbol]
      (along z-axis)

  (7) S   --> *

Without loss of generality, let's assume at the beginning, S is at (0,0,0). [surrounded by infinite number of #'s] Notice that, the neighborhood of

(0,0,0) is (0,0,0),(1,0,0),(0.1,0),(0,0,1),(-1,0,0),(0,-1,0),(0,0,-1)

In general, the neighborhood of (i,j,k) is

(i,j,k),(i+1,j,k),(i,j+1,k),(i,j,k+1),(i-1,j,k),(i,j-1,k),(i,j,k-1)

Consider * as a unit cube * = ⬠

Gu works as a "universal 3-d array grammar" extended from 2-d universal array grammar introduced in [16]. In the conventional syntactic pattern recognition, each pattern is characterized by a grammar. When the number of classes under consideration is very large, the grammar becomes very big involving many grammar symbols and production rules evolved from combining all classes of patterns, each represented by its individual grammar. Therefore parsing a given input pattern is very time consuming. This in turn makes classification, clustering and recognition very difficult, if not impossible.

The 3-d universal array grammar introduced in this paper can overcome such difficulty. Each 3-d object can be represented by a 1-d string (parsing sequence). Patterns of similar properties or characteristics are represented by the same or similar parsing sequence, derived from the following parallel parsing algorithm:

Here is the basic spirit of the algorithm. From each pixel under scanning, there are 6 directions to follow, namely 1 to 6 according to the 6 grammatical rules. We assign a precedence for the next choice: namely $1 > 2 > 3 > 4 > 5 > 6$, i.e., from the beginning (starting location), we try the x-direction neighbor. If there is a black pixel, then out put a "1", and move scanner one unit toward the x-axis. Continuing this way till there is no black pixel to the positive x-axis direction, then continue with the positive y-axis direction and so on and so on.

Example 2.2:

$$S \xrightarrow{1^2} \xrightarrow{2^3} \xrightarrow{5^4} \xrightarrow{7} 3*4*5 \text{ brick}$$
(Solid)

In general: m*n*p solid brick sequence is $1^{m-1}2^{n-1}5^{p-1}7$, where m,n,p => 1.

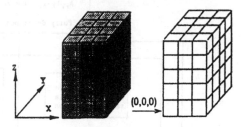

As a special case, when m=n=p in the above example, we will obtain a *cube* represented by the following sequence: $1^{n-1}2^{n-1}5^{n-1}7$.

Here, the sequence shows the sequential/parallel nature of the object generation and parsing. For instance, at the beginning, a dot moves sequentially and its trajectory forms a straight line segment. Then this line segment moves in parallel (every pixel forming this line move at the same time) to form an area (plane). And then this plane moves in parallel (every pixel in this plane moves at the same time) to form a volume.

Notice that in dealing with 3-dimensional patterns by the conventional sequential methods, it normally takes $O(n ** 3)$ cubic time, whereas here it only takes $O(n)$ linear time.

Note: if no positions (locations) specified during derivation process, by default, all locations wherever applicable are applied.

For a detailed parsing algorithm and more complicated examples, please see the following:

308

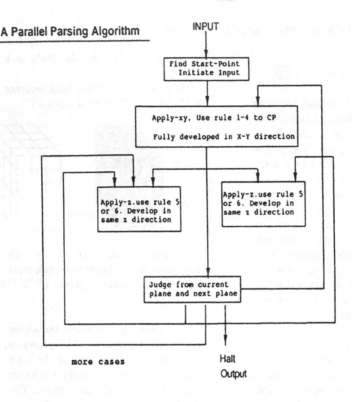

A Parallel Parsing Algorithm

INPUT

```
Find Start-Point
Initiate Input
```

```
Apply-xy, Use rule 1-4 to CP

Fully developed in X-Y direction
```

```
Apply-z.use rule 5
or 6. Develop in
same z direction
```

```
Apply-z.use rule 5
or 6. Develop in
same z direction
```

```
Judge from current
plane and next plane
```

more cases

Halt
Output

Main procedure:

1. Apply_xy(B)
   From the Curent Plane B, repeat to apply Rule 1..4, until Rule 1..4 all can not be applied,
   in other words, fully develop in X_Y plane. Return the clanged B.

2. Check_xy(rule#)
   Check one rule and apply this rule if it can be applied(rule 1..4). Return the applied
   locations, also return changed B

3. Check-apply-z(B, direction, Z-count)
   Main procedure in the Algorithm. From the Current Plane, try to apply Rule 5 or Rule6
   (according to dirction) along Z, then call Apply-xy to develop in new current plane, if
   the new current plane is in the old current plane then recursive call in the same direction,
   otherwise, first, let the current plane be the difference between the new current plane and
   old current plane, recursive all in the other direction, second, recover the current plane,
   and recursive call in the old direction.

Example 2.3:

A standard vessel (hollow up brick)

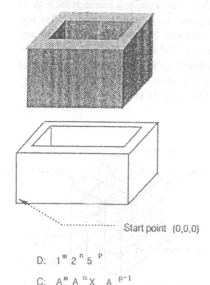

Start point (0,0,0)

D: $1^m \, 2^n \, 5^{\;P}$

C: $A^m \, A^n X \; A^{\;p-1}$

*D is the parsing sequence. C is the apply locations.
X = {[[(0.. m), (0..0)] [(0..m), (n..n)][(m..m) (0..n)]
A means all nonterminals in Current Plane.

Example 2.4:

A reversed vessel

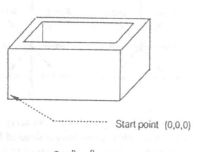

Start point (0,0,0)

D: $1^m \, 2^n \, 5^{\;P}$

C: $A^m \, A^n X \; A^{\;p-1}$

Example 2.5:

A vessel with four feet

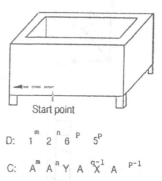

Start point

D: $1^m \; 2^n \; 6^{\;P} \; 5^P$

C: $A^m \, A^n Y \, A \; \overset{q-1}{X} \, A^{\;p-1}$

As shown, Gu can be considered as a "universal" AG (3-D), which can generate a rather large class of interesting 3-D objects, each represented by a parsing (derivation) sequence.

Notice that this parsing sequence can be generated actually in a sequential/parallel mode, which serves as a compromise between purely sequential methods which take too much time, and purely parallel methods, which are faster but take too much hardware.

This idea is also good for representing a special kind of 3-d objects known as "wire-like' objects, which have been widely used in industrial parts inspection [9,10,12]. The following are some examples. Notice that although these examples are all straight line segments with 90 degrees angles, other more complicated examples involving curved lines or line segments of arbitrary turning angles can be approximated by these basic line segments and 45 degrees or 90 degrees angles.

Example 2.6: Wire-like objects :

For other shapes and line segments with *arbitrary* degrees, they can be simulated and approximated by "26-nbd", or even "6-nbd". For example, a line segment with 10 degrees can be approximated by line segments with 0 degrees, followed by line segments of 90 degrees, followed by 0 degrees, followed by 90 degrees etc. A curved line can be approximated by tangent line segments at 0, 45 and 90 degrees.[17]

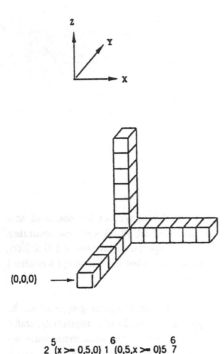

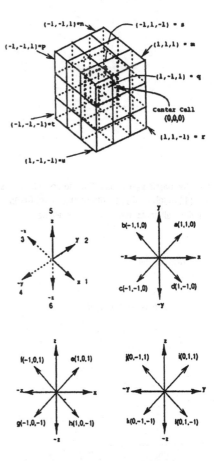

Figure 2.1 26-neighborhood of an array in 3d space, each neighborhood is denoted by an alphanumerical ranging from 1 to 6, and a to u (excluding "oh")

Notice that because of the definition of 6 neighborhood in this paper, these wire-like objects are restricted to line segments of "90 degrees" of changes only. This does not have to be the case if we relax the neighborhood to a 3 x 3 cube, (hence known as 26- neighborhood). By doing so, it is more flexible in that it can also describe line-drawing segments with "45 degrees" of change of directions, and is closer to the realistic world of 3-d wire-like objects, which can be digitized with a quantization of a "45 degree" thresholding. Such a "26-nbd" is shown in Figure 2.1.

3 Discussions and Future Research

We have introduced a formal model known as "3-d universal array grammar" (3duag) for 3-dimensional object representation. It is extended from 2-d universal array grammar (2duag) [16]. But the difference here is not just the dimensionality, though. The types of production rules are different and it is parallel. The 2duag in [16] uses regular (type 3) rules while here the 3duag Gu uses "context-free" (type 2) rules. Please note that here by "context-free" we mean to borrow the terminology from Chomsky [1]. Because its dimensions and the use of blanks (#) in the context, it is still more or less sensitive to the # symbols.

Nevertheless it is interesting to see that 3duag Gu is more powerful (in terms of generative capability) than 2duag, in that the following "multi-branch wire like" patterns can be generated by Gu (with 6-neighborhood and 26-neighborhood) but not by any 2duag (with 4 neighborhood and 8-neighborhood) even in 2d space [16]:

Figure 3

(a)multi-branch
wire-like pattern
(a noise-free example,
using 6-nbd)symbolizing
Chinese word "center"

(b)cross-like
diagonal pattern
(using 26-nbd)
symbolizing
word "man"

Further, neither Gu nor 2duag can generate any diagonal "cross-like" patterns shown in Figure 3 (b). This is because the limitation of "6-neighborhood". But with "26- neighborhood" it can.

Therefore there is certain hierarchy in the patterns depending on not only its production rules but also neighborhood definitions, It would be interesting to investigate such 3-dimensional pattern hierarchy.

The idea introduced in this paper not only can generate many interesting 3-D objects, but also can be used for 3-D object learning, understanding and description. For example, according to the sequence of rules (universal array grammar), Figure 2.7 can be described as: a wire-like object with 5 units of segments stretching toward x-axis, followed by a 6 units of line segments stretching toward y-axis, *and* by a 6 units of line segments stretching toward z-axis. It is the author's hope that this ground work can also pave a road for further studies of 3-d formal model for object pattern recognition and to stimulate research in 3-d object clustering analysis involving noisy and distorted patterns.

Acknowledgement

This work was supported in part by the College of Computer Science, Northeastern University and University of Paris VII and XIII. Part of this paper was written when the author was visiting at the LITP center of University of Paris VII. The excellent environment and facility provided by LITP and encouragement of Professors M. Nivat and A. Saoudi are appreciated.

4 Bibliography

1. C.Cook and P.S.P.Wang, "A chomsky hierarchy of isotonic array grammars and languages", *Computer Graphics and Image Processing*, v8, 144-152 (1978)

2. S.Edelman, H. Bulthoff, D.Weinshall, *Stimulus familiarity determines recognition strategy for novel 3-D*, MIT AI Lab. Memo. 1138, July 1989.

3. K.S.Fu, *Syntactic pattern recognition and applications*, Englewood Cliffs, N.J., Prentice-Hall, 1982

4. W.I.Grosky and P.S.P.Wang, "The relation between uniformly structured tessellation automata and parallel array grammars", *Proc. IEEE ISUSAL 75*, Tokyo, 97-102 (1975)

5. T. Marill, *Computer perception of 3D objects*. MIT AI Lab Memo. 1136, August 1989.

6. A. Rosenfeld, " Preface ", *Array grammars, patterns and recognizers* , P.S.P. Wang(Ed.), *World Scientific Publishing Co. (WSP)*, 1989.

7. A. Rosenfeld, *Picture languages: formal models for picture recognition*, Academic Press, New York, 1979

8. R. N. Shepard and J.Metzler, "Mental rotation of 3-D objects" *Science* 171, pp.701-703(1971).

9. R.Siromoney, "Array language and Lindenmayer systems- a survey", *The Book of L*, G. Rozenberg and A. Salomma (ed), Springer Verlag, 1986

10. S.Ullman, *An approach to object recognition: aligning pictorial descriptions*. MIT AI Lab. Memo. 931, Dec. 1986.

11. P.S.P.Wang, *3D object analysis by sequential/parallel array grammars*, Computer Science and Mathematics Dept. TR-91-7, University of Paris XIII, 1991

12. P.S.P.Wang, "Hierarchical structures and complexities of isometric patterns", *IEEE Trans. PAMI*, v5, n1, 92-99 (1983)

13. P.S.P.Wang, "An application of array grammars to clustering analysis for syntactic patterns", *Pattern Recognition*, v17, n4, 441-451 (1984)

14. P.S.P.Wang, "3D object representation by array grammars", *IJPRAI*, v7 n2, 1992, to appear

15. P.S.P. Wang(Ed.), *Array grammars, patterns and recognizers World Scientific Publishing Co. (WSP)*, 1989.

16. P.S.P.Wang, "Parallel object pattern analysis - a syntactic approach", *Proc. ICARCV'92*, Singapore, 1992, to appear

17. P.S.P.Wang,"A Formal Parallel Model for 3-D Object Pattern Analysis", *Handbook of Pattern Recognition and Computer Vision* (ed by C.H.Chen, L.Pau and P.Wang), WSP, (1993) to appear

Lecture Notes in Computer Science

For information about Vols. 1–570
please contact your bookseller or Springer-Verlag

Vol. 610: F. von Martial, Coordinating Plans of Autonomous Agents. XII, 246 pages. 1992. (Subseries LNAI).

Vol. 611: M. P. Papazoglou, J. Zeleznikow (Eds.), The Next Generation of Information Systems: From Data to Knowledge. VIII, 310 pages. 1992. (Subseries LNAI).

Vol. 612: M. Tokoro, O. Nierstrasz, P. Wegner (Eds.), Object-Based Concurrent Computing. Proceedings, 1991. X, 265 pages. 1992.

Vol. 613: J. P. Myers, Jr., M. J. O'Donnell (Eds.), Constructivity in Computer Science. Proceedings, 1991. X, 247 pages. 1992.

Vol. 614: R. G. Herrtwich (Ed.), Network and Operating System Support for Digital Audio and Video. Proceedings, 1991. XII, 403 pages. 1992.

Vol. 615: O. Lehrmann Madsen (Ed.), ECOOP '92. European Conference on Object Oriented Programming. Proceedings. X, 426 pages. 1992.

Vol. 616: K. Jensen (Ed.), Application and Theory of Petri Nets 1992. Proceedings, 1992. VIII, 398 pages. 1992.

Vol. 617: V. Mařík, O. Štěpánková, R. Trappl (Eds.), Advanced Topics in Artificial Intelligence. Proceedings, 1992. IX, 484 pages. 1992. (Subseries LNAI).

Vol. 618: P. M. D. Gray, R. J. Lucas (Eds.), Advanced Database Systems. Proceedings, 1992. X, 260 pages. 1992.

Vol. 619: D. Pearce, H. Wansing (Eds.), Nonclassical Logics and Information Proceedings. Proceedings, 1990. VII, 171 pages. 1992. (Subseries LNAI).

Vol. 620: A. Nerode, M. Taitslin (Eds.), Logical Foundations of Computer Science – Tver '92. Proceedings. IX, 514 pages. 1992.

Vol. 621: O. Nurmi, E. Ukkonen (Eds.), Algorithm Theory – SWAT '92. Proceedings. VIII, 434 pages. 1992.

Vol. 622: F. Schmalhofer, G. Strube, Th. Wetter (Eds.), Contemporary Knowledge Engineering and Cognition. Proceedings, 1991. XII, 258 pages. 1992. (Subseries LNAI).

Vol. 623: W. Kuich (Ed.), Automata, Languages and Programming. Proceedings, 1992. XII, 721 pages. 1992.

Vol. 624: A. Voronkov (Ed.), Logic Programming and Automated Reasoning. Proceedings, 1992. XIV, 509 pages. 1992. (Subseries LNAI).

Vol. 625: W. Vogler, Modular Construction and Partial Order Semantics of Petri Nets. IX, 252 pages. 1992.

Vol. 626: E. Börger, G. Jäger, H. Kleine Büning, M. M. Richter (Eds.), Computer Science Logic. Proceedings, 1991. VIII, 428 pages. 1992.

Vol. 628: G. Vosselman, Relational Matching. IX, 190 pages. 1992.

Vol. 629: I. M. Havel, V. Koubek (Eds.), Mathematical Foundations of Computer Science 1992. Proceedings. IX, 521 pages. 1992.

Vol. 630: W. R. Cleaveland (Ed.), CONCUR '92. Proceedings. X, 580 pages. 1992.

Vol. 631: M. Bruynooghe, M. Wirsing (Eds.), Programming Language Implementation and Logic Programming. Proceedings, 1992. XI, 492 pages. 1992.

Vol. 632: H. Kirchner, G. Levi (Eds.), Algebraic and Logic Programming. Proceedings, 1992. IX, 457 pages. 1992.

Vol. 633: D. Pearce, G. Wagner (Eds.), Logics in AI. Proceedings. VIII, 410 pages. 1992. (Subseries LNAI).

Vol. 634: L. Bougé, M. Cosnard, Y. Robert, D. Trystram (Eds.), Parallel Processing: CONPAR 92 – VAPP V. Proceedings. XVII, 853 pages. 1992.

Vol. 635: J. C. Derniame (Ed.), Software Process Technology. Proceedings, 1992. VIII, 253 pages. 1992.

Vol. 636: G. Comyn, N. E. Fuchs, M. J. Ratcliffe (Eds.), Logic Programming in Action. Proceedings, 1992. X, 324 pages. 1992. (Subseries LNAI).

Vol. 637: Y. Bekkers, J. Cohen (Eds.), Memory Management. Proceedings, 1992. XI, 525 pages. 1992.

Vol. 639: A. U. Frank, I. Campari, U. Formentini (Eds.), Theories and Methods of Spatio-Temporal Reasoning in Geographic Space. Proceedings, 1992. XI, 431 pages. 1992.

Vol. 640: C. Sledge (Ed.), Software Engineering Education. Proceedings, 1992. X, 451 pages. 1992.

Vol. 641: U. Kastens, P. Pfahler (Eds.), Compiler Construction. Proceedings, 1992. VIII, 320 pages. 1992.

Vol. 642: K. P. Jantke (Ed.), Analogical and Inductive Inference. Proceedings, 1992. VIII, 319 pages. 1992. (Subseries LNAI).

Vol. 643: A. Habel, Hyperedge Replacement: Grammars and Languages. X, 214 pages. 1992.

Vol. 644: A. Apostolico, M. Crochemore, Z. Galil, U. Manber (Eds.), Combinatorial Pattern Matching. Proceedings, 1992. X, 287 pages. 1992.

Vol. 645: G. Pernul, A M. Tjoa (Eds.), Entity-Relationship Approach – ER '92. Proceedings, 1992. XI, 439 pages, 1992.

Vol. 646: J. Biskup, R. Hull (Eds.), Database Theory – ICDT '92. Proceedings, 1992. IX, 449 pages. 1992.

Vol. 647: A. Segall, S. Zaks (Eds.), Distributed Algorithms. X, 380 pages. 1992.

Vol. 648: Y. Deswarte, G. Eizenberg, J.-J. Quisquater (Eds.), Computer Security – ESORICS 92. Proceedings. XI, 451 pages. 1992.

Vol. 649: A. Pettorossi (Ed.), Meta-Programming in Logic. Proceedings, 1992. XII, 535 pages. 1992.

Vol. 650: T. Ibaraki, Y. Inagaki, K. Iwama, T. Nishizeki, M. Yamashita (Eds.), Algorithms and Computation. Proceedings, 1992. XI, 510 pages. 1992.

Vol. 652: R. Shyamasundar (Ed.), Foundations of Software Technology and Theoretical Computer Science. Proceedings, 1992. XIII, 405 pages. 1992.

Vol. 653: A. Bensoussan, J.-P. Verjus (Eds.), Future Tendencies in Computer Science, Control and Applied Mathematics. Proceedings, 1992. XV, 371 pages. 1992.

Vol. 654: A. Nakamura, M. Nivat, A. Saoudi, P. S. P. Wang, K. Inoue (Eds.), Prallel Image Analysis. Proceedings, 1992. VIII, 312 pages. 1992.